Studies in Computational Intelligence

Volume 496

Series Editor

Janusz Kacprzyk, Warsaw, Poland

For further volumes:
http://www.springer.com/series/7092

Roger Lee
Editor

Software Engineering Research, Management and Applications

Editor

Roger Lee
Software Engineering and Information Technology Institute
Central Michigan University
Michigan
USA

ISSN 1860-949X ISSN 1860-9503 (electronic)
ISBN 978-3-319-03270-2 ISBN 978-3-319-00948-3 (eBook)
DOI 10.1007/978-3-319-00948-3
Springer Cham Heidelberg New York Dordrecht London

Printed on acid-free paper

Springer is part of Springer Science+Business Media (www.springer.com)

Preface

The purpose of the 11th International Conference on Software Engineering Research, Management and Applications (SERA 2012) held on August 7 – August 9, 2013 in Prague, Czech Republic, was to bring together scientists, engineers, computer users, and students to share their experiences and exchange new ideas and research results about all aspects (theory, applications and tools) of Software Engineering Research, Management and Applications, and to discuss the practical challenges encountered along the way and the solutions adopted to solve them. The conference organizers selected the best 17 papers from those papers accepted for presentation at the conference in order to publish them in this volume. The papers were chosen based on review scores submitted by members of the program committee and underwent further rigorous rounds of review.

In chapter 1, Haeng-Kon Kim proposes the XML security model for mobile commerce services based electronic commerce system to guarantee the secure exchange of trading information. To accomplish the security of XML, the differences of XML signature, XML encryption and XML key management scheme respect to the conventional system should be provided. The new architecture is proposed based on unique characteristics of mobile commerce on XML

In chapter 2, Benjamin Aziz presents a formal model of VOs using the Event-B specification language. Grid computing is a global-computing paradigm focusing on the effective sharing and coordination of heterogeneous services and resources in dynamic, multi-institutional Virtual Organisations (VOs). They have followed a refinement approach to develop goal-oriented VOs by incrementally adding their main elements: goals, organisations and services.

In chapter 3, Omar Badreddin, Andrew Forward, and Timothy C. Lethbridge present modeling characteristics of attributes from first principles and investigate how attributes are handled in several open-source systems. They look at code-generation of attributes by various UML tools and present their own Umple language along with its code generation patterns for attributes, using Java as the target language.

In chapter 4, Cagla Atagoren and Oumout Chouseinoglou conduct a case study in one of the leading, medium sized software companies of Turkey by utilizing the root cause analysis (RCA) method. The collected defect data has been analyzed with Pareto charts and the root causes for outstanding defect categories have been identified with the use of fishbone diagrams and expert grading, demonstrating that these techniques can be effectively used in RCA. The main root causes of the investigated defect items have been identified as lack of knowledge and extenuation of the undertaken task, and corrective actions have been proposed to upper management.

In chapter 5, Amina Magdich, Yessine Hadj Kacem, and Adel Mahfoudhi propose through their paper an extension of MARTE/GRM sub-profile to consider the modeling of information needed for the half-partitioned and global scheduling step. The recent extension of Unified Modeling Language (UML) profile for Modeling and Analysis of Real-Time Embedded systems (MARTE) is enclosing a lot of stereotypes and sub-profiles providing support for designers to beat the shortcomings of complex systems development. In particular, the MARTE/GRM (Generic Resource Modeling) package offers stereotypes for annotating class diagrams with the needed information which will be extracted to fulfill a scheduling phase.

In chapter 6, Iakovos Ouranos and Petros Stefaneas sketch some first steps towards the definition of a protocol algebra based on the framework of behavioural algebraic specification. Following the tradition of representing protocols as state machines, we use the notion of Observational Transition System to express them in an executable algebraic specification language such as CafeOBJ.

In chapter 7, Sébastien Salva and Tien-Dung Cao propose a model-based testing approach which combines two monitoring methods, runtime verification and passive testing. Starting from ioSTS (input output Symbolic Transition System) models, this approach generates monitors to check whether an implementation is conforming to its specification and meets safety properties. This paper also tackles the trace extraction problem by reusing the notion of proxy to collect traces from environments whose access rights are restricted.

In chapter 8, Donghwoon Kwon,Young Jik Kwon, Yeong-Tae Song, and Roger Lee investigate how user characteristics affect quality factors for an effective Shopping mall websites implementation. User characteristics consist of gender, age, school year, department, experience, and purchasing experience during a specified period. They also selected a total of 14 quality factors from the literature review such as design, customer satisfaction, etc. As a proof of their hypothesis to investigate how those user characteristics and quality factors are interrelated, they have used 6 hypotheses. To verify them, the results have analyzed the SAS 9.2 statistic package tool and they have asked 519 participants to fill out a questionnaire for 5 Chinese and 8 Korean websites.

In chapter 9, Omar Badreddin, Andrew Forward, and Timothy C. Lethbridge introduce a syntax for describing associations using a model-oriented language called Umple. They show source code from existing code-generation tools and highlight how the issues above are not adequately addressed. They outline code generation patterns currently available in Umple that resolve these difficulties and address the issues of multiplicity constraints and referential integrity.

In chapter 10, Damla Aslan, Ayça Tarhan, and Onur Demirörs report a case study that aimed to investigate the effect of process enactment data on product defectiveness in a small software organization. They carried out the study by defining and following a methodology that included the application of Goal-Question-Metric (GQM) approach to direct analysis, the utilization of a questionnaire to assess usability of metrics, and the application of machine learning methods to predict product defectiveness. The results of the case study showed that the accuracy of predictions varied according to the machine learning method used, but in the overall, about 3% accuracy improvement was achieved by including process enactment data in the analysis.

In chapter 11, Javier Berrocal, José García-Alonso and Juan Manuel Murillo study implicit relationships that often exist between different types of elements that subsequently have to be identified and explicitly represented during the design of the system. This requires an in-depth analysis of the generated models on behalf of the architect in order to interpret their content. Misunderstandings that take place during this stage can lead to an incorrect design and difficult compliance with the business goals. They present a series of profiles that explicitly represent these relationships during the initial development phases, and which are derived to the system design. They are reusable by the architect, thereby decreasing the risk of their misinterpretation.

In chapter 12, Oumout Chouseinoglou and Semih Bilgen analyze traditional approaches in software engineering education (SEEd), which are mostly inadequate in equipping students with these unusual and diverse skills. Their study, as part of a larger study aiming to develop a model for assessing organizational learning capabilities of software development organizations and teams, proposes and implements a novel educational approach to SEEd combining different methodologies, namely lecturing, project development and critical thinking. The theoretical background and studies on each approach employed in this study are provided, together with the rationales of applying them in SEEd.

In chapter 13, Étienne André, Christine Choppy, and Gianna Reggio propose activity diagram patterns for modeling business processes, devise a modular mechanism to compose diagram fragments into a UML activity diagram, and propose semantics for the produced activity diagrams, formalized by colored Petri nets. Our approach guides the modeler task (helping to avoid common mistakes), and allows for automated verification.

In chapter 14, Barbara Gallina, Karthik Raja Pitchai and Kristina Lundqvist propose S-TunExSPEM, an extension of Software & Systems

Process Engineering MetaModel 2.0 (SPEM 2.0) to allow users to specify safety-oriented processes for the development of safety-critical systems in the context of safety standards according to the required safety level. Moreover, to enable exchange for simulation, monitoring, execution purposes, S-TunExSPEM concepts are mapped onto XML Process Definition Language 2.2 (XPDL 2.2) concepts. Finally, a case-study from the avionics domain illustrates the usage and effectiveness of the proposed extension.

In chapter 15, Martin Babka, Tomáš Balyo, and Jaroslav Keznikl describe an application in the code performance modeling domain, which requires SMT-solving with a costly decision procedure. Then they focus on the problem of finding minimum-size satisfying partial truth assignments. they describe and experimentally evaluate several methods how to solve this problem. These include reduction to partial maximum satisfiability – PMAXSAT, PMINSAT, pseudo-Boolean optimization and iterated SAT solving. They examine the methods experimentally on existing benchmark formulas as well as on a new benchmark set based on the performance modeling scenario.

In chapter 16, Jacob Geisel, Brahim Hamid, and Jean-Michel Bruel deal with a specification language for development methodologies centered around a model-based repository, by defining both a metamodel enabling process engineers to represent repository management and interaction and an architecture for development tools. The modeling language they propose has been successfully evaluated by the TERESA project for specifying development processes for trusted applications centered around a model-based repository of security and dependability (S&D) patterns.

In chapter 17, Haeng-Kon Kim and Roger Lee discuss some of the problems of the current mobile service applications development and show how the introduction of CBD (Component Based Development) provides flexible and extensible solutions to it. Mobile service applications resources become encapsulated as components, with well defined interfaces through which all interactions occur. Builders of components can inherit the interfaces and their implementations, and methods (operations) can be redefined to better suit the component. New characteristics, such as concurrency control and persistence, can be obtained by inheriting from suitable base classes, without necessarily requiring any changes to users of these resources. They describe the MSA (Mobile Service Applications) component model, which we have developed, based upon these ideas, and show, through a prototype implementation, how we have used the model to address the problems of referential integrity and transparent component (resource) migration.

It is our sincere hope that this volume provides stimulation and inspiration, and that it will be used as a foundation for works to come.

August 2013

Petr Hnetynka
Program Chair

List of Contributors

José García-Alonso
University Of Extremadura,
Spain
jgaralo@unex.es

Étienne André
Université Paris 13,
France
Etienne.Andre@lipn.univ-paris13.fr

Cagla Atagoren
Başkent University,
Turkey
caglaatagoren@gmail.com

Benjamin Aziz
University of Portsmouth,
United Kingdom
benjamin.aziz@port.ac.uk

Damla Aslan
Simsoft Computer Technologies Co.,
Turkey
damla.sivrioglu@simsoft.com.tr

Martin Babka
Charles University, Czech Republic
babka@ktiml.mff.cuni.cz

Omar Badreddin
School of Electrical Engineering and Computer Science,
Canada
obadr024@eecs.uottawa.ca

Tomáš Balyo
Charles University,
Czech Republic
balyo@ktiml.mff.cuni.cz

Javier Berrocal
University of Extremadura,
Spain
jberolm@unex.es

Semih Bilgen
Middle East Technical University,
Turkey
semih-bilgen@metu.edu.tr

Jean-Michel Bruel
IRIT, University of Toulouse,
France
bruel@irit.fr

Tien-Dung Cao
Tan Tao University,
Vietnam
dung.cao@ttu.edu.vn

Christine Choppy
Université Paris 13,
France
Christine.Choppy@lipn.univ-paris13.fr

Oumout Chouseinoglou
Başkent University,
Turkey
umuth@baskent.edu.tr

Onur Demirörs
Informatics Institute, METU,
Turkey
demirors@ii.metu.edu.tr

Andrew Forward
School of Electrical Engineering and Computer Science, Canada
aforward@eecs.uottawa.ca

Barbara Gallina
Mälardalen University,
Sweden
barbara.gallina@mdh.se

Jacob Geisel
IRIT, University of Toulouse,
France
geisel@irit.fr

Brahim Hamid
IRIT, University of Toulouse,
France
hamid@irit.fr

Yessine Hadj Kacem
CES Laboratory, ENIS,
Tunisia
yessine.hadjkacem@ceslab.org

Jaroslav Keznikl
Charles University,
Czech Republic
Academy of Sciences of the Czech Republic,
Czech Republic
keznikl@d3s.mff.cuni.cz
keznikl@cs.cas.cz

Haeng-Kon Kim
Catholic University of Deagu,
Korea
hangkon@cu.ac.kr

Donghwoon Kwon
Towson University, USA
dkwon3@students.towson.edu

Young Jik Kwon
Daegu University, Korea
yjkwon@daegu.ac.kr

Roger Lee
Central Michigan University, USA
lee@cps.cmich.edu

Timothy C. Lethbridge
School of Electrical Engineering and Computer Science, Canada
tcl@eecs.uottawa.ca

Kristina Lundqvist
Mälardalen University,
Sweden
kristina.lundqvist@mdh.se

Amina Magdich
CES Laboratory, ENIS,
Tunisia
amina.magdich@ceslab.org

Adel Mahfoudhi
CES Laboratory, ENIS,
Tunisia
adel.mahfoudhi@ceslab.org

Juan Manuel Murillo
University Of Extremadura,
Spain
juanmamu@unex.es

Iakovos Ouranos
Hellenic Civil Aviation Authority,
Heraklion Airport,
Greece
iouranos@central.ntua.gr

Karthik Raja Pitchai
Mälardalen University,
Sweden
kpi10001@student.mdh.se

Gianna Reggio
DIBRIS, Italy
gianna.reggio@unige.it

Sébastien Salva
University of Auvergne, France
sebastien.salva@udamail.fr

Yeong-Tae Song
Towson University, USA
ysong@towson.edu

Petros Stefaneas
National Technical University of Athens, Greece
petros@math.ntua.gr

Ayça Tarhan
Hacettepe University,
Turkey
atarhan@cs.hacettepe.edu.tr

Contents

Security Certification Model for Mobile-Commerce

Haeng-Kon Kim

Abstract. The most important technology in the mobile commerce based on mobile applications is to guarantee the certification and security of trading information exchange. Many technologies are proposed as a standard to support this security problem. M(Mobile)-commerce is a new area arising from the marriage of electronic commerce with emerging mobile and pervasive computing technology. The newness of this area and the rapidness with which it is emerging makes it difficult to analyze the technological problems that m-commerce introduces and, in particular, the security and privacy issues. This situation is not good, since history has shown that security is very difficult to retro-fit into deployed technology, and pervasive m-commerce promises to permeate and transform even more aspects of life than e-commerce and the Internet has. One of them is an XML (eXtensible Markup Language). This is used in various applications as the document standard for electronic commerce system. The XML security has become very important topic. In this paper, we propose the XML security model for mobile commerce services based electronic commerce system to guarantee the secure exchange of trading information. To accomplish the security of XML, the differences of XML signature, XML encryption and XML key management scheme respect to the conventional system should be provided. The new architecture is proposed based on unique characteristics of mobile commerce on XML. Especially the method to integrate the process management system need to the electronic commerce is proposed.

Keywords: Mobile commerce, Mobile security, XML management, Mobile, networks.

Haeng-Kon Kim
School of Information Technology, Catholic University of Deagu, Korea
e-mail: hangkon@cu.ac.kr

R. Lee (Ed.): *SERA*, SCI 496, pp. 1–15.
DOI: 10.1007/978-3-319-00948-3_1

1 Introduction

Mobile commerce is an interesting and challenging area of research and development. It presents many issues that cover many disciplines and may best be addressed by an active participation of computer and telecommunications experts, social scientists, economists and business strategists. M-commerce introduced several new classes of applications, reviewed networking requirements, and discussed application development support. Since the area of mobile commerce is very new and still emerging, several interesting research problems are currently being addressed or should be addressed by the research and development community. It is believed that user trust will play a crucial role in acceptance and widespread deployment of mobile commerce applications. Regarding m-payment, some systems are under development or already operational. One of the main future challenges will be to unify payment solutions, providing the highest possible level of security.

Much of information is propagated by Internet. Internet that is an open communication system provides browsers based on easy protocols and various tools for information handling. Therefore m-commerce is proliferated. This m-Commerce is based on the standards for document processing in Internet

In the last few years, advances in and widespread deployment of information technology have triggered rapid progress in m-commerce. This includes automation of traditional commercial transactions (electronic retailing, etc.) as well as the creation of new transaction paradigms that were infeasible without the means of widely deployed information technology. New paradigms include electronic auctioning of purchase orders, as well as novel, with less transaction models such as Napster [1]. M-commerce has heightened the focus on security both of systems and also for messaging and transactions [2].

The enterprises perform not only the internal activities but also the interactive businesses with other companies to secure the competitive power of them. In general, the trading business between enterprises is performed typically according to the pre-defined business process by exchanging the contracted documents.

The purpose of this paper is to propose a business model for B2B environment for m-commerce. This model is based on the business process management system which manages the conventional internal processes of enterprises. This model also analyzes the key elements needed to m-commerce for inter-enterprises. Especially, the documents and data exchanged between companies is formalized by using the XML messages that are approved as the standard tools for information exchanges. The business processes exchange the XML messages. During all processes, therefore, the efficient business integration may be possible. This model ensures the secure information exchange which is an essential factor in m-commerce as in figure 1.

The m-commerce should be based on the public key encryption system to authenticate the valid users. The method to ensure the reliability and security of user's public keys is required. Public key infrastructure (PKI)

provides secure and reliable method to open the user's public keys to the public [3]. Public key infrastructure has very important roles in Internet E-Commerce. It opens the user's public keys to public in secure and reliable manner. Since the XML technology is used as the format of message exchange in Internet e-Business, the security for XML documents becoming essential and XML digital signature should be supported for secure m-commerce [4,5].

In this paper, the security application of m-commerce is designed which is reliable by using X.509 certificate based on PKI. A web service is designed to implement the PKI-based security application for mutual authentication. The digital signature protocol based on PKI and XML is also designed to solve the security and repudiation problem of message exchange in B2B on m-commerce.

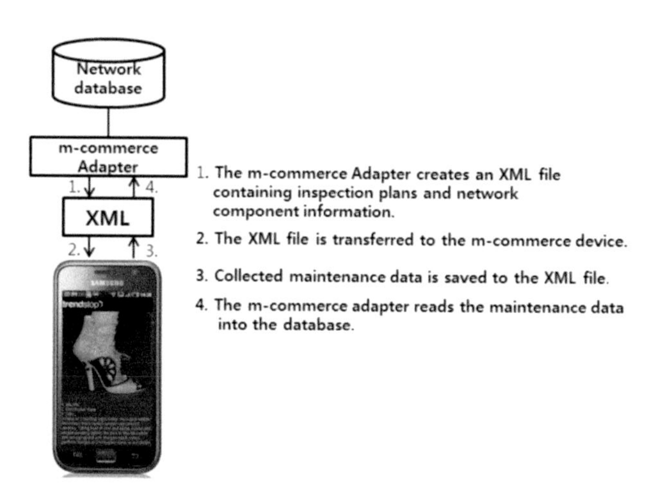

Fig. 1 Secure Information Exchange in M- Commerce

2 Background Study

2.1 Security for Mobile Commerce Applications

Security for mobile Commerce application is a crucial issue. Without secure commercial information exchange and safe electronic financial transactions over mobile networks, neither service providers nor potential customers will trust mobile commerce systems. From a technical point of view, mobile commerce over wireless networks is inherently insecure compared to electronic commerce over the Internet. The reasons are as follows:

- Reliability and integrity: Interference and fading make the wireless channel error-prone. Frequent handoffs and disconnections also degrade the security services.

- Confidentiality/privacy: The broadcast nature of the radio channel makes it easier to tap. Thus, communication can be intercepted and interpreted without difficulty if no security mechanisms such as cryptographic encryption are employed.
- Identification and authentication: The mobility of wireless devices introduces an additional difficulty in identifying and authenticating mobile terminals.
- Capability: Wireless devices usually have limited computation capability, memory size, communication bandwidth and battery power. This will make it difficult to utilize high-level security schemes such as 256-bit encryption.

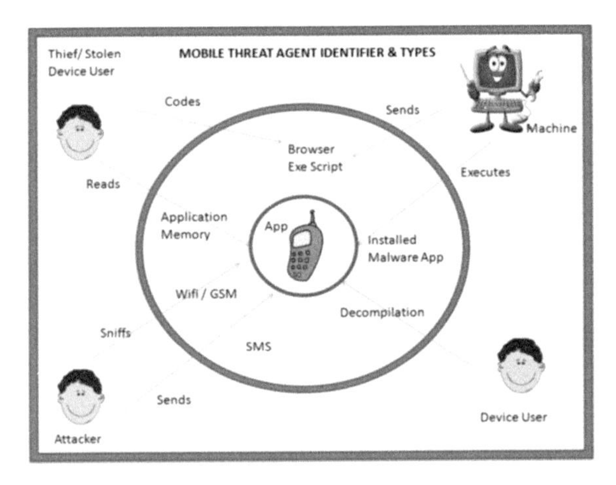

Fig. 2 Securities Mechanisms and Systems

Mobile commerce security is tightly coupled with network security. The security issues span the whole mobile commerce system, from one end to the other, from the top to the bottom network protocol stack, from machines to humans. Therefore, many securities mechanisms and systems used in the mobile application and commerce may be involved as in figure 2.

Public key encryption system is an asymmetric system which is based on mathematical functions. It has the pair of keys one is opened to public and the other is saved securely instead of private key encryption system. Then the key is opened is called public key, the other is called private key. The majority security systems for E-Commerce based on public key algorithm because the key management and distribution are difficult. It also resolves the anonymous and user authentication problems.

Public key infrastructure should be constructed based on public key certificates. The certification authority(CA) authenticates the trading subjects. The certification authority creates digital signature by using their own private key and attaches them to the certificate for proving the subject users are valid. The certificate includes the public key of certificate's users and information of subject users.

2.2 M-Commerce Framework

This emerging area of m-commerce creates new security and privacy challenges because of new technology, novel applications, and increased pervasiveness. Mobile applications will differ from standard e-commerce applications, because the underlying technology has fundamental differences:

- **Limitations of Client Devices.** Current (and looming) PDAs are limited in memory, computational power, cryptographic ability, and (for the time being) human I/O. As a consequence, the user cannot carry his entire state along with him, cannot carry out sophisticated cryptographic protocols, and cannot engage in rich GUI interaction.
- **Portability of Client Device.** PDAs have the potential to accompany users on all activity, even traditionally offline actions away from the desk-top. Besides creating the potential for broader permeation of e-transactions, this fact also makes theft, loss, and damage of client devices much more likely.
- **Hidden and Unconscious Computing.** Both to compensate for limited PDA storage, as well as to provide new ways to adapt a user's computing environment to her current physical environment, pervasive computing often permits client devices to transparently interact with the infrastructure without the user's direct interaction. This unconscious interaction can include downloading executable content.
- **Location Aware Devices.** When the user is mobile, the infrastructure can potentially be aware of the location of the user (e.g., in a particular telephone cell). This knowledge introduces a wide range of applications which have no analogue in the stationary user model.
- **Merchant Machines.** In the e-commerce world, the merchant (i.e., the party that is not the user) has powerful machines, with ample storage and computation, usually in a physically safe place. However, to fully exploit the potential interacting with mobile, PDA equipped users, merchant machines may move out into the physical world. This move brings with its own challenges of increased physical exposure, limited computation and state, and limited interconnection.

The most threatened factor to the m-commerce is the security problems. The messages exchanged by an XML message via Internet is not secure because the user authentication is not guaranteed as shown in Fig. 3 [6].

We are aware that consensus within business and industry of future applications is still in its infancy. However, we are interested in examining those future applications and technologies that will form the next frontier of electronic commerce. To help future applications and to allow designers,

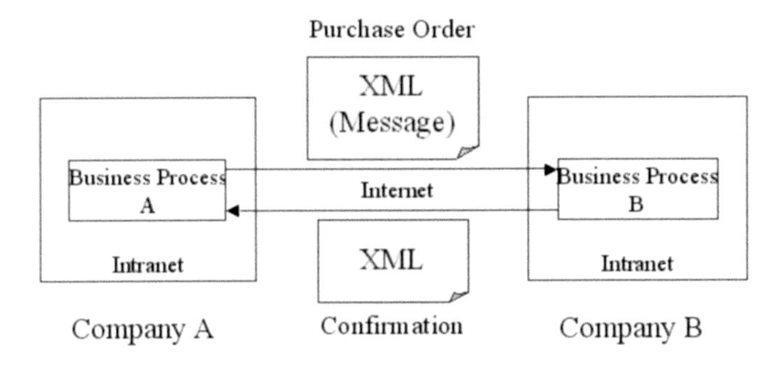

Fig. 3 Unsecured message exchange

developers and researchers to strategize and create mobile commerce applications, a four level integrated framework is proposed as in figure 4.

These four levels are as follows: m-commerce applications, user infrastructure, middleware and network infrastructure which simplifies the design and development. By following this framework a single entity is not forced to do everything to build m-commerce systems, rather they can build on the functionalities provided by others. The framework also provides a developer and provider plane to address the different needs and roles of application developers, content providers and service providers.

Service providers can also act as content aggregators, but are unlikely to act as either an application or content provider due to their focus on the network and service aspects of m-commerce. Content provider can build its service using applications from multiple application developers and also can aggregate content from other content providers and can supply the aggregated content to a network operator or service provider.

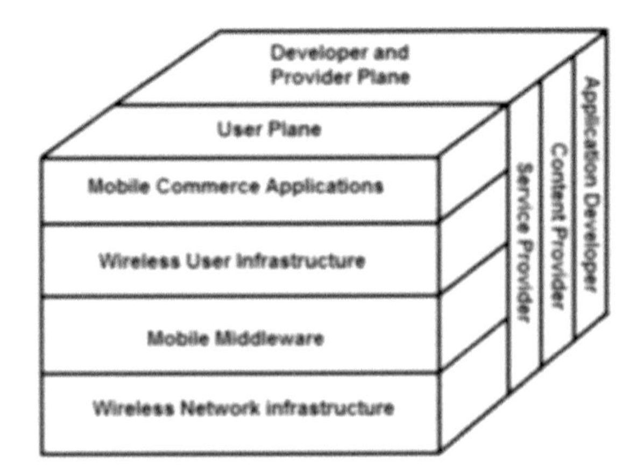

Fig. 4 Framework of M-commerce

2.3 Mobile Applications and XML

M-commerec services are software interface which can be found and called by another programs on the web regardless of location and platforms. M-commerec service is independent on platforms, devices and location. M-commerec service provides dynamic functionality. M-commerec service can be also applied to the conventional systems by low cost. The mobile applications service in m-commerce is a standardized software technology which combines conventional computer system programs between businesses on Internet. This standard technology enables all business functionalities and services. The M-commerec services by using Internet overcome the differences of communications among the heterogeneous operation systems and programming languages. So to speak, the web services are software components which conform e-Business standard and have business logics of Internet. XML standard describes the classes of data objects for XML documents. It also describe the operations of computer programs which process these XML documents. XML is an application of SGML (Standard Generalized Markup Language).

XML documents consist of entities which are storage units. The entity contains parsed data or un-parsed data. The parsed data consists of characters. Some of these characters are character data, the others are markups. The markups encode the arrangement plan of physical storage and the description of logical structure. XML provides a mechanism which enforces the arrangement plan of storage and logical structure. The software module as it called XML processor reads XML document and accesses the content and structure of that. XML is a standard for organizing the data, XSL (eXtensible Stylesheet Language) is a standard for method to output this data. XSL is a translation technology. XSL is a language to translate each field of XML to relevant tags of HTML and represent to web browser. XML schema is the term for file to define the structure and content of XML documents. DTD (Document Type Definition) is also a kind of schema, but it has some defects. DTD should be described by E-BNF and so difficult. On the other hand, XML schema can be desctibed just using XML itself. Moreover, XML schema can use various data types that are not supported in DTD. In XML schema the elements can be reused. So to peak, XML schema extended model of DTD. XML schema can define precisely the types of XML documents and the relationships of elements. XML documents should be parsed to make a tree structure from XML elements. DOM (Document Object Model) is a model to store parsed data as a tree structure and permits accessing particular element. According to DOM, XML documents are analyzed to client and server structure as in figure 5.

Recently, XML is in the spotlight as a technology applicable to various applications like B2B and B2C. The importance of security is increased in E-Commerce because the most businesses are processed in electronically. Especially, the standards for security in documents exchanging using XML in m-commerce have been established. The XML-Signature Group of IETF and

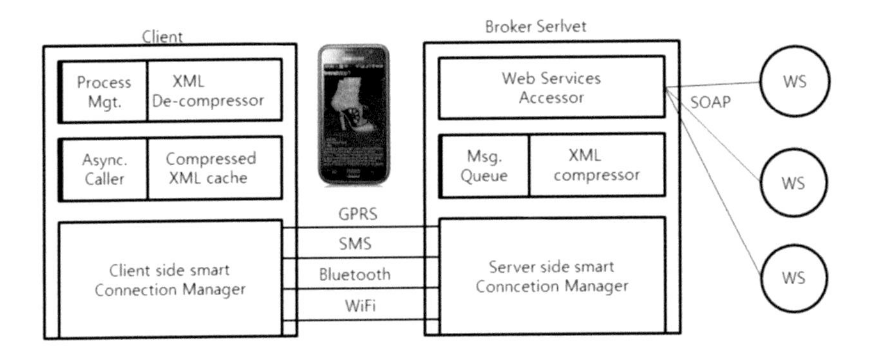

Fig. 5 Client-Server Structure for M-commerce on XML

W3C recommended the specification for "XML-Signature Syntax and Processing". This specification describes the syntax and processes for XML digital signature.

The following should be considered for security of XML digital signature.

- Confidentiality
- Integrity
- Authentication
- Authorization
- Non-Repudiation

3 Mobile Commerce Security Model

3.1 M-Commerce Security Issues

As mentioned earlier, m-commerce is not possible without a secure environment, especially for those transactions involving monetary value. Depending on the point of views of the different participants in an m-commerce scenario, there are different security challenges . These security challenges relate to:

- **The mobile device -** Confidential user data on the mobile device as well as the device itself should be protected from unauthorized use. The security mechanisms employed here include user authentication (e.g. PIN or password authentication), secure storage of confidential data (e.g. SIM card in mobile phones) and security of the operating system.
- **The network operator infrastructure -** Security mechanisms for the end user often terminate in the access network. This raises questions regarding the security of the users data within and beyond the access network. Moreover, the user receives certain services for which he/she has to pay. This often involves the network operator and he/she will want to be assured about correct charging and billing.

- **The kind of m-commerce application -** M-commerce applications, especially those involving payment, need to be secured to assure customers, merchants, and network operators. For example, in a payment scenario both sides will want to authenticate each other before committing to a payment. Also, the customer will want assurance about the delivery of goods or services. In addition to the authenticity, confidentiality and integrity of sent payment information, non-repudiation is important.

The figure 6 shows the security issues for m-commerce in the view of stakeholders as application developers, contents provider, wireless service provider, equipment vendors and other service provider in this paper.

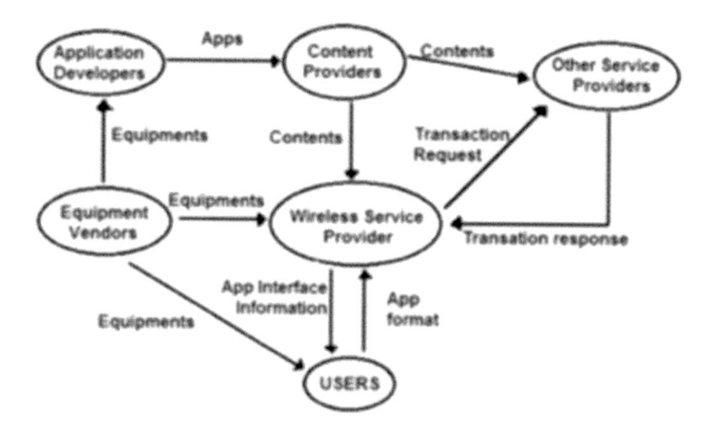

Fig. 6 M-commerce Security Issues

3.2 Mobile Applications Signature

The syntax of XML signature is a complicated standard to provide various functionalities. It can be applied any signatures because it is designed to have high-level extensibility and flexibility. W3C recommendation defined XML signature syntax and processing rules for them. Traditionally, middle-ware unites different applications, tools, networks and technologies; allowing user access via a common interface. Mobile middle-ware can be defined as an enabling layer of software that is used by the applications development to connect the m-commerce applications with different networks and operating systems without introducing mobility awareness in the applications. To allow for web content to be accessible from everywhere, from PCs to TVs to palm devices to cellular phones, the World Wide Web consortium (W3C) had developed several recommendations. These recommendations include the Extensible Makeup Language (XML) for richer semantic information, improved Cascading Style Sheets (CSS) and Extensible Style Sheet Language (XSL) to further separate content from presentation, and a Document Object Model (DOM) which defines a language independent application programming interface that applications can use to access and modify the structure, content and style of HTML and XML documents. Fig.7 shows the Mobile middleware for Certification Model for Mobile-Commerce [7].

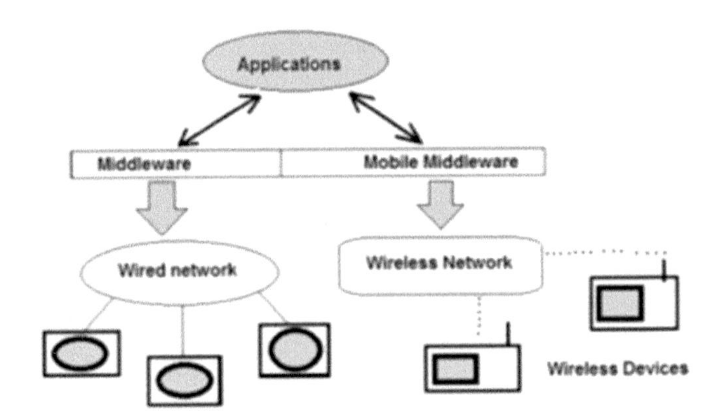

Fig. 7 Mobile middleware Certification Model for Mobile-Commerce in this paper

XML signature starts with an element <Signature>. The element <Signature>is an important one that consists of signature and identifying the signatures. The element <SignedInfo>lists "the signed information" which are the objects to sign by us. The particular data streams for Digest is represented by the element <References>. The URI (Uniform Resource Identifier) syntax is used to prescribe these streams. The element <KeyInfo>may be used efficiently in automation of XML signature processing because it provides identifying mechanism for verification keys. The element <Object>is a container which can retain any types of data objects. Two elements for <SignatureProperties>and <Manifest>are defined that should be contained in the element <Object>. The element <SignatureProperties>is a pre-defined container to verify signatures. It retains the assertions for signatures. These assertions may be used to verify the signatures and integrity. The element <Manifest>is used to verify references for application domains. It also provide a convenient method for multiple-signers to sign multiple documents. If the element <Manifest>does not used, the results of signature increase in volume and the performance may be depreciated. The creation information for certificates and the issued certificates are exchanged in the form of XML documents. The important information is encrypted as a unit of XML element.

3.3 Structure for XML Security

In this paper, the security system is designed based on the web service platform. This system executes and verifies XML signatures independent from the conventional applications. Consider the Purchase Order is submitted by Company A via Internet and is confirmed by Company B as shown in Fig. 7. Company A executes digital signature before transmission and Company B confirms after reception. So, the secure SOAP message exchanges are possible. In this process the Proxy has a role to check the digital signatures under surveillance of delivered messages. The real object to execute and to confirm the digital signature is implemented as a web service. The following is the procedures for. Mobile Secure exchange of XML messages as in figure 8.

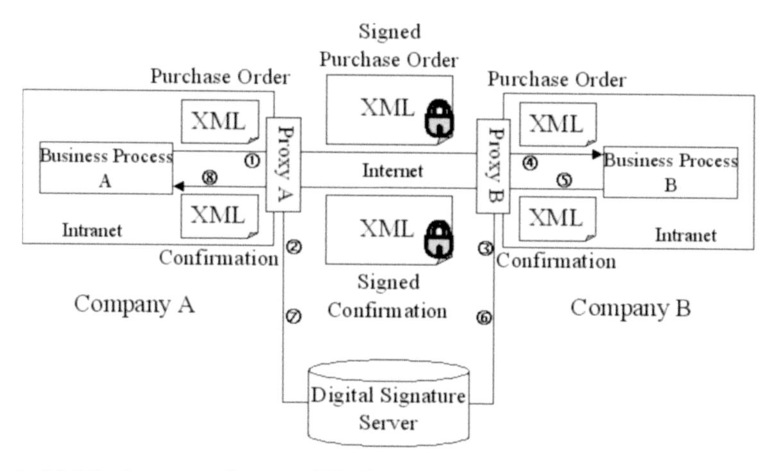

Fig. 8 Mobile Secure exchange of XML messages

Step 1. The business process A of company A transmits the message for Purchase Order to business process B of company B.

Step 2. When the purchase is passing proxy A, the digital signature is executed by sending the message to digital signature server.

Step 3. The proxy B of company B receives the signed message and sends it to the confirmation server. The confirmation server verifies the signed message.

Step 4. The verification results are sent to proxy B. If the signature is valid, proxy B removes the signature and sends it to business process B. The information of signer may be preserved.

Step 5. The business process B transacts the message for Purchase Order. The business process B makes a reply message and transmits it to company A.

Step 6. When the reply message is passing proxy B, the digital signature is executed using the private key of company B by sending the message to digital signature server.

Step 7. The company A sends the message from Proxy A to the confirmation server.

Step 8. If the digital signature is valid, the signature is removed from the message and the message is sent to the business process A.

The proxy determines whether it executes digital signature or not by checking the XML messages on network. Consequently, the workflow A and B do not concern the execution and confirmation of signatures. It is a forte that the conventional applications may not be changed.

The content verifier of the proxy server determines whether it needs a digital signature or not by checking the existence of an element <Signature>in XML schema. If it needs, two modules are required. One is to translate the XML message to the form of SOAP message, the other is reverse.

3.4 *Execution of Digital Signature*

Figure 7 shows an example of the message for Purchase Order with digital signature. The procedure to execute the message in Fig. 9. by digital signature web service is as follows:

Step 1. Determine the object for digital signature. This is given as the form of URI in general.

Step 2. Calculate the value of Digest for each object for signature. The object for signature is defined in the element <Reference>and each

```
<?xml version="1.0" encoding="UTF-8"?>
<Signature xmlns="http://www.w3.org/2000/09/xmldsig#">
<SignedInfo Id="foobar">
<CanonicalizationMethod Algorithm="http://www.w3.org/TR/2001/REC-xml-c14n-20010315"/>
<SignatureMethod Algorithm="http://www.w3.org/2000/09/xmldsig#dsa-sha1"/>
<Reference URI="http://www.acompany.com/news/2000/03_27_00.htm">
<DigestMethod Algorithm="http://www.w3.org/2000/09/xmldsig#sha1"/>
<DigestValue>j6lwx3rvEPO0vKtMup4NbeVu8nk=</DigestValue> </Reference>
<Reference URI="http://www.w3.org/TR/2000/WD-xmldsig-core-20000228/signature-sample.xml">
<DigestMethod Algorithm="http://www.w3.org/2000/09/xmldsig#sha1"/>
<DigestValue>UrXLDLBIta6skoV5/A8Q38GEw44=</DigestValue> </Reference>
</SignedInfo>
<SignatureValue>MC0E~LE=</SignatureValue>
<KeyInfo>
<X509Data>
<X509SubjectName>CN=Ed Simon, O=XML Security Inc., ST=OTTAWA, C=CA</X509SubjectName>
<X509Certificate> MIID5jCCA0+gA...lVN </X509Certificate>
</X509Data>
</KeyInfo>
</Signature>
```

Fig. 9 XML Digital Signature

Digest is stored in the element <DigestValue>. The element <DigestMethod>defines the algorithm.

Step 3. The element <SignedInfo>contains the elements <Reference>of each objects for signature. The element <CanonicalizationMethod> designates the algorithm that normalizes the element <SignedInfo>.

Step 4. The Digest of the elements <SignedInfo>is calculated and signed, then stored in the element <SignatureValue>.

Step 5. If the information of public key is required, it is stored in the element <KeyInfo>. This is a certificate of X.509 for sender and needed to confirm the digital signature. The procedure for confirmation is shown in Fig. 10.

Step 6. Finally, the XML digital signature is generated by including all created elements to the element <Signature>.

Fig. 8 shows the procedures to confirm the reliability of digital signature.

The information of certificates is extracted from the element <KeyInfo>to confirm the generated digital signature. It is compared to the certificate stored in the root certificate registry. Then the reliability is ensured.

In our works, we applied our model to mobile financial applications are likely to be one of the most important components of m-commerce as in figure 11. They could involve a variety of applications such as mobile banking and brokerage service, mobile money transfer, and mobile payments as shown in the figure 11. One interesting mobile financial application is micro payment involving small purchases such as vending and other items. A mobile device can communicate with a vending machine using a local wireless network to purchase desired items. Micropayments can be implemented in a variety of ways. One way is that the user could make a call to a certain number where per minute charges equal the cost of the vending item.

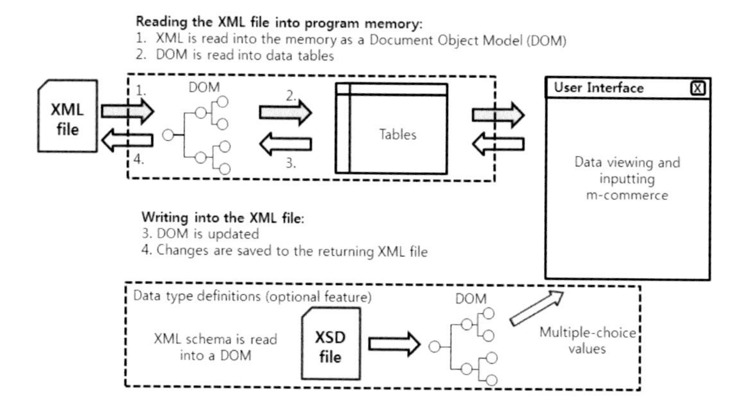

Fig. 10 Confirmation of Reliability of Mobile Security Model

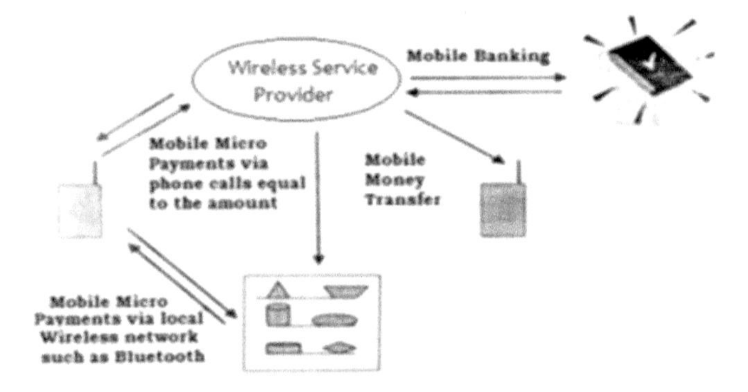

Fig. 11 One Execution example of M-commerce Services

4 Conclusion and Further Study

M-commerce introduced several new classes of applications, reviewed networking requirements, and discussed application development support. Since the area of mobile commerce is very new and still emerging, several interesting research problems are currently being addressed or should be addressed by the research and development community. It is believed that user trust will play a crucial role in acceptance and widespread deployment of mobile commerce applications.

Regarding m-payment, some systems are under development or already operational. One of the main future challenges will be to unify payment solutions, providing the highest possible level of security.

In this paper, PKI-based digital signature is designed based on XML and web services. It ensures the secure trading and non-repudiation in E-Commerce. The XML digital signature is designed and the operation structure is also proposed when two companies exchange the trading information as the form of XML messages. By using the concepts of proxy and web service, the conventional application programs can be operated without change. All information for document exchange is represented in XML. Only the secret information of XML document is encrypted. Because the digital signature is executed whole document, the security of trading and non-repudiation are guaranteed.

In the future, we will research for connecting to the CA, distribution of CRL (Certificate Revocation List) and key renewal for CA for improvement our model.

References

1. The Napster.com home page, `http://www.napster.com`
2. Chari, S., Kermani, P., Smith, S., Tassiulas, L.: Security Issues in M-Commerce: A Usage-Based Taxonomy. In: Liu, J., Ye, Y. (eds.) E-Commerce Agents. LNCS (LNAI), vol. 2033, pp. 264–282. Springer, Heidelberg (2001)

3. RFC: 2560 X.509 Internet Public Key Infrastructure Online Certificate Status Protocol - OCSP (1996)
4. W3C, Extensible Markup Language (XML) (1998), http://www.w3c.org/XML
5. XML Signature Requirements WD, W3C Working Draft (October 1999), http://www.w3.org
6. Cho, K.M.: Framework of Content Distribution in Mobile Network Environment. In: Proc. the 2003 International Conference on Internet Computing (IC 2003), pp. 429–434 (2003)
7. http://www.roseindia.net/services/m-commerce/mobile-commerce.shtml
8. XML-Signature Syntax and Processing, W3C Recommendation (February 2002), http://www.w3c.org
9. XML Encryption Syntax and Processing, W3C Working Draft (October 2001), http://www.w3c.org
10. Decryption Transform for XML Signature, W3C Working Draft (October 2001), http://www.w3c.org
11. Takase, T., et al.: XML Digital Signature System Independent Existing Applications. In: Proc. the 2002 Symposium on Application and the Internet, pp. 150–157 (2002)
12. Xavier, E.: XML based Security for E-Commerce Applications. In: Eighth Annual IEEE International Conference and Workshop on the Engineering of Computer Based Systems, pp. 10–17 (2001)
13. Cho, K.M.: Packaging Strategies of Multimedia Content in DRM. In: Proc. the 2003 International Conference on Internet Computing (IC 2003), pp. 243–248 (2003)
14. Cho, K.M.: Web Services based XML Security Model for Secure Information Exchange in Electronic Commerce. The Journal of Korean Association of Computer Education 7(5), 93–99 (2004)

On Formalising Policy Refinement in Grid Virtual Organisations

Benjamin Aziz

Abstract. Grid computing is a global-computing paradigm focusing on the effective sharing and coordination of heterogeneous services and resources in dynamic, multi-institutional Virtual Organisations (VOs). This paper presents a formal model of VOs using the Event-B specification language. We have followed a refinement approach to develop goal-oriented VOs by incrementally adding their main elements: goals, organisations and services. Our main interest is in the problem of policy refinement in VOs, so policies are represented as invariants that should be maintained throughout the refinement process. As an illustration, we show how a VO resource-usage policy is represented at different levels of abstraction.

1 Introduction

Grid computing is a global-computing paradigm focusing on the effective sharing and coordination of heterogeneous services and resources in dynamic, multi-institutional Virtual Organisations (VOs) [13]. A Grid VO can be seen as a temporary or permanent coalition of geographically dispersed organisations that pool services and resources in order to achieve common goals. This paper presents a formal model of VOs using the Event-B specification language [3]. We have followed a refinement approach to develop goal-oriented VOs by incrementally adding their main elements: goals, organisations and services. Our main interest is in the problem of policy refinement in VOs.

Policy refinement is the process of transforming a high-level abstract policy specification into a low-level concrete one [16]. Current approaches to policy refinement in distributed and dynamic systems suppose that the refinement of the abstract system entities into the concrete objects/devices is done as a previous phase to the refinement of policies, by assuming there exist pre-defined hierarchies of concrete

Benjamin Aziz
University of Portsmouth, Portsmouth, United Kingdom
e-mail: benjamin.aziz@port.ac.uk

R. Lee (Ed.): *SERA*, SCI 496, pp. 17–31.
DOI: 10.1007/978-3-319-00948-3_2

objects/devices [22] or by taking the concrete system architecture as an input [8]. Here, we use the stepwise refinement approach [6] to develop simultaneously both the system entities and their policies. In our case, policies are represented as invariants that should be maintained throughout the refinement process. We illustrate this approach by analysing the case of a resource-usage policy, the so-called *cost-balancing policy*, where the cost of achieving a goal in a VO is divided equally among the VO members. This is a particular case of the $1/N$ policy [23], a representative Grid policy indicating that all resource utilisation is equally distributed among the VO-member resources.

The work presented here has a twofold aim; on one hand, we would like to gain a more formal understanding of VOs and their lifecycle, especially in the presence of policy contraints. On the other hand, we would like to experiment with the process of designing VOs following the refinement process paying particular attention to non-functional properties such as resource usage and security. In recent years, the need for adopting rigorous approaches for designing distributed systems such as VOs has risen due to the various challenges posed by the use of such systems in safety and security critical collaborative environments such as collaborative engineering in the aerospace domain [14], Grid-based operating systems [17] and others.

The structure of the paper is the following. Section 2 introduces the main elements of a VO and the VO life cycle. Next, Section 3 gives a brief overview of Event-B. Section 4 presents a motivating case scenario involving cost-balancing policies. Section 5 presents our abstract model of VOs; a model containing only goals and representing the VO lifecycle. An intermediate refinement is described in Section 6, which includes goals and organisations. Our concrete model is presented in Section 7, including goals, organisations and services. Section 8 presents related work and finally, Section 9 concludes the paper and highlights future work.

2 On Virtual Organisations and Their Lifecycle

The entities that form a VO are drawn from a "club of potential collaborators" called a *Virtual Breeding Environment* (VBE) [10]. A VBE can be defined as an association of organisations subscribing to a base long term cooperation agreement, adopting common operating principles and infrastructure with the objective of participating in future potential VOs. In this paper, we take the view that potential partners in a VO are selected from a VBE. We are interested in goal-oriented VOs, so organisations willing to participate in a VO will join the VBE, advertising the goals they can achieve and the services provided to fulfill such goals.

For the management of a VO, we are following a VO life-cycle adopted by other projects such as ECOLEAD [11] and TrustCoM [5]. The life-cycle includes the following phases:

- VO Identification: In this phase, the VO Administrator sets up the VO by selecting potential partners from the VBE, using search engines or registries. In our model, we will be looking for partners that can achieve the goals identified in the

VO. The identification phase ends with a list of candidates that potentially could perform the goals needed for the current VO.
- VO Formation: In the formation phase, the initial set of candidates is reduced to a set of VO members. This process may involve a negotiation between potential partners. After this has been completed, the VO is configured and can be considered to be ready to enter the operation phase.
- VO Operation: The operation phase could be considered the main life-cycle phase of a VO. During this phase the VO members contribute to the VOs task(s) by executing pre-defined business processes (e.g. service orchestration) to achieve the VO goals. Membership and structure of VOs may evolve over time in response to changes of objectives or to adapt to new opportunities in the business environment; this is a feature we are not considering in the current version of our model.
- VO Dissolution: During dissolution, the VO structure is dissolved and final operations are performed to annul all contractual binding of the partners.

Figure 1 illustrates the VO lifecycle. As part of our model, we show in the paper a formalisation of the VO lifecycle, where each VO phase is modelled as an event. In our view, a VO policy is a property that should be respected across the VO phases. We model the initial actions needed to enable the integration of organisations into a VO as an additional phase called Initialisation.

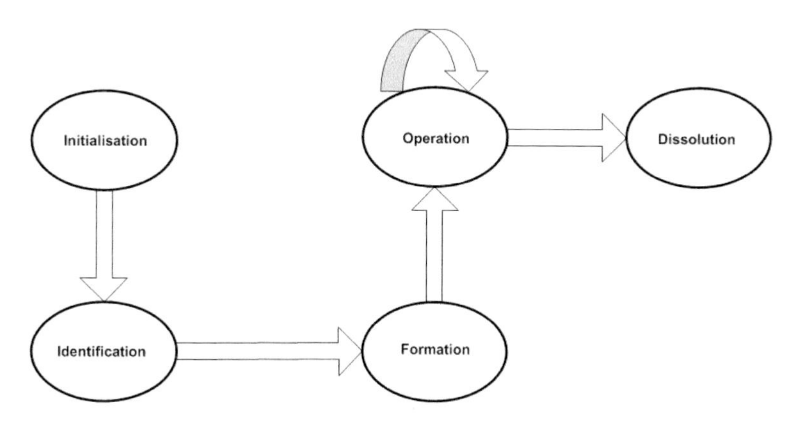

Fig. 1 The VO Lifecycle

3 A Brief Overview of Event-B

Event-B [2] is an extension of Abrial's B method [1] for modelling distributed systems. This section presents a brief overview of Event-B; we refer the reader to [2, 12] for a more complete description of this formal method. Modularity is central to the Event-B method and this is achieved by structuring specifications and development into *Machines*. A *machine* encapsulates a local state and provides operations on the state, as shown in Figure 2.

	MACHINE AM
	SEES AC
CONTEXT AC	**VARIABLES** v
SETS T	**INVARIANT** I
CONSTANTS c	**INITIALISATION** *Init*
AXIOMS A	**EVENTS**
END	E_1 = **WHEN** G **THEN** S **END**
	...
	$E_n = \cdots$
	END

Fig. 2 Abstract Machine Notation in Event-B

The **CONTEXT** component specifies the types and constants that can be used by a machine. It is uniquely identified by its name AC and includes clauses **SETS**, defining carrier sets (types); **CONSTANTS**, declaring constants; and **AXIOMS**, defining some restrictions for the sets and including typing constraints for the constants in the way of set membership.

A machine is introduced by the **MACHINE** component, which is uniquely identified by its name AM. A machine may reference a context, represented by clause **SEES**, indicating that all carrier sets and constants defined in the context can be used by the machine. Clause **VARIABLES** represents the variables (state) of the model, which are initialised in *Init* as defined in the **INITIALISATION** clause. The **INVARIANT** clause describes the invariant properties of the variables, denoting usually typing information and general properties. These properties shall remain true in the whole model and in further refinements. The **EVENTS** clause defines all the events (operations) describing the behaviour of the system. Each event is composed of a guard G (a predicate) and an action S, which is a statement, such that if G is enabled, then S can be executed. If several guards are enabled at the same time then the triggered event is chosen in a nondeterministic way.

Statements in the bodies of events have the following syntax:

$$
\begin{aligned}
S == \; & x := e \;\mid \\
& \textbf{IF } cond \textbf{ THEN } S1 \textbf{ ELSE } S2 \textbf{ END} \;\mid \\
& x :\in T \;\mid \\
& \textbf{ANY } z \textbf{ WHERE } P \textbf{ THEN } S \textbf{ END} \;\mid \\
& S1 \parallel S2
\end{aligned}
$$

Assignment and conditional statements have the standard meaning. The non-deterministic assignment $x :\in T$ assigns to variable x an arbitrary value from the given set (type) T. The non-deterministic block **ANY** z **WHERE** P **THEN** S **END** introduces the new local variable z that is initialised non-deterministically according to the predicate P and then used in statement S. Finally, $S1 \parallel S2$ models parallel (simultaneous) execution of $S1$ and $S2$ provided they do not have conflict on state variables. Statements are formally defined using a weakest precondition semantics.

In order to be able to ensure the correctness of a system, a machine should be consistent and feasible. This is assured by proving the initialisation is feasible and establishes the invariant, and then each event is feasible and preserves the invariant. Proof obligations are generated automatically and verified using the RODIN toolkit [20]. Proof obligations are generated via before-after predicates denoting the relation between the variable values before and after the execution of a statement.

Event-B supports stepwise refinement, the process of transforming an abstract, non-deterministic specification into a concrete, deterministic, system that preserves the functionality of the original specification. We use a particular refinement method, *superposition refinement* [7], where the state space is extended while preserving the old variables. During the refinement process, new features that are suggested by the requirements are represented by new variables added to the system. Simultaneously, events are refined to take the new features into account. This is performed by strengthening their guards and adding substitutions on the new variables.

3.1 Our Approach

The general approach we adopt in this paper involves the following steps:

- First, we use Event-B to model, at an abstractb level, a specific system. This will be in our case the system of goal-oriented VOs.
- Second, we use the refinement mechanism supported by Event-B to add more detail gradually to the original abstract model, until one arrives at the required level of detail. In this case, this will be realised by refining our abstract goal-oriented VOs to VOs with organisations and goal costs, then again refine further to VOs with service sets.
- Finally, we express any policy constraints we need (in our case, the cost balancing constraint we discuss in the next section) in terms of the machine invariants starting from some level of detail in the refinement chain. This could either start at the abstract level, or at any level of the refined machines. We then show that the same policy (invariant) is respected and upheld by the lower levels of refinement.

This approach is general and can be applied to any domain and with any policy requirements. The rest of the paper considers only one example of the application of this approach.

4 Case Study: $1/N$ Cost-Balancing Policy in Auction-Based Routing VOs

The case study that motivated this paper is based on an auctioning VO that allows transportation customers in a supply chain scenario to place their requests for transport on an online auctioning system. Transportation companies can then bid for these requests through a transporters' portal at the backend of the auctioning system. The collection of the customers and the service providers forms one VO called the *Auctioning VO*.

At the same time, each transportation company can form a second VO called a *Routing VO*, which will involve along with the transportation company all the necessary computational resources needed for computing the routing calculation resulting in the bid offer. The highly complex computations could be outsourced to other organisations, which is why the Routing VO is needed. In both VOs the manager is the Transporter Association Portal (TAPortal), through which the administrator creates and populates the two VOs. This scenario is depicted in Figure 3.

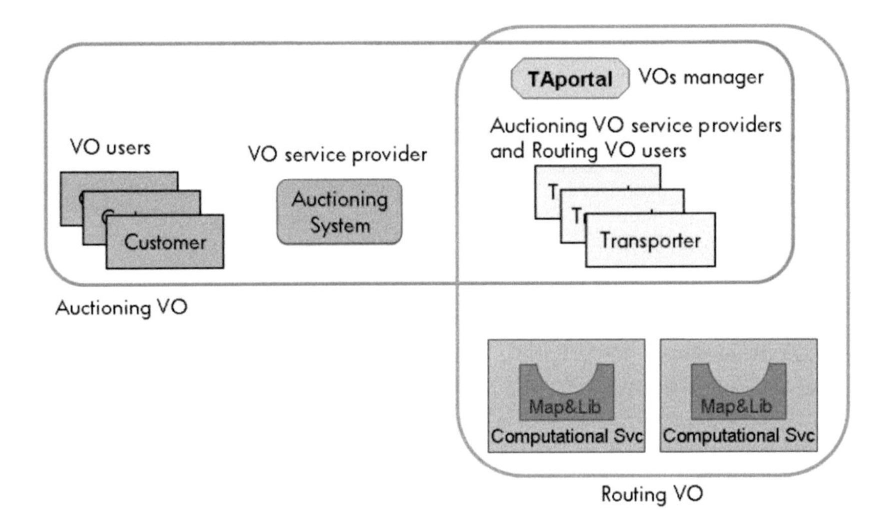

Fig. 3 Auctioning-based Routing VOs

In the above case study, a *cost balancing* policy would be desireable in the Auctioning VO, in the event that a customer of the VO is planning to divide their transportation task among N number of service providers, while determning what the cost associated with each transportation stage (service) would be. The bid calculated by each transporter is then compared to the budget advertised by the customer in their request, and the winning bid is the one with the best cost estimate.

Such a policy is known as a $1/N$ *cost-balancing* policy, and it is one example of VO-wide policies that are typical in Grid systems [23], which deal with the problem of managing VO resources by dividing equally the resource utilisation among the member organisations. Such policies are useful in critical applications [14] and Grid-based operating systems [17], since they facilitate the regulation of resource usage.

Informally, our version of the policy states that the cost of achieving a goal is divided equally among the VO members (organisations) that are collaborating toward achieving that goal. This then implies that the cost of services employed by each organisation toward the goal will be equal to the cost of services employed by any of its sibling organisations. Ideally, this cost must not exceed the budget allocated to

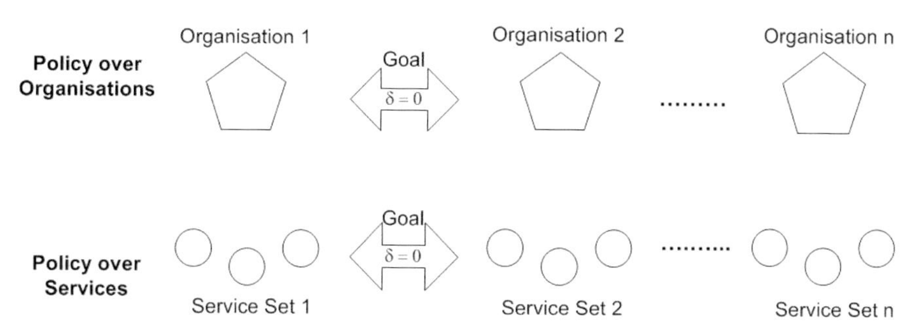

Fig. 4 The $1/N$ cost-balancing policy

the organisation. Figure 4 illustrates this policy across the two layers of abstraction (organisations and services).

The policy is formalised in terms of a cost distance variable, $\delta \in \mathbb{N}$, which measures the difference between any two entities (organisations or sets of services) working on the same goal. When delta returns zero, then the policy becomes a $1/N$ cost-balancing policy, where N is the cardinality of the set of entities sharing the cost. On the other hand, if δ is set to some non-zero value, then this will imply that any two organisations are allowed to have some difference in their cost associated with achieving the main goal of the VO. It is outside the scope of this paper to determine what the value of δ should be, this will be largely dependant on each specific case of the auctioning problem.

The top layer in Figure 4 shows this policy ($\delta = 0$ for some Goal) among the various Organisations $1 \ldots n$, whereas in the lower more refined layer, we see the same policy this time on Service Sets $1 \ldots n$, where each set is the representation (refinement) of its corresponding organisation.

5 An Abstract Model of Goal-Oriented VOs

The first model of a VO is goal-oriented; it captures the idea that a VO is driven by the aim to achieve a set of goals that some VBE makes possible. The model defines a machine, which represents the VO lifecycle as discussed in Section 2 based on this idea of goal-driven VOs. The machine and its context are shown in Figure 5.

The VBE is modelled as a context that introduces a carrier set (type) called Goals. Goals form a non-empty finite set. The context also includes the type Status, which is a flag representing the different phases of the VO lifecycle. The machine has four events corresponding to the four phases of the VO lifecycle as described in Section 2. The VO machine contains variables that represent the *status* (or VO lifecycle phase) of the machine, the *goals* of the VO and the *completed goals* of the VO. The machine is initialised such that the goals variable is assigned some non-empty value from the VBE goals and so that the first event at which the machine commences is the Identification event.

```
MACHINE VO SEES VBE

VARIABLES
status, goals, completedGoals

INVARIANTS
/*Here we define the types of goals and completedGoals*/
status ∈ Status ∧ goals ∈ ℙ1(Goals) ∧ completedGoals ⊆ goals

INITIALISATION
goals :∈ ℙ1(Goals) || completedGoals := ∅ || delta := 0 || status := Id
END

Identification
/*Nothing to identify*/
WHEN status = Id THEN status := Fr END

Formation
/*Nothing to form*/
WHEN status = Fr THEN status := Op END

Operation
/*Pick an uncompleted goal and achieve it*/
WHEN status = Op ∧ (completed_goals ≠ goals) THEN
ANY aGoal WHERE aGoal ∈ (goals\ completedGoals) THEN
completedGoals := completedGoals ∪ {aGoal} END

Dissolution
/*No more uncompleted goals, therefore stop*/
WHEN status = Op ∧ goals = completedGoals THEN status := Stop END

END
```

```
CONTEXT VBE

SETS
Goals, Status

CONSTANTS
Id, Fr ,Op, Stop

AXIOMS
Status = { Id, Fr, Op, Stop }
ℙ1(Goals) ≠ ∅ ∧ finite(Goals)

END
```

Fig. 5 The abstract machine, VO, and its abstract context, VBE

The Identification event only changes the status flag to the next event (Formation). At this level of abstraction, there is no concept of organisations and therefore it is impossible to identify potential VO candidates. In the following event, Formation, again the only update to the machine's state is to change the status flag to indicate to the Operation event, also since there is no concept of organisations at this stage and hence, it is impossible to model VO membership formation. The Operation event is triggered as long as the set of completed goals has not yet reached the set of VO goals. When this is the case, a goal (aGoal) is chosen non-deterministically from the set of incomplete goals and added to the set of completed goals. Note that for simplicity, we do not model operational failure here. Finally, once the set of completed goals reaches the set of VO goals, the Dissolution event is triggered, which in turn sets the status flag to the Stop value indicating the end of the VO lifecycle.

This machine is too abstract to represent our cost-balancing policy, which refers to goals and organisations. Nevertheless, we have included it to show the modelling style we follow in the rest of the paper. The machine also demonstrates in an abstract manner that the aim of a VO lifecycle is to start and finish some specific goal.

6 Goal-Oriented VOs with Organisations

In the first refinement, we introduce the concepts of *organisations* and *goal cost*. The refined machine and its context are shown in Figure 6. The context VBERef1 is the refined VBE which introduces the type Organisations. The context also introduces two new constants, GoalCandidates and GoalCost. The former models the possible groups (sets) of organisations that when collaborating together can achieve a particular goal. The fact that GoalCandidates is a relation and not a function implies that there could be more than one such set of organisations per goal. The latter is a function that reflects the cost of achieving a goal as advertised by a set of organisations. Here we assume that cost is a stable value, which leads to GoalCost being a function rather than a relation.

The VORef1 machine consists again of the four VO lifecycle events; Identification, Formation, Operation and Dissolution. In the Identification event, the set of organisations that are candidates to join the VO are identified using the relation goalCandidates, which restricts the domain of GoalCandidates defined in the VBERef1 context to the set of VO goals. The next event is Formation, in which the VO members defined by the function, goalMembers, and their budget defined by the function, memberBudget, are updated. The goalMembers function is defined as being a functional subset of the more general goalCandidates relation. On the other hand, memberBudget is selected such that for an organisation operating towards achieving a goal, then the member budget assigned to that organisation is equal to the total cost of the goal divided by the cardinality of the set of organisations working towards that goal. In other words, a member receives $1/N$ of the cost of the goal:

$$(\forall\ g,o.\ g \in goals \wedge o \in goalMembers0(g) \wedge card(goalMembers(g)) \neq 0 \wedge finite(goalMembers(g)) \Rightarrow$$
$$memberBudget(o) = GoalCost(g \mapsto goalMembers(g)) \div card(goalMembers0(g)))$$

As this division is carried over integers, we know that each member will receive equal share of the cost and that due to the remainder, the total cost is less than the sum of the individual member budgets. However, this error remains in practice small since the goal cost will be much larger than the number of participants. This can be forced even in cases of small goal costs by adjusting the measurement unit (e.g. the cost in Euros to the cost in Cents).

In the next event, Operation, an uncompleted goal, aGoal, is chosen as well as a set of member organisations such that this set is capable of achieving the goal

MACHINE VORef1 **REFINES** VO **SEES** VBERef1

VARIABLES
status, goals, completedGoals, goalCandidates, goalMembers, delta, memberBudget

INVARIANTS
/***Type of goalCandidates***/
goalCandidates $\in$ goals $\leftrightarrow \mathbb{P}1$(Organisations) $\wedge$

/***Type of goalMembers***/
goalMembers $\in$ goals $\rightarrow \mathbb{P}1$(Organisations) $\wedge$

/***Type of delta***/
delta $\in \mathbb{N}$

/***Type of memberBudget***/
memberBudget $\subseteq$ Organisations $\rightarrow \mathbb{N}1$ $\wedge$

/***The 1/N cost-balancing policy invariant: VO members have equal budgets***/
$\forall g, o1, o2. g \in$ goals $\wedge$ o1$\in$goalMembers(g) $\wedge$ o2$\in$goalMembers$(g) \Rightarrow$
(memberBudget$(o1)$ $-$ memberBudget$(o2)$ $=$ delta) $\vee$
(memberBudget$(o1)$ $-$ memberBudget$(o2)$ $=$ 0-delta)

INITIALISATION
goals :$\in \mathbb{P}1$(Goals) $\|$ completedGoals := $\emptyset$ $\|$ goalCandidates := $\emptyset$ $\|$ goalMembers := $\emptyset$
$\|$ delta := 0 $\|$ memberBudget := $\emptyset$ $\|$ status := Id END

Identification REFINES Identification
/***Identify potential candidates***/
WHEN status = Id THEN goalCandidates := goals $\lhd$ GoalCandidates $\|$ status := Fr END

Formation REFINES Formation
/***Form the VO organisation membership***/
ANY goalMembers0, memberBudget0 WHERE status = Fr $\wedge$

/***The definition of goalMember0***/
goalMembers0 $\in$ goals $\rightarrow \mathbb{P}1$(Organisations) $\wedge$ goalMembers0 $\subseteq$ goalCandidates $\wedge$

/***The definition of memberBudget0***/
memberBudget0 $\in$ Organisations $\rightarrow \mathbb{N}1$ $\wedge$

/***The 1/N cost-balancing policy condition***/
($\forall g, o. g \in goals \wedge o \in$ goalMembers0(g) $\wedge$ card(goalMembers0(g)) $\neq 0 \wedge$ finite(goalMembers0(g)) $\Rightarrow$
memberBudget0(o) = GoalCost$(g \mapsto$ goalMembers0$(g))\div$card(goalMembers0(g)))

THEN goalMembers := goalMembers0 $\|$ memberBudget := memberBudget0 $\|$ status := Op END

Operation REFINES Operation
/***Operate on an uncompleted goal with the right member set***/
ANY aGoal, memberSet WHERE status = Op $\wedge$ completedGoals $\neq$ goals $\wedge$
aGoal $\in$ (goals$\setminus$ completedGoals) $\wedge$ memberSet = goalMembers(aGoal) THEN
completedGoals := completedGoals $\cup$ {aGoal} END

Dissolution REFINES Dissolution
/***No more uncompleted goals therefore stop the VO lifecycle***/
WHEN status = Op $\wedge$ goals = completedGoals THEN status := Stop END

END

CONTEXT VBERef1 **REFINES** VBE

SETS
Organisations

CONSTANTS
GoalCandidates, GoalCost

AXIOMS
GoalCandidates $\in$ Goals $\leftrightarrow \mathbb{P}1$(Organisations) $\wedge$ GoalCost $\in$ Goals $\times \mathbb{P}1$(Organisations) $\rightarrow \mathbb{N}1$

END

Fig. 6 The first refinement of the VO model

(as defined by the goalMembers function). This goal is then added to the set of completed goals of the VO. Once the set of completed goals reaches the set of VO goals, the Dissolution event is triggered, which ends the VO lifecycle by setting the status goal to Stop.

At this level, we can define the following invariant, which expresses the $1/N$ cost-balancing policy using the delta distance measure.

Invariant 1 (All VO members have equal goal budgets). $\forall g, o1, o2. g \in goals \wedge$
$o1 \in goalMembers(g) \wedge o2 \in goalMembers(g) \Rightarrow$
$(memberBudget(o1) - memberBudget(o2) = delta) \vee$
$(memberBudget(o1) - memberBudget(o2) = 0\text{-}delta)$ □

This invariant states that the difference in member budget between any two organisations working on the same goal is only *delta* (or *–delta*) units away, where delta is a variable measuring the cost distance. The machine sets this variable to zero in order to implement the $1/N$ cost-balancing policy. However, other values are also possible, which would reflect incremental cost-sharing policies (similar to salary systems). As we mentioned above, the invariant is enforced thanks to the condition stating that each member will receive $1/N$ of the cost of a goal among N organisations working on that goal.

7 Goals, Organisations and Services

The second refinement, which represents our concrete model, is based on the concept of *services* and their relation to goals and organisations. The concrete machine and its context are shown in Figures 7. Based on this context, the second refinement, VORef2, of the VO machine is defined as in Figure 7.

The context, VBERef2, defines a new type called Services. These are the services advertised in a VBE. In addition to these, the context defines three relational valued constants. These are Requires, which models the set of services that a goal requires, Offers, which models a set services offered by an organisation in a VBE and finally ServiceCost, which models the price of a set of services as advertised by an organisation in a VBE.

The concrete machine resembles the previous refinement except that an extra variable, memberServices, is introduced. This variable represents the service currently offered by the member organisations and used by the VO. The memberServices is given a value in the Formation event as a function from organisations to sets of services such that for any particular organisation, o, working towards a goal, g, then memberServices(o) is the set of services both required by g and offered by o. Our cost balancing policy imposes the same condition as in the previous refinement, which is that the budget received by each member organistion is equal to the total goal cost divided by the cardinality of the set of organisations working on that goal.

Now, we can state the following policy invariant at the level of services.

Invariant 2 (Sets of member services have equal costs). $\forall g, o1, o2.\ g \in goals \wedge$
$o1 \in goalMembers(g) \wedge o2 \in goalMembers(g) \Rightarrow$

MACHINE VORef2 **REFINES** VORef1 **SEES** VBERef2

VARIABLES
status, goals, completedGoals, goalCandidates, goalMembers, delta, memberBudget, memberServices

INVARIANTS
/∗**Define the type of memberServices**∗/
memberServices $\in$ Organisations $\rightarrow \mathbb{P}1$(Services) $\wedge$

/∗**The 1/N cost-balancing policy invariant: sets of member services have equal costs**∗/
$\forall g, o1, o2.$ g $\in$ goals $\wedge$ o1 $\in$ goalMembers(g) $\wedge$ o2 $\in$ goalMembers(g) $\Rightarrow$
(ServiceCost($o1$)(memberServices($o1$)) $-$ ServiceCost($o2$)(memberServices($o2$)) $=$ delta) $\vee$
(ServiceCost($o1$)(memberServices($o1$)) $-$ ServiceCost($o2$)(memberServices($o2$)) $=$ 0-delta)

INITIALISATION
goals :$\in \mathbb{P}1$(Goals) $\|$ completedGoals := $\emptyset$ $\|$ goalCandidates := $\emptyset$ $\|$ goalMembers := $\emptyset$ $\|$
delta := 0 $\|$ memberBudget := $\emptyset$ $\|$ memberServices := $\emptyset$ $\|$ status := Id END

Identification REFINES Identification
/∗**Identify potential candidates**∗/
WHEN status = Id THEN goalCandidates := goals $\lhd$ GoalCandidates $\|$ status := Fr END

Formation REFINES Formation
/∗**Form the VO membership**∗/
ANY goalMembers0, memberBudget0, memberServices0 WHERE status = Fr $\wedge$

/∗**Type of goalMembers0**∗/
goalMembers0 $\in$ goals $\rightarrow \mathbb{P}1$(Organisations) $\wedge$ goalMembers0 $\subseteq$ goalCandidates $\wedge$

/∗**Type of goalBudget0**∗/
memberBudget0 $\in$ Organisations $\rightarrow \mathbb{N}1$ $\wedge$

/∗**Type of memberServices0**∗/
memberService0 $\in$ Organisations $\rightarrow \mathbb{P}1$(Services) $\wedge$

/∗**The definition of memberServices0**∗/
($\forall$g,o g $\in$ goals $\wedge$ o $\in$ goalMembers0(g) $\Rightarrow$ memberServices0(o) $=$ Requires(g) $\cap$ Offers(o)) $\wedge$

/∗**An extra condition on memberServices0:**
The cost of member services is $\leq$ their member budget∗/
$\forall$ g,o. g $\in$ goal $\wedge$ o $\in$ goalMembers(g) $\Rightarrow$ ServiceCost(o)(memberService0(o)) $\leq$ memberBudget0(o) $\wedge$

/∗**The 1/N cost-balancing policy condition**∗/
($\forall g, o. g \in goals \wedge o \in$ goalMembers0(g) $\wedge$ card(goalMembers0(g)) $\neq 0 \wedge$ finite(goalMembers0(g)) $\Rightarrow$
memberBudget0(o) $=$ GoalCost($g \mapsto$ goalMembers0(g))$\div$card(goalMembers0(g)))
THEN
goalMembers := goalMembers0 $\|$ memberBudget := memberBudget0 $\|$ status := Op END

Operation REFINES Operation
/∗**Pick an uncompleted goal and achieve it**∗/
ANY aGoal, memberSet WHERE status = Op $\wedge$ completedGoals $\neq$ goals $\wedge$
aGoal $\in$ (goals$\setminus$ completedGoals) $\wedge$ memberSet = goalMembers(aGoal) THEN
completedGoals := completedGoals $\cup$ {aGoal} END

Dissolution REFINES Dissolution
/∗**When no more uncompleted goals, stop the VO**∗/
WHEN status = Op $\wedge$ goals = completedGoals THEN status := Stop END

END

CONTEXT VBERef2 **REFINES** VBERef1

SETS
Services

CONSTANTS
Requires, Offers, ServiceCost

AXIOMS
Requires $\in$ Goals $\rightarrow \mathbb{P}1$(Services) $\wedge$ Offers $\in$ Goals $\rightarrow \mathbb{P}1$(Services) $\wedge$
ServiceCost $\in$ Organisations $\rightarrow$ ($\mathbb{P}1$(Services) $\rightarrow \mathbb{N}1$)

END

Fig. 7 The second refinement of the VO model

$(ServiceCost(o1)(memberServices(o1)) - ServiceCost(o2)(memberServices(o2))$
$= delta) \vee$
$(ServiceCost(o1)(memberServices(o1)) - ServiceCost(o2)(memberServices(o2))$
$= 0\text{-}delta)$ □

The invariant states that the distance among sets of services belonging to one member is delta from the sets of services employed by another member towards the same goal. This invariant constitutes a more refined version of the invariant mentioned for the previous machine in the sense that equality among member budgets for achieving a goal is now propagated to the level of services resulting in the cost of all services offered by a member towards that goal being equal to the cost of all services offered by any other member working on the same goal.

8 Related Work

There is a fresh interest in the problem of policy refinement, given the complexity of dynamic distributed systems as envisaged in global computing. Bandara et al [8] uses a goal-oriented technique for policy refinement. In their work, a formal representation of a system, based on the Event Calculus [19], is used in conjunction with adductive reasoning techniques to derive the sequence of operations that will allow a given system to achieve a desired policy. An abstract policy is represented as a goal, and goal-oriented techniques are used to refine a policy into more concrete ones. Their approach differs from ours in that they assume the existence of a concrete architecture, which is expressed in UML and then translated to the Event Calculus.

In [22], Chadwick et al. propose a refinement approach for access control policies in Grids. Central to their approach is the existence of a hierarchy representing resources at different layer of abstractions. A policy is represented at the most abstract layer, which is then refined into more concrete policies following the resource hierarchy. The hierarchy and the policies are specified using the ontological language OWL, so semantic-web reasoning is exploited to infer the concrete policies. Our work can be seen as a generalisation of their techniques in which the resource hierarchy and policies are generated simultaneously, exploiting the stepwise refinement approach.

Another line of work related to our is the formal modelling of Grids and VOs. Németh and Sunderam [18] define an operational model of grids and VOs based on the theory of ASMs [21]. They start first by defining a generic model that can be used to describe both distributed and Grid computing. This generic model consists of the universes of applications, processes, users, resources, nodes and tasks. These universes are related to one another through multiple mappings, which define the structure of systems. Our formal models can be seen as abstractions of their models.

The work of Janowski et al. [15] identifies two combinations of real-world enterprises that lead to the achievement of common goals. These are the extended enterprise and the virtual enterprise (which corresponds to the notion of a VO in our terminology). In the former, members of an extended enterprise satisfy one another's needs by matching the output of one member to the input of another. On the other hand, a virtual enterprise allows member organisations to cooperate and coordinate their resources and infrastructures in order to achieve the common goal. Hence, a virtual enterprise is a tighter coalition than an extended enterprise, which operates beyond the business interface of its members.

9 Conclusions

VOs are examples of distributed systems in which participants offer different kind of capabilities and resources in order to achieve common goals. Given the complex nature and rich state of this kind of systems, an incremental approach to build VOs is necessary. Here we have shown how to develop VOs and their policies using the refinement approach. We have also developed a similar model for refining other security-related policies [4], such as the Chinese Wall policy [9].

A key characteristic in our approach is to express system entities and their policies at the same level of abstraction. Then both components (i.e. entities and policies) are refined simultaneouly. The stepwise refinement approach allows one to build a system in an incremental way, adding at each step more concrete/operational detail. The refinement theory [6] guarantees the correctness of the whole approach and the existence of automatic tools [20] facilitates the verification process.

The use of the Rodin toolkit in discharging proofs and animating the models has been helpful in improving our understanding of the problem we are tackling (cost-balancing policy refinement) and its domain of application (VOs) in the sense that some assumptions made about the problem and/or its domain proved not to be valid.

As future work we plan to include in our model failure in achieving a goal. This would trigger the evolution sub-phase of the operational phase of the VO lifecycle, which is used to represent more dynamic behaviour in a VO.

References

1. Abrial, J.R.: The B Book. Cambridge University Press (1996)
2. Abrial, J.R. (ed.): Modeling in Event-B: System and Software Engineering. Cambridge University Press (2010)
3. Abrial, J.-R., Mussat, L.: Introducing dynamic constraints in B. In: Bert, D. (ed.) B 1998. LNCS, vol. 1393, pp. 83–128. Springer, Heidelberg (1998)
4. Arenas, A., Aziz, B., Bicarregui, J., Matthews, B.: Managing conflicts of interest in virtual organisations. Electron. Notes Theor. Comput. Sci. 197(2), 45–56 (2008), `http://dx.doi.org/10.1016/j.entcs.2007.12.016`, doi:10.1016/j.entcs.2007.12.016
5. Arenas, A.E., Djordjevic, I., Dimitrakos, T., Titkov, L., Claessens, J., Geuer-Pollmann, C., Lupu, E.C., Tiptuk, N., Wesner, S., Schubert, L.: Toward web services profiles for trust and security in virtual organisations. In: Collaborative Networks and their Breeding Environments (PRO-VE 2005). Springer (2005)
6. Back, R.J., Wright, J.V.: Refinement Calculus: A Systematic Introduction. Springer (1998)

7. Back, R.J., Sere, K.: Superposition refinement of reactive systems. Formal Aspects of Computing 8(3), 324–346 (1996)
8. Bandara, A.K., Lupu, E.C., Moffett, J., Russo, A.: A goal-based approach to policy refinement. In: Fifth IEEE International Workshop on Policies for Distributed Systems and Networks, POLICY 2004. IEEE (2004)
9. Brewer, D., Nash, M.: The chinese wall policy. In: IEEE Symposium on Research in Security and Privacy. IEEE (1989)
10. Camarihna-Matos, L.M., Afsarmanesh, H.: Elements of a ve infrastructure. Journal of Computers in Industry 51(2), 139–163 (2003)
11. Camarihna-Matos, L.M., Afsarmanesh, H., Ollus, M.: Ecolead: A holistic approach to creation and management of dynamic virtual organizations. In: Collaborative Networks and their Breeding Environments (PRO-VE 2005). Springer (2005)
12. Consortium, R.: Event B Language. Technical Report, Deliverable D7 (2005), `http://rodin.cs.ncl.ac.uk/deliverables/rodinD10.pdf`
13. Foster, I., Kesselman, C., Tuecke, S.: The anatomy of the grid: Enabling scalable virtual organizations. International Journal of Supercomputer Applications 15(3) (2001)
14. Golby, D., Wilson, M., Schubert, L., Geuer-Pollmann, C.: An assured environment for collaborative engineering using web services. In: Proceedings of the 2006 Conference on Leading the Web in Concurrent Engineering: Next Generation Concurrent Engineering, pp. 111–119. IOS Press, Amsterdam (2006), `http://dl.acm.org/citation.cfm?id=1566652.1566674`
15. Janowski, T., Lugo, G.G., Hongjun, Z.: Composing enterprise models: The extended and the virtual enterprise. In: Proceedings of the Third IEEE/IFIP International Conference on Intelligent Systems for Manufacturing: Multi-Agent Systems and Virtual Organizations. IEEE (1998)
16. Moffett, J.D., Sloman, M.S.: Policy hierarchies for distributed system management. IEEE Journal of Selected Areas in Communications, Special Issue on Network Management 11(9) (1993)
17. Morin, C.: Xtreemos: A grid operating sytem making your computer ready for participating in virtual organizations. In: 10th IEEE International Symposium on Object/Component/Service-Oriented Real-Time Distributed Computing (ISORC 2007). IEEE (2007)
18. Németh, Z., Sunderam, V.S.: Characterizing grids: Attributes, definitions, and formalisms. Characterizing Grids: Attributes, Definitions, and Formalisms 1(1), 9–23 (2003)
19. Kowalski, R.A., Sergot, M.J.: A logic-based calculus of events. New Generation Computing 4, 67–95 (1986)
20. RODIN Consortium: Specification of basic tools and platforms. Technical Report, Deliverable D10 (2005), `http://rodin.cs.ncl.ac.uk/deliverables/rodinD10.pdf`
21. Dexter, S., Doyle, P., Gurevich, Y.: Abstract state machines and schoenhage storage modification machines. Journal of Universal Computer Science 3(4), 279–303 (1997)
22. Su, L., Chadwick, D.W., Basden, A., Cunningham, J.A.: Automated decomposition of access control policies. In: Sixth IEEE International Workshop on Policies for Distributed Systems and Networks, POLICY 2005. IEEE (2005)
23. Wasson, G., Marty, H.: Toward explicit policy management for virtual organisations. In: 4th IEEE International Workshop on Policies for Distributed Systems and Networks, POLICY 2003 (2003)

Exploring a Model-Oriented and Executable Syntax for UML Attributes

Omar Badreddin, Andrew Forward, and Timothy C. Lethbridge

Abstract. Implementing UML attributes directly in an object-oriented language may not appear to be complex, since such languages already support member variables. The distinction arises when considering the differences between modelling a class and implementing it. In addition to representing attributes, member variables can also represent association ends and internal data including counters, caching, or sharing of local data. Attributes in models also support additional characteristics such as being unique, immutable, or subject to lazy instantiation. In this paper we present modeling characteristics of attributes from first principles and investigate how attributes are handled in several open-source systems. We look code-generation of attributes by various UML tools. Finally, we present our own Umple language along with its code generation patterns for attributes, using Java as the target language.

Keywords: Attributes, UML, Model Driven Design, Code Generation, Umple, Model-Oriented Programming Language.

1 Introduction

A UML attribute is a simple property of an object. For example, a Student object might have a *studentNumber* and a *name*. Attributes should be contrasted with associations (and association ends), which represent relationships among objects.

Constraints can be applied to attributes; for example, they can be immutable or have a limited range. In translating UML attributes into languages like Java it is common to generate accessor (get and set) methods to manage access.

Omar Badreddin · Andrew Forward · Timothy C. Lethbridge
School of Electrical Engineering and Computer Science,
University of Ottawa, Canada K1N 6N5
e-mail: {obadr024,aforward,tcl}@eecs.uottawa.ca

R. Lee (Ed.): *SERA*, SCI 496, pp. 33–53.
DOI: 10.1007/978-3-319-00948-3_3

In this paper, we study the use of attributes in several systems and discuss how to represent attributes in a model-oriented language called Umple. Umple allows models to be described textually as an extension to Java, PHP, Ruby or C++. We present code-generation patterns for attributes as used by Umple for the Java language.

1.1 A Quick Look at Umple

Umple is a set of extensions to existing object-oriented languages that provides a concrete syntax for UML abstractions like attributes, associations, state machines. To distinguish between Umple and Java, the Umple examples use dashed borders in shading, and Java examples use solid-line borders with no shading.

Figure 1 is a snippet of Umple on the left, with its corresponding UML diagram on the right. Methods have been left out of this example; this illustrates one of the features of Umple, the ability to use it incrementally, first to create high level models, and later on to add more and more implementation detail until the system is complete.

```
class Student {}
class CourseSection {}
class Registration {
  String grade;
  * -- 1 Student;
  * -- 1 CourseSection;
}
```

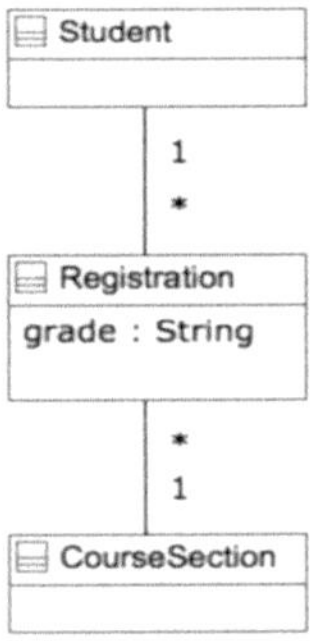

Fig. 1 Umple class diagram for part of the student registration system

Figure 1 shows two associations and an attribute so that the reader can see how they are defined in Umple. The remainder of the paper focuses exclusively on attributes.

One of our motivations is our previous research [1] indicating that most developers remain steadfastly code-centric; hence visual modeling tools are not being adopted as widely as might be desired. Another motivation is that there is much repetitive code in object-oriented programs. We wish to incorporate abstractions to promote understandability and reduce code volume [2].

An Umple program contains algorithmic methods that look the same as their Java counterparts. Constructors, instance variables and code for manipulating attributes, associations, and state machines are generated.

Umple is intended to be simple from the programmer's perspective because there is less code to write and there are fewer degrees of freedom than in Java or UML. Despite the restrictions in Umple, it is designed to have ample power to program most kinds of object-oriented systems. The current version of Umple is written in itself.

Please refer to [3] for full details about Umple. The Umple language can be explored in an online editor [3], which includes many examples.

2 Attributes in Practice: A Study of Seven Systems

To ground our work in the pragmatics of industrial software development, we analyzed how real projects implement attributes. This will help us identify code-generation patterns and areas where Umple could be improved.

Key goals of our empirical analysis of software attributes are to determine how attributes are defined, accessed and used in practice, and also to find attribute patterns that can enhance the vocabulary with which attributes are defined in Umple.

For our research, we considered seven open-source software projects. The criteria by which the projects were selected are described below, followed by a review of the results and the implications for building a model-oriented syntax to describe attributes.

We sampled existing software systems by selecting a random sample of projects from selected repositories. A candidate repository contained at least 1000 full projects in Java or C#. We analyzed 33 repositories, and selected three that met our criteria.

Candidate projects were selected by randomly picking a repository, then randomly selecting a language (Java, or C#), and finally randomly selecting one of the first 1000 projects. The 7 projects analysed include: from GoogleCode: fizzbuzz, ExcelLibrary, ndependencyinjection and Java Bug Reporting Tool; from SourceForge: jEdit and Freemaker; and from Freecode (formerly Freshmeat): Java Financial Library.

We documented all member variables. For each we recorded the project, namespace, object type, variable name, and other characteristics presented in Table 1.

2.1 *Analysis and Results*

We used reverse engineering tools to extract member variables from source code, and manually inspected each attribute. We identified 1831 member variables in 469 classes. Of the member variables identified, 620 were static (class variables) and 1211 were instance variables. Table 2 gives a distribution of the types of static variables.

Table 1 Categorizing member variables

Category	Values	Description
Set in Constructor	No, Yes	Is the member variable set in the object's constructor?
Set Method	None, Simple, Custom	Is the variable public, or does it have a setter method? If so, is there custom behavior apart from setting the variable (such as validating constraints, managing a cache or filtering the input)
Get Method	None, Simple, Custom	Is the variable public, or does the variable have a getter method? If so, does it have any custom behavior like returning defaulted values, using cached values or filtering the output.
Notes	Free Text	Other characteristics such as whether the variable is static, read-only, or derived.

Table 2 Distribution of static (class) variables

Object Type	Frequency	%	Description
Integer	431	69%	All whole number types including primitive integers, unsigned, and signed numbers.
String	53	9%	All string and string builder objects.
Boolean	29	5%	All True/False object types.
Other	107	17%	All other object and data types
Total	620	100%	

Out of the 620 static members analyzed, 90% were read-only constants, 69% were publically visible, and 83% were written in ALL_CAPS, a common style for static variables. From this point onwards, we will focus on the instance variables.

Table 3 gives the distribution of all instance members (i.e. non-static variables) for the five basic attribute types. The 'other' includes custom data types, plus types corresponding to classes like *Address*. Member variables consist of attributes, associations and internal data. To help determine which variables are most likely attributes; we used a two-phased approach. First, we analyzed whether the variables were included in the object's constructor and whether the member variable had get and set accessor methods. This analysis is shown in Table 4.

Only 3% of the variables were initialized during construction, could be overwritten in set method, and accessed in a get method. The most common occurrence was no access to a variable at all (not in constructor, and also no accessor methods). The second most common was a variable whose value was set only after construction.

To filter out potential internal data (local variables), we removed from our potential attributes list all variables that did not have get. We also visually inspected the list and observed that most no-getter variables were cached objects and results (i.e. size or length), or user-interface controls. In total, 637 member variables were removed during this process. We also filtered out five member variables with the word *cache*, or *internal* in their name; as they most likely also refer to internal data.

Table 3 Distribution of instance variable types

Object Type	# of Variables	%	Description
Integer	326	27%	All whole number types including primitive integers, unsigned, and signed numbers.
String	169	14%	All string and string builder objects.
Boolean	121	10%	All True/False object types.
Double	12	1%	All decimal object types like doubles, and floats
Date / Time	9	1%	All date, time, calendar object types.
Other	574	47%	All other data types
Total	1211	100%	

To find variables representing attributes, as opposed to associations, we worked recursively. An attribute is considered to have as its type either: a) a simple data type identified in the first five rows of Table 3, or b) a class that only itself contains instance variables meeting conditions a and b, with the proviso that in this recursive search process, if a cycle is found, then the variable is deemed an association. This approach was partially automated (identifying and removing 12 association member variables) where both ends of the association were defined within the system. The remaining variables were inspected manually, and subjective judgments were made to categorize the variable type as entity or complex. An entity class is one that is comprised of only primitive data types, or associations to other entity classes. A complex class is comprised of primitive data, as well as associations to other complex classes. Table 5 was used to help distinguish class categories.

Table 4 Analyzing all instance variables for presence in the constructor and get/set methods

Constructor	Setter	Getter	Freq	%	Likelihood of being an attribute (High, Medium, Low)
Yes	Yes	Yes	32	3%	High, full variable access
Yes	Yes	No	8	1%	Low, no access to variable
Yes	No	Yes	44	4%	High, potential immutable variable
Yes	No	No	160	13%	Low, more likely an internal configuration
No	Yes	Yes	318	26%	High, postpone setting variable
No	Yes	No	41	3%	Low, no access to variable
No	No	Yes	179	15%	Medium, no access to set the variable
No	No	No	429	35%	Low, no access at all to set/get variable
Total	1211	100%			

Table 5 Entity versus complex object type criteria hinds

Entity Class	Complex Class
Properties, Formats, Types and Data	Nodes, Worksheets
Files, Records, and Directories	Writers, Readers
Colors, Fonts, and Measurements	Engines, Factories and Strategies
Indices, Offsets, Keys and Names	Proxies, Wrappers, and Generic Objects
	Actions, Listeners, and Handlers
	Views, Panes and Containers

This process identified internal, attribute and association variables. Once complete, we were left with 457 potential attributes. The distribution of attribute types is shown in Table 6. As expected, most potential attributes are integers, strings and Booleans.

Table 6 Distribution of attribute types

Object Type	Freq.	%	Description
Integer	200	44%	All whole number types (e.g. integers, signed, and unsigned).
String	102	22%	All string and string builder objects.
Boolean	67	15%	All True/False object types.
Double	6	1%	All decimal object types like doubles, and floats
Date / Time	5	1%	All date, time, calendar object types.
Other	77	17%	All other data types
Total	457	100%	

Table 7 divides attributes into 4 categories. Only 29 attributes (6%) had immutable-like qualities (available in the constructor, with no setter). About 31% of the attributes were managed internally with no setter and not available in the constructor. Finally, only about 11% of the attributes were available in the object's constructor.

Table 7 Constructor and Access Method Patterns (all attributes have a get method)

Constructor	Setter	Frequency	%	Probable Intention
Yes	Yes	23	5%	Fully editable
Yes	No	29	6%	Immutable
No	Yes	262	57%	Lazy / postponed initialization
No	No	143	31%	Derived or calculated attribute
Total	457	100%		

Implementation of Set and Get Methods: As described in Table 1, a set or get method, if present, can be simple or custom. Table 8 illustrates the frequency of the various combinations of attribute set and get methods.

Table 8 Distribution of attribute properties based on type of setters and getters

Setter	Getter	Frequency	%
Simple	Simple	250	55%
Simple	Custom	1	0%
Custom	Simple	9	2%
Custom	Custom	25	5%
None	Simple	46	10%
None	Custom	126	28%
Total	457	100%	

Over 55% of attributes had simple set / get mechanisms, 10% had simple get methods with no set method, and the remaining 35% had at least some custom set or get method.

Attribute Multiplicities: We distinguished between *one* (0..1 or 1) and *many* (*) based on the attribute type. List structures and object types with a plural noun (e.g. *Properties*) were identified as *many*, all other structures were identified as *one*.

Overall 93% of attributes had a multiplicity of *one*, leaving only 7% with a *many* multiplicity. To more finely categorize the multiplicity types would be too subjective, as the multiplicity constraints are programmed in diverse ways.

Characteristics of Custom Access Methods: The following custom set method implementations were observed: having a caching mechanism, lazy loading, updating multiple member variables at once, and deriving the underlying member variable's value based on the provided input.

The following custom *get* method implementations were observed: constant values returned, default values returned if the attribute had not been set yet, lazy loading of attribute data, attribute values derived from other member variable(s), and the attribute value returned from a previously cached value. A summary of the implementation types for set and get methods is in Table 9.

Table 9 Distribution Set and Get Method implementations

Method Implementation	Description	Freq.	%
Derived Set	Input filtered prior to setting variable's value	4	1%
Other Custom Set	Caching / updating multiple members at once	30	7%
Derived Get	Based on a cache, or other member variables	105	23%
Other Custom Get	Custom constraints applied to variable	28	6%
Constant Get	Always returns the same value	19	4%

The frequencies in Table 9 are based on the total number of attributes and not simply those attributes with custom set or get methods. The most interesting observation from this table is that almost a quarter of all attributes were somehow derived from other data of the class.

2.2 *Key Findings*

Key findings based on the results above include

- **Simple set and get methods:** Many attributes follow a simple member variable get and set approach, suggesting that such behavior could be the default, helping to reduce the need for explicit setters and getters.
- **Immutable attributes:** Few attributes are set during construction, implying a separation between building objects and populating their attributes. Despite this, we believe it is still important to allow attributes to be immutable and,

hence, set only in the constructor. Immutability helps ensure the proper implementation of hash codes and equality; for example, to allow consistent storage and retrieval from hash tables. It is also important for asynchronous and distributed processing where tests need to be done to see if one object is the same as another.

- **Attribute multiplicities:** Attribute multiplicities are almost always 'one' (93%). Based on this, Umple only supports the generic 'many' multiplicity and not specific numeric multiplicities as found in associations.
- **Static attributes:** Class level attributes (i.e. static) were mostly written in ALL_CAPS (83%); a convention that could be added directly to a language, removing the need for the 'static' keyword.

By analyzing existing projects we were able to align our model-oriented language Umple with the observed trends in representative software projects. This alignment will be expanded upon in the next section. We were also able to provide code generation that is aligned to industry practice – in order to help make the quality of the generated code similar in style and quality to code that a software developer would write him or herself.

3 Umple Syntax for Attributes

In this section we show how the Umple language (introduced in Section 1) allows the programmer to specify attributes, with common characteristics found in practice as presented in the last section. In UML, attributes represent a special subset of semantics of UML associations, although pragmatically we have found it more useful in Umple to consider them as entirely separate entities.

The main features of Umple's syntax for attributes, and its code generation, result from answering the following three questions.

- Q1: Is the attribute value required upon construction?
- Q2: Can the attribute value change throughout the lifecycle of the object?
- Q3: What traits / constraints limit the value and accessibility of the attribute?

As we discuss in Section 4, most current code generators provide the most liberal answers to the questions above: no, the value is not required upon construction, yes the attribute value can change, and no there are no constraints on or special traits of the attribute. In UML, you can add OCL constraints to answer Q3, but there is no straight-forward way to specify answers to Q1 and Q2.

As observed in the previous section (see Table 7), the answer to Q1 is usually 'no' (89%), and the answer to Q2 is split between 'yes' (62%) and 'no' (38%).

The answer to Q3 is *none* half the time (55%) – in other words most attributes have straightforward set and get behavior. The other half, there are a large number of possible characteristics to consider, since each project has unique constraints under which an attribute much conform. Two of the characteristics observe reasonably frequently are uniqueness and default values; we discuss these in Section 3.3

In the work below, we show that these answers above could be reflected in a model-oriented syntax, and in generated code. We also determined which scenarios do not make semantic or pragmatic sense; to further simplify the attribute syntax. Further discussion of code generation in Umple is in Section 5.

Is the Attribute Specified in the Constructor (Q1)?: First, let us consider attributes that are available in the constructor (Q1.Yes). By default an attribute's value is required at construction, and the syntax to describe this scenario is to declare the attribute with no extra adornment. E.g.

```
String x;
Integer y;
```

For attributes that are not to be set during construction (Q1.No) the Umple syntax is to provide an initial value (which can be null) to the attribute, as shown below.

```
String x = "Open";
Integer y = 1;
String z = null;
String p = nextValue();
```

The initialized value follows the semantics of the target language (e.g. Java or PHP). It can either be a constant as we see for x and y, uninitialized as we see for z or an arbitrary method call (that the developer must define) as in the case of p.

Can the Attribute Change After Construction (Q2)? By default in Umple, an attribute's value can change after construction (Q2.Yes), requiring no additional syntax to describe this scenario. A set method is generated in this case.

Attributes that cannot change after construction (Q2.No) are marked 'immutable'; the value set in the constructor cannot then be changed. No set method is generated.

```
  immutable String x;
```

As we discussed in Section 2, immutability is very useful to provide consistent semantics for equality and hashing, although not many attributes exhibited the immutable property. Part of the issue being the difficulty in specifying immutable attributes in the languages we studied (Java and C#).

There are instances where an attribute should be immutable, but it might be the case where the value is not available at the time of construction. Examples of this include application frameworks where the creation of an object is controlled by the framework and is outside the developer's control. In these cases, an initially empty object is provided to the application, to be immediately populated with the attribute data that cannot then be changed. Therefore, to support this case in Umple, we allow immutable attributes to delay instantiation by using the *lazy* keyword.

```
lazy immutable y;
```

As in Section 3.1 the use of this syntax means that no argument is created in the constructor for the attribute. The generated code will contain a flag to track whether the object has been set yet, only allowing a single set to occur. We elaborate on immutability and the underlying executable implementation in Section 5.2.

What Other Characteristics Does the Attribute Possess (Q3)? The potential characteristics are limitless. In our analysis of existing software we found three somewhat-common patterns that are incorporated into Umple.

Before we consider these patterns, we should recognize that many attributes have no explicit constraints. In general, a property like a name (or jobTitle) has no constraints apart from those enforced by the underlying language (i.e. type checking).

The first characteristic we considered is uniqueness. In databases, guaranteeing uniqueness allows for efficient searching and equality assertions; many domains also have data that is unique by design (e.g. flight numbers in an airline). In some cases, objects are automatically assigned a unique identifier upon creation, whereas in others uniqueness is checked whenever the attribute is set.

In UML, an attribute's uniqueness can be specified with a qualifier, which is really a special type of attribute. Below is an example Airline that has many RegularFlights.

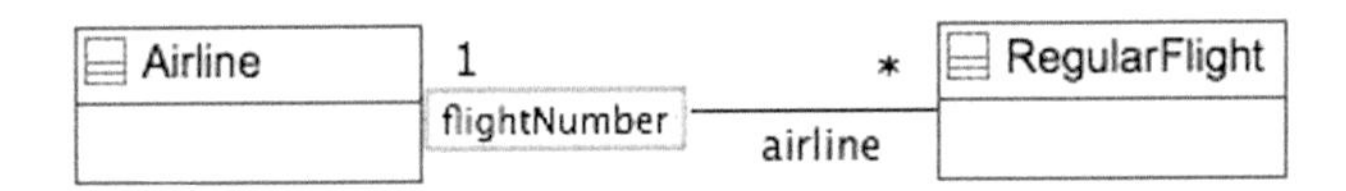

Fig. 2 Unique flightNumber on the airline association

Two RegularFlights of the same Airline should not have the same flightNumber. It is also possible to allow for global uniqueness within a system, for example an ipAddress attribute should perhaps be unique throughout the entire application.

In the cases above, the developer must define unique attributes. The example below provides a mechanism to allow the underlying system to manage the generation of valid and unique identifiers, within or outside the context of an association. The Umple syntax to describe the constraints above is shown below.

```
unique Integer flightNumber on airline;
unique String ipAddress;
```

Uniqueness for integer attributes can also be managed automatically in Umple using the *autounique* keyword.

```
autounique Integer flightNumber on airline;
```

A *defaulted* value ensures an attribute is never *null*. Any time the internal value of the attribute is null the get method returns the default value.

```
defaulted type = "Long";
```

Not all attributes conform to a standard set/get semantics. In addition, many member variables are not attributes, but are support variables used internally [4]. In Umple, the syntax for *internal* attributes is shown below. Internal attributes do not form part of the constructor and do not have accessor methods, allowing developers to manage this data in the way they see fit.

```
internal Integer cachedSize = 0;
```

Finally, let us consider a *many* multiplicity. Using square bracket [] syntax, attributes can also be represented as multiple instances of the attribute type.

```
String[] names;
```

4 Generating Code for Attributes: Existing Tools

After investigating how attributes are used in practice, we studied the code generation patterns of existing tools to see how they implement attributes. The UML modeling tools considered were identified from a Gartner report [5] and an online list of UML tools [6]. We selected four open source projects and one closed source application to analyze. ArgoUML and StarUML are two of the most active open source UML modeling tools and RSA has the largest market share; using popular tools helps to ensure that our study is relevant [5, 7].

Table 10 UML code generation tools

Tool	Version	Source
ArgoUML	0.26.2	argouml.tigris.org
StarUML	5.0.2.1570	staruml.sourceforge.net
BOUML	4.11	bouml.free.fr
Green	3.1.0	green.sourceforge.net
RSA	7.5	ibm.com/software/awdtools/architect/swarchitect

Table 11 lists tools not considered either because they did not provide code generation (at all, or of class diagrams), or did not run in our environment.

For the tools listed in Table 10, we used a Student class with two attributes, an integer representing an id, and a list of names (represented as simple strings).

```
class Student {
  Integer id;
  String[] names
}
```

```
Student
id : Integer
names : String[]
```

Fig. 3 Student class with a simple id attribute and a list attribute

Table 11 Additional UML tools not considered for our case study

Tool	Version	Source
Acceleo	2.5.1	acceleo.org
Jink UML	0.745	code.google.com/p/jink-uml
Hugo	0.51	pst.ifi.lmu.de/projekte/hugo
Umbrello	2.0	uml.sourceforge.net
Umlet	9.1	umlet.com
Fujaba	5.0.1	wwwcs.upb.de/cs/fujaba/
Modelio	1.0.0	modeliosoft.com
Topcased	1.2.0	topcased.org
NetBeans UML Modeling	6.7	netbeans.org
Papyrus	1.11.0	papyrusuml.org

ArgoUML: An open source modeling platform that provides code generation for Java, C++, C#, PHP4 and PHP5. Below is the code generated from Figure 3.

```
import java.util.Vector;
public class Student {
  public Integer id;
  public Vector names; }
```

The generated code provides public access to set and get the attributes; a pattern not often used in practice as it breaks encapsulation by fully exposing the internals of the class. The attributes are made directly available without the inclusion of accessor methods.

StarUML: This is an open source modeling tool. Its generated code is almost identical to that of ArgoUML, except that the import clause was missing, and names is declared erroneously as a String. StarUML's generated code hence does not account for the many multiplicity, resulting in unusable generated code.

Bouml: This is another free tool based on UML 2 that provides source code generation for C++, Java, Idl, PHP and Python. Its code is very similar to the two systems above, with the primary difference being that the attributes are declared as private and genericity is used:

```
class Student {
  private int id;
  private List<String> names;
}
```

The Bouml source code does not provide any mechanism to set or get the attributes; this code must be written by hand after code generation.

Green UML: This is another UML editor that can generate source code from a class diagram.

```
import java.util.List;
public class Student {
  int id;
  List<String> courses;
  public Student() {} }
```

Although Green UML does attempt to provide some additional source code generation to manage the various types of association multiplicities available; the results provide little if any added benefit in representing the model's intentions.

Rational Software Architect (RSA) and RSA Real-Time: IBM's Rational Software Architect is a full-fledged development environment that supports model-driven development including source code generation from UML diagrams. Its code for Figure 3 is as follows:

```
import java.util.Set;
public class Student {
  public Integer id;
  public Integer getId() { return id; }
  public void setId(Integer id) { this.id = id; }
  private Set<String> courses;
  public Set<Student> getCourses() { return courses; }
  public void setCourses(Set<Student> courses)
   { this. courses = courses; }
}
```

RSA's model transformation into Java provided flexibility regarding the template patterns including (a) which Java collection to use, and (b) whether or not to include get/set methods. RSA Real Time generated the same code as RSA for class diagrams.

After reviewing several code generators it became apparent that an opportunity exists to provide better support for aligning the modeling's world of access and multiplicity constraints directly in the generated code, alleviating the modeller/developer from code duplication to ensure consistency between the model and the code. As an Umple model *is* the Umple code, such consistency is easily achieved with adequate code-generation capabilities.

5 Generating Code for Attributes Using Umple

As seen earlier, existing code generating tools do not consider the many complicated facets of implementing attributes in a target language like Java. In this section, we demonstrate code generation patterns based on Umple attributes and generated into Java code.

The following example shows how one would declare attributes in the first steps of modeling a system using Umple. For conciseness, we have omitted the code comments and some additional methods not related to the attributes in the generated Java.

5.1 Basic Attributes

At the core of an Umple attribute is a name. The implications on code generation include a parameter in the constructor, a default type of String and a simple set and get method to manage access to the attribute. The attribute code in Umple is shown below, and code generated in Java follows.

```
class Student { name; }
```

```
public class Student {
  private String name;
  public Student(String aName) { name = aName; }
  public boolean setName(String aName) {
    name = aName;
    return true; }
  public String getName() { return name; } }
```

The syntax is similar to RSA generated code, and to the *simple* cases observed in the open source projects. As seen Section 2, few attributes are set in the constructor. In Umple, this can be achieved by specifying an initial value as shown below. The generated code in Java would only differ in the constructor, and follows.

```
  class Student { name = "Unknown"; }
```

```
  public Student() { name = "Unknown"; }
```

Please note the initial value can be null, or some user defined function written in the underlying target language (i.e. Java).

5.2 *Immutable Attributes*

If a Student variable was declared immutable, as presented in Section 3, the resulting Java code would be the same as the basic attribute implementation, except that there would be no setName method.

By default, immutable attributes must be specified on the constructor, and no setter method is provided. But, Umple also supports lazy instantiation of immutable objects as shown below and discussed in Section 3.

```
class Student { lazy immutable name;}
```

By declaring a lazy immutable attribute we follow the same convention whereby the *name* attribute will not appear in the constructor; but we also provide a set method that can only be called once.

```
public class Student {
  private String name;
  private boolean canNameBeSet;
  public Student() { canNameBeSet = true; }
  public boolean setName(String aName) {
    if (!canNameBeSet) { return false; }
    canNameBeSet = false;
    name = aName;
    return true; }
  public String getName() { return name; } }
```

The implementation above includes an additional check *canNameBeSet* to ensure that the variable is only set once, but should be used with caution in threaded access to avoid issues from parallel processing conflicts.

5.3 *Defaulted Attributes*

As discussed in Section 3, a defaulted attribute provides an object with a default configuration that can be overwritten. The code generated for Java follows.

```
class Student { defaulted name = "Unknown"; }
```

```
public class Student {
  private String name;
  public Student(String aName) { name = aName; }
  public boolean setName(String aName) {
    name = aName;
    return true; }
  public String getName() {
    if (name == null) { return "Unknown"; }
    return name; } }
```

Below are the subtle differences between initialized and defaulted attributes. First, a defaulted attribute is specified in the constructor, an initialized attribute is not. Second, a defaulted value is guaranteed tp be non-*null*, an initialized attribute only guarantees an attribute in set to particular value in the constructor (and can change afterwards).

5.4 Unique Attributes

The Umple language currently only supports code generation for autounique attributes as shown below. The code generated for Java follows.

```
class Student { autounique id;}
```

```
public class Student {
  private static int nextId = 1;
  private int id;
  public Student() { id = nextId++; }
  public int getId() { return id; } }
```

The implementation is constrained to non-distributed systems; but allows for a simple mechanism to uniquely identify an object.

5.5 Constant Class Attributes

A constant class level attribute is identified using the convention of ALL_CAPS. The UML modeling standard is to underline; a convention that is difficult to achieve in a development environment as most developer code is written in plain text. The code generated for Java follows.

```
class Student { Integer MAX_PER_GROUP = 10; }
```

```
public class Student { public static final int MAX_PER_GROUP = 10; }
```

5.6 Injecting Constraints Using Before/After

To support vast array of other types of custom implementations of set and get methods, as well as provide a generic mechanism for managing pre and post-conditions of an attribute, we introduce the before and after keywords available in the Umple language. Let us begin with a simple example.

```
class Operation {
  name;
  before getName {
    if (name == null) { /* long calculation and store value */ }
  }
  after getName {
    if (name == null) { throw new RuntimeException("Error"); }
  }
}
```

In the code above, we are caching the derivation of the complex process to determine the value of *name*. The code is also verifying that the *getName* method always returns a value (never null). The code provided in the *before* block will be run prior to desired operation (i.e. getName) and the code block provided in the *after* block runs after (or just before returning) from the desired operation. The code generated for Java for the *getName* method is shown below.

```
public String getName()
{
  if (name == null) { /* long calculation and store value */ }
  String aName = name;
  if (name == null) { throw new RuntimeException("Error"); }
  return aName;
}
```

The before and after mechanisms can be used with any Attribute A summary of the operations is described below.

Before and after can be applied to associations, and constructors as well. This mechanism can, for example, provide additional constraints to a class, or to initialize several internal variables.

Table 12 Applying before and after operations to Attributes

Operation	**Applies To (UB = Upper Bound)**
setX	Attributes (UB <= 1)
getX	Attributes
addX	List Attributes (UB > 1)
removeX	List Attributes (UB > 1)
getXs	List Attributes (UB > 1)
numberOfXs	List Attributes (UB > 1)
indexOfX	List Attributes (UB > 1)

An operation can have several before and after invocations. This chaining effect allows each statement to focus on a particular aspect of the system such as a precondition check of inputs, or a post-condition verification of the state of the system.

It should be noted that the syntax of Umple's before and after mechanism is purposely generic with a relatively fine-grained level of control. The intent of this mechanism is to act as a building block to include additional constraint-like syntaxes for common conditions such as non-nullable, boundary constraints and access restrictions. By including before and after code injections at the model level, additional code injection facilities are possible at the model level, without having to modify the underlying code generators. For example, the immutable property discussed is implemented internally (i.e. Umple is built using Umple) using before conditions on the set methods.

6 Related Work

There is literature on code generation from UML [8-11]. In [8], an abstract class is generated for the set and get methods and an instantiable class implements operations. This adds a layer of complexity. Umple provides a more direct approach, and the generated code more closely resembles that which would be written by hand. Whereas the approach above seems guided more by the limitations of using UML.

Jifeng, Liu, and Qin [12] present an object-oriented language that supports a number of features like subtypes, visibility, inheritance, and dynamic binding. Their textual object-oriented language is an extension of standard predicate logic [13]. The approach to Umple was not to create a *new language*, but rather to enhance existing ones with a more model-oriented syntax and behaviour.

Reverse engineering tools tend to generate a UML attribute when they encounter a member variable, a practice widely adopted by software modeling tools. Sutton and Maletic [14] advocate that attributes reflect a facet of the class interface that can be read or written rather than representing the implementation details of a member variable. They present their findings on the number of class entities, attributes and relationships that were recovered using several reverse engineering tools, revealing the inconsistencies in the reverse engineering approaches. They present their prototype tool, *pilfer*, that creates UML models that reflect the abstract design rather than recreating the structure of the program.

Gueheneuc [15] has analyzed existing technology and tools in reverse engineering of Java programs, and highlights their inability to abstract relationships that must be inferred from both the static and dynamic models of the Java programs. They developed PADL (Pattern and Abstract-level Description Language) to describe programs in class diagrams. However, their proposed approach requires the availability and analysis of both static and dynamic models to build the class diagrams. In another study [16], two commercial reverse engineering tools (Together and Rose) are compared to research prototypes (Fujaba and Idea); they note that different tools resulted in significantly different elements recovered from the source code.

Lange and Chaudron [17] conducted an empirical analysis of three software systems and identified violations to a number of well-formedness rules. In one of the systems, 67% of attributes were declared as public without using setters and getters.

Experimentation with Umple [18] users reveals evidence that software developer comprehension of the code is enhanced when compared to traditional object oriented code [19-21]. Umple was deployed and evaluated in open source projects [22]

In most of the cases above, automated analysis done by reverse engineering tools resulted in vastly different perceptions about the systems being studied. Our approach, although subjective at times and error prone due to several manual steps throughout the process, attempts to provide a structured approach to reviewing, categorizing and understanding how attributes are used in practice.

7 Threats to Validity

Our empirical investigation of existing implementation of attributes has two main threats of validity; Firstly, to what extent are the seven selected projects representative of typical uses of attributes; and secondly, to what extent are attribute patterns affected by the capabilities provided by the existing programming languages.

To mitigate the risks of non-representation we were diligent to select projects in a random fashion from a large group of projects written in different languages (yet languages that we were experienced in). The process to select projects was well documented and can easily be repeated for future studies into this subject.

The second threat is to what extent the capabilities of the underlying programming language affects the types of patterns that can be observed. This threat is somewhat of an extension to our first threat, and is more difficult to mitigate, as we cannot understand what we do not know. One way to better deal with this would be to repeat the study using different programming languages with different attribute semantics.

8 Conclusion

This paper analyzed the syntax, semantics and pragmatics of attributes. We studied how attributes are used in practice; and discovered the difficulty in extracting modeling abstractions from analyzing source code. Our approach used manual inspection, which, although subject to human error is probably comparable to analysis by automated tools since there are so many special cases to be considered.

We demonstrated how attributes are represented in the Umple model-oriented language and showed the code-generation patterns used to translate Umple into Java. When compared to the code generated for attributes by existing tools, we believe our patterns have a lot to offer.

References

1. Forward, A., Lethbridge, T.C.: Problems and opportunities for model-centric versus code-centric software development: A survey of software professionals. In: International Workshop on Models in Software Engineering, MiSE, pp. 27–32 (2008)
2. Forward, A., Lethbridge, T.C., Brestovansky, D.: Improving program comprehension by enhancing program constructs: An analysis of the umple language. In: International Conference on Program Comprehension (ICPC), pp. 311–312 (2009)
3. Umple language online, `http://www.try.umple.org` (accessed 2013)
4. Sutton, A., Maletic, J.I.: Recovering UML class models from C++: A detailed explanation. Inf. and SW Tech. 49, 212–229 (2007)
5. Norton, D.: Open-Source Modeling Tools Maturing, but Need Time to Reach Full Potential, Gartner, Inc., Tech. Rep. G00146580 (April 20, 2007)
6. Wikipedia Listing of UML modeling tools, `http://en.wikipedia.org/wiki/List_of_UML_tools` (accessed 2013)
7. Blechar, M.J.: Magic Quadrant for OOA&D Tools, 2H06 to 1H07, Gartner Inc., Tech. Rep. G00140111 (May 30, 2006)
8. Harrison, W., Barton, C., Raghavachari, M.: Mapping UML designs to Java. ACM SIGPLAN Notices 35, 178–187 (2000)
9. Long, Q., Liu, Z., Li, X., Jifeng, H.: Consistent code generation from uml models. In: Australian Software Engineering Conference, pp. 23–30 (2005)
10. Brisolara, L.B., Oliveira, M.F.S., Redin, R., Lamb, L.C., Carro, L., Wagner, F.: Using UML as front-end for heterogeneous software code generation strategies. In: Design, Automation and Test in Europe, pp. 504–509 (2008)
11. Xi, C., JianHua, L., ZuCheng, Z., YaoHui, S.: Modeling SystemC design in UML and automatic code generation. In: Conference on Asia South Pacific Design Automation, pp. 932–935 (2005)
12. Jifeng, H., Liu, Z., Li, X., Qin, S.: A relational model for object-oriented designs. In: Chin, W.-N. (ed.) APLAS 2004. LNCS, vol. 3302, pp. 415–436. Springer, Heidelberg (2004)
13. Unifying Theories of Programming. Prentice Hall (1998)
14. Sutton, A., Maletic, J.I.: Recovering UML class models from C++: A detailed explanation. Inf. and SW Tech. 49, 212–229 (2007)
15. Gueheneuc, Y.: A reverse engineering tool for precise class diagrams. In: CASCON, pp. 28–41. ACM and IBM (2004)
16. Kollman, R., Selonen, P., Stroulia, E., Systa, T., Zundorf, A.: A study on the current state of the art in tool-supported UML-based static reverse engineering. In: Ninth Working Conference on Reverse Engineering (WCRE 2002), pp. 22–30 (2002)
17. Lange, C.F.J., Chaudron, M.R.V.: An empirical assessment of completeness in UML designs. In: EASE 2004, pp. 111–121 (2004)
18. Badreddin, O.: Umple: a model-oriented programming language. In: 2010 ACM/IEEE 32nd International Conference on Software Engineering, vol. 2. IEEE (2010)
19. Badreddin, O.: Empirical Evaluation of Research Prototypes at Variable Stages of Maturity. In: ICSE Workshop on User Evaluation for Software Engineering Researchers, USER (to appear, 2013)

20. Badreddin, O., Lethbridge, T.C.: Combining experiments and grounded theory to evaluate a research prototype: Lessons from the umple model-oriented programming technology. In: User Evaluation for Software Engineering Researchers (USER). IEEE (2012)
21. Badreddin, O., Forward, A., Lethbridge, T.C.: Model oriented programming: an empirical study of comprehension. In: CASCON. ASM and IBM (2012)
22. Badreddin, O., Lethbridge, T.C., Elassar, M.: Modeling Practices in Open Source Software. In: 9th International Conference on Open Source Systems, OSS 2013 (to appear, 2013)

A Case Study in Defect Measurement and Root Cause Analysis in a Turkish Software Organization

Cagla Atagoren and Oumout Chouseinoglou

Abstract. In software projects, final products aim to meet customer needs and concurrently to have the least number of defects. Defect identification and removal processes offer valuable insights regarding all stages of software development. Therefore, defects are recorded during the software development process with the intentions of not only fixing them before the product is delivered to the customer, but also accumulating data that can be researched upon. That data can later be used for software process improvement. One of the techniques for analyzing defects is the root cause analysis (RCA). A case study is conducted in one of the leading, medium sized software companies of Turkey by utilizing the RCA method. The collected defect data has been analyzed with Pareto charts and the root causes for outstanding defect categories have been identified with the use of fishbone diagrams and expert grading, demonstrating that these techniques can be effectively used in RCA. The main root causes of the investigated defect items have been identified as lack of knowledge and extenuation of the undertaken task, and corrective actions have been proposed to upper management. The case study is formulated in a way to provide a basis for software development organizations that aim to conduct defect analysis and obtain meaningful results. All stages of the research and the case study are explained in detail and the efforts spent are given.

Keywords: Defect measurement analysis in software projects, cause and effect charts, root cause analysis, fishbone diagrams.

Cagla Atagoren
Information Technology and System Management, Başkent University,
06810, Ankara, Turkey
e-mail: caglaatagoren@gmail.com

Oumout Chouseinoglou
Statistics and Computer Science Department, Başkent University,
06810, Ankara, Turkey
e-mail: umuth@baskent.edu.tr

R. Lee (Ed.): *SERA*, SCI 496, pp. 55–72.
DOI: 10.1007/978-3-319-00948-3_4

1 Introduction

One of the major goals of software engineering is to develop high quality products with least defects. Software quality not only focuses on the final product but also on the artifacts of the software development processes. In that respect the concepts of software process management and software process improvement (SPI) continuously gain importance [1]. Software industry is a developing and emerging industry in Turkey with a very high growing potential [2], thus it is critical for Turkish software organizations to follow the best practices with respect to software development processes. The continuous increase in demand for software makes the use of a well-defined and structured software process management approach a necessity for successful completion, especially in complex or large defense projects. On the other hand, developing and operationalizing quality software is a subtle process [3]. Within the quality viewpoint, one of the aims in the software industry is having the least defects on the product being delivered to the customer; where software defect (or fault or bug) is defined as any flaw or imperfection in a software work product or software process [4]. Similarly, every point in the software that does not meet user requirements can be considered as a defect [5] whereas for a software engineer anything that needs to be changed is a defect [6].

As a result of the SPI approach, all defects identified from the first day of the development till the delivery of the product to the customer and the finalization of the project are being recorded in order to be analyzed and used for future SPI. It is important that the recorded defect data is reliable and of high quality, otherwise misleading results and interpretations may emerge from the analyses conducted on that data. Furthermore, a poorly conducted analysis may yield also to misrepresentative results. Therefore, both the defect data to be used and also the techniques and methodologies to be employed in defect analysis are of crucial importance for SPI [7][8]. Moreover, the Causal Analysis and Resolution activity in CMMI level 5 is expressed by two Specific Goals, namely Determine Cause of Defects and Address Cause of Defects. The specific practices of the first goal are selecting the defect data for analysis and analyzing causes by requiring the following: "Analyze selected defects and other problems to determine their root causes" [1]. The defects recorded in software projects can be categorized based on different approaches, and the defect categories may vary based on the software being developed, the research conducted or the characteristics of the software development organization [9]. One such categorization is with respect to the events that have caused the defect. Based on such a categorization, the source of the defect can be identified and the deficiencies of the processes can be removed, thus preventing the origination of similar defects in future. As a result of this, SPI is realized [10].

Root-cause analysis (RCA) and fishbone (cause and effect) diagrams are used to identify the possible reasons (root causes) of specific problems or situations, focusing on the belief that defects are best solved by attempting to correct or eliminate root causes, instead of merely addressing the immediately obvious symptoms [11]. RCAs have been identified as one of the tools used for defect removal within the people review profiles in software engineering practices [12]. As it is possible to reach the causes of a problem by starting from its results and using statistical methods, it is important that the cross relationship between the causes and the effects is identified and displayed with graphical tools, such as fishbone diagrams, fault tree diagrams, causal maps, matrix diagrams, scatter charts, logic trees, or causal factor charts [9] but this is best and most easily accomplished with fishbone diagrams [1]. The use of such diagrams allows the structuring of the problem solving process, the delivery and exhibition of all the information regarding the problem, the undertaking of a systematic approach while moving from the already known towards the unknown and the utilization of all experts that do have some prior experience with the problem at hand. As stated in [7], identifying root causes is part of SPI activities regarding defect prevention. The common steps of RCA methods and related work practices used in software developing organizations are detailed in [9].

The aim of this research is to display and interpret the results of an RCA case study that has been conducted in one of the major software development organizations of Turkey. Within the scope of the case study, defect data collected from 13 software projects developed from 2006 to 2012 have been investigated. The results of this research have been used by the assessed software development organization for the formation of defect prevention plans and SPI activities. The rest of the paper is organized as follows: in Section 2 a brief literature survey regarding defects, defect classifications, RCA in software projects and similar case studies is provided. Section 3 presents the details and gives the results of the case study. The last section concludes the paper with a discussion of the findings and future research.

2 Related Work

Defect cause analyses and RCA has been employed in the area of software engineering since the 80s, a survey of existing RCA methods and the theoretical background of RCA are provided in detail in [9] and some of the best examples are given in [13]. Among these, the work of Yu [14] follows a similar approach to the RCA conducted in this study, as the team in question uses a fishbone analysis to identify the root causes, which is also the case in [1], where fishbone diagrams are used for the causal analysis of the defects and to provide a basis for SPI. The efficient use of this tool in the software engineering domain has provided the grounds of employing it in our research.

Leszak et al. [15], present several defect analyses conducted retrospectively on a network element software product. The research is based on different data sets from the same project context, where the input data used is defect numbers and defect classification data from a sample from all defects, which constitutes the basis for RCA. To conduct RCA, a cross-functional team was constituted with members from software and hardware domains, the independent integration and certification department, quality support group and other departments. The aims of the conducted RCA are summarized as analyzing sample defects to find systematic root causes of defects; analyzing major customer-reported modification requests during the maintenance release and proposing improvement actions as inputs for other development projects, in order to reduce number of critical defects and rework cost. This retrospective approach has provided an important insight as the defect data in our study is similarly historical and has been collected from different completed projects. Similarly cross-functional teams to conduct the RCAs were formed and the RCA results obtained were used in similar fashion.

Lehtinen et al. [9] present a "lightweight" RCA method, namely ARCA, specifically focusing on medium-sized software companies. ARCA consists of four steps and does not require defect reports as it focuses on group meetings to detect the target problem. Lehtinen and Mntyl [16] apply ARCA in order to target defects in four medium-sized software companies by creating a two-dimensional classification, where the first dimension is based on common software engineering process areas and the second dimension describes the type of causes detailing them under four major categories, namely people, tasks, methods and environment. The target problem causes have been detected through anonymous e-mail inquiries, followed by cause analysis workshops. The authors have spotted 14 types of causes in 6 process areas and their results indicate that development work and software testing are the most common process areas, whereas lack of instructions and experiences, insufficient work practices, low quality task output, task difficulty, and challenging existing product are the most common types of the causes. Pinpointing both the process areas that require improvements and the improvements required, the authors argue that the proposed classification is useful for understanding problem causes. Within the defect classification approaches, Orthogonal Defect Classification (ODC) [17] is an approach for categorizing defects into classes that collectively point to the part of the process which needs attention, with the requirement that defects can be classified semantically and that the set of all values of defect attributes must form a spanning set over the process sub-space. ODC's extension is to assist defect prevention by providing data that can be used in RCAs. Buglione and Abran [1], working within the perspective of CMMI and by introducing the concepts of cause analysis, RCA and ODC at lower levels of CMMI, propose a quantitative approach to RCA and fishbone by overcoming some of the limitations noted in ODC and arguing that the use of RCA would assist the organization in its measurement ability and at the rating of other processes at lower maturity levels. Kumaresh and Baskaran [10] combine methodologies such as ODC, iteration defect reduction and capturing defects at earlier stages to generate a defect prevention cycle that they argue it can be used for continuous improvement of quality and defect prevention. The cycle consists of the stages of defect identification, classification, analysis, RCA, prevention, and SPI. The study uses 637 defects gathered from 5 different projects. Instead of analyzing all defects, the authors try to focus on the majority of the defects by using the Pareto principle and RCA. RCA is realized with the use of fishbone diagrams. The Pareto approach has been used in our study too, to focus in the majority of defects, to lessen the effort spent and to efficiently use the teams conducting RCA. A research utilizing ODC in a RCA is [18], where 4,372 software defects gathered from a software project developed by the Ministry of Health, Turkey have been categorized with the use of ODC and then have been investigated with RCAs.

Apart from the aforementioned studies, several more works have been investigated in the area of RCAs on computer bugs and defects. Yin et al. [19] give a methodology for classifying network software bugs based on a combination of static analysis and manual classification. The bugs have been classified in five dimensions, namely root cause, trigger condition, effect, code location and operational issues. Although not directly related to software defects, Schroeder and Gibson [20] utilize a root cause approach to study failures in high-performance computing systems: root causes are grouped under main categories which are further detailed to root causes. Similarly, such categorizations and groupings of defects have been used in our study.

3 Case Study

This case study has been conducted with the use of defect data gathered from 13 software projects developed from 2006 to 2012 in a middle-sized software organization in Turkey, having CMMi capability level 2 but showing indicators of being on level 3 in some areas. The organizational and software development maturity of the assessed organization are above average among software development organizations in Turkey. Due to the request of the software organization, the name of the organization is not revealed in this study, and thus it will be referred to as Organization X. Moreover, the software projects of Organization X are being developed within a data confidentiality perspective as they are proprietary projects and require high levels of security clearance. In accordance with these confidentiality considerations, the data gathered and used in this research has been altered with the use of a black-box algorithm provided by Organization X. Therefore, the data given in this case study does not reflect the actual data; however, all data has been altered with the same black-box algorithm for consistency considerations.

The software projects developed by Organization X are mainly grouped into two major categories based on project characteristics. With respect to the aforementioned confidentiality considerations these project groups will be briefly summarized as **Type-1 projects**, which have stricter, more controlled development procedures and require to follow several software development standards and **Type-2 projects**, where the software development process is more relaxed with no predefined standards to be followed. All projects investigated in this research by Organization X have been following the waterfall development lifecycle.

The empirical case study in this paper follows the field study methodology [21][22]. The defect data was collected from the company defect database and was analyzed in two stages. Firstly, the inaccurate, incomplete and inconsistent defect items were either removed or corrected before analysis by using interviews. Then the defect data were analyzed with respect to their characteristics and were grouped accordingly. The RCA meetings were conducted with the use of query forms. All the effort spent in the analysis was measured to provide a feedback for any other similar undertaking. The methodology is further detailed in the following sections.

3.1 Application of the Case Study

The data used has been retrieved from the defect database of Organization X in May 30, 2012. The defect data consists of 36,424 defect items and belongs to 13 projects conducted from 2006 to 2012, of which 8 are Type-1 and 5 are Type-2 projects. All projects have followed the waterfall lifecycle and the defect items have been recorded by project members at any stage of the software development lifecycle of the specific project, with the use of a predefined defect record interface. The fields of a defect item being recorded and the explanation of each field is given in Table 1. The development phase that the defect was recorded and the phase that it originated are given in Table 2, and the defect type and the phase that it originated are given in Table 3. The tables 2 and 3 are generated with data from all 13 projects.

As can be seen from Table 2, 7,499 defects have originated from requirements phase and 1,931 of them were found in next phases. As finding and fixing a defect in future phases is much more expensive than finding and fixing it during the requirements and early design phases [23], the aforementioned high number is of critical importance for Organization X. Table 3 shows that 27,920 of 36,424 total defects were corrected before the delivery of the product to the customer, a fact that shows that peer-review effectiveness is within acceptable levels with respect to internal goals of Organization X.

Before applying RCA, the organization had considered the application of the ODC approach for the analysis of the defect. However, as important information such as "defect triggers" were missing from the defect data recorded and as such information is not possible to be post-inserted, the organization shifted the defect analysis approach to only RCA, which suited best the defect data in hand. Prior to the RCAs, every defect item has been analyzed with respect to its accuracy, integrity and consistency by a measurement team consisting of 7 members. Any problematic defect item or missing fields have been resolved by conducting interviews either with the personnel who has recorded the defect or with project members of the specific software project. A total of 315 problematic defect items have been corrected. This analysis has shown that the most common defect recording problem that has been assessed is the confusion between the fields of "defect origin" and "when found". Moreover, due to the fact that defect categories have not been used at the early stages of the defect recording process, a number of defect items have not been categorized while being recorded. However, as the rest of the data in these uncategorized defect items has been considered to yield meaningful and important information, they have not been excluded from the overall analysis.

When the defect data in hand is analyzed with respect to the levels of severity, it is seen that 22.2% are uncategorized, 0.6% are critical, 20% are major and 56.6% are minor defects. Moreover, the major and critical defects (in total 20.6% of all defects) are labeled as Software Change Requests (SCR). As SCRs are defects that have been identified post to customer delivery, the effect of these defects to the product are more severe. When the analysis results were investigated, the first prevention plan that was decided to be undertaken by the organization has been identified as the minimization of major and critical defects within the SCR defect type.

Table 1 Defect Fields and Their Explanations

Field	Explanation
Project Name	The abbreviation of project name
Defect Type:	Can have three values: • Software Change Request (SCR): change request and/or defect notice that has been communicated after the submission of the deliverable to the customer. • Peer Review (PR): change request and/or defect notice that has been identified prior to the submission of the deliverable to the customer. • Informal Action Item (INFAI): change request and/or defect notice identified during product development.
Defect Number	An unique identifier used for identifying defects
When Found	The development lifecycle phase that was being conducted when the defect was discovered: Requirements (REQs), Design, Code, Test, Field
Defect Origin	The development lifecycle phase where the defect originated: REQs, Design, Code, Test, Field. The defect origin is determined by the person(s) who resolves the defect after the closure of the defect. Defect reporter is not responsible for entering this data.
Occurrence Number of Defect	Number of times this defect is found
Defect Category	Defect categories are based on the characteristic of the defect and mainly developed based on the ODC. The categories due to the aforementioned confidentiality considerations are concealed and in this research they are given as follows: A,B,C,D,E,F,G,H,I,J,K. However, after the conclusion of this research, Organization X has accepted to reveal the names of the defect categories that were analyzed in RCA. Namely these are: • A: Behavior/ Missing/ Traceability • D: Conflict/ Consistency/ Verifiability • E: Documentation/ Understanding The K category consists of uncategorized defects thus they have been excluded from the RCA undertaken in this research.
Defect Severity	The effect of the defect on both the project and product with respect to budget and calendar considerations that were analyzed in RCA. Namely these are: • Critical: Defects that halt the functioning software • Major: Defects that cause software to produce incorrect or erroneous outputs • Minor: Defects that effect the functioning software but can easily be fixed.

Table 2 Origins and Detection Stages of Defect Data

		Defect Origin					
		REQs	***Design***	***Code***	***Test***	***Field***	***Total***
When Found	***REQs***	5,568					5,568
	Design	134	1,159				1,293
	Code	646	319	7,127			8092
	Test	1,151	570	7,592	11995		21,308
	Field	0	2	149	8	4	163
	Total	7,499	2,050	14,868	12,003	4	36,424

Table 3 Origins and Types of Defect Data

		Defect Type			
		INFAI	***PR***	***SCR***	***Total***
Defect Origin	***REQs***	598	4,747	2.154	7,499
	Design	360	939	751	2,050
	Code	5,876	6.106	2,886	14,868
	Test	176	9.114	2,713	12,003
	Field	0	4	0	4
	Total	7,010	20,910	8,504	36,424

As the defects display variations with respect to project types and as the number of defects for Type-1 projects is higher than the defects in Type-2 projects, the defect categorization analyses have been conducted and assessed separately for the two project types. The defect data for Type-1 and Type-2 projects is given respectively in figures 1 and 2, where it can be seen that the most common defects in Type-1 projects are in categories D, A and E; whereas in Type-2 projects they are in

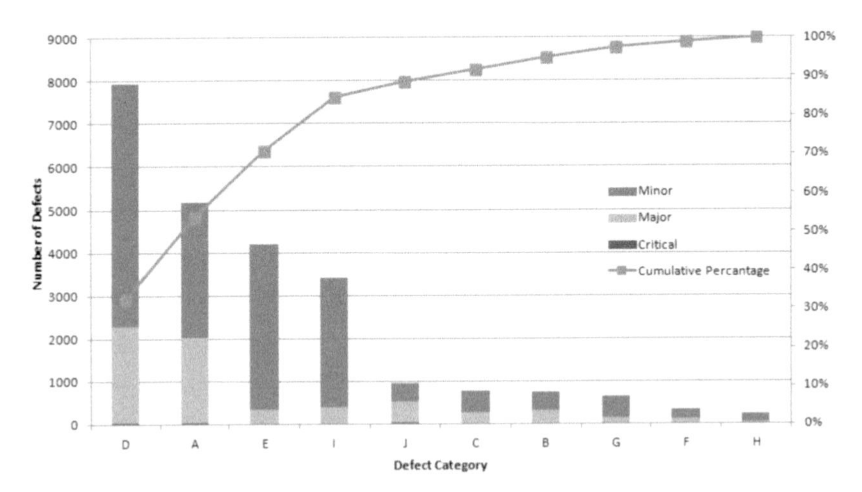

Fig. 1 Defect Categories for Type-1 Projects

categories K, D and E. Based on expert opinions, in order to obtain more efficient results it has been decided to conduct RCA on categories where the defects densify, similar to [10]. Therefore, to select the categories for the RCA, the Pareto principle and Pareto histograms have been employed. The Pareto principle briefly states that 20% of the causes result to 80% of the problems and can be used to break a big problem down into smaller pieces to identify the most significant factors, and to direct where to focus efforts. Pareto histograms are drawn based on the occurrence rate of an event, are especially suitable for the identification of the most important cause in a problem or situation and can be used by software teams to focus on the major problem areas rapidly [24].

Type-1 and Type-2 projects are analyzed separately and the defects identified between the years 2006 to 2012 are grouped in categories as Pareto histograms and are given in figures 1 and 2. In Figure 1, 70% of the defects belong to A, D and E categories. On the other hand, in Figure 2, 90% of the defects belong to K , A, D and E categories. Based on the results obtained from the Pareto histograms, the organization decided to conduct RCA on A, D, and E defect categories, where the 74% of all defects resides. Accordingly, three separate RCA teams have been formed consisting of members from different departments whose skills and expertise are in accordance with the defect category being assessed, similar to [15]. RCAs have been conducted for each of these three categories separately with the aim of identifying the root causes of these defect categories, followed by prevention plans for each root cause. In order to identify the root causes the teams conducted problem cause brainstorming sessions (utilizing brainwriting) and root cause identification sessions (utilizing the Five Whys approach), as described in [25]. The details of the formed teams and the RCAs are given in detail in the following section.

Prior to the conduction of RCAs, the measurement team of the organization identified 5 main groups to categorize each root cause similar to approaches used

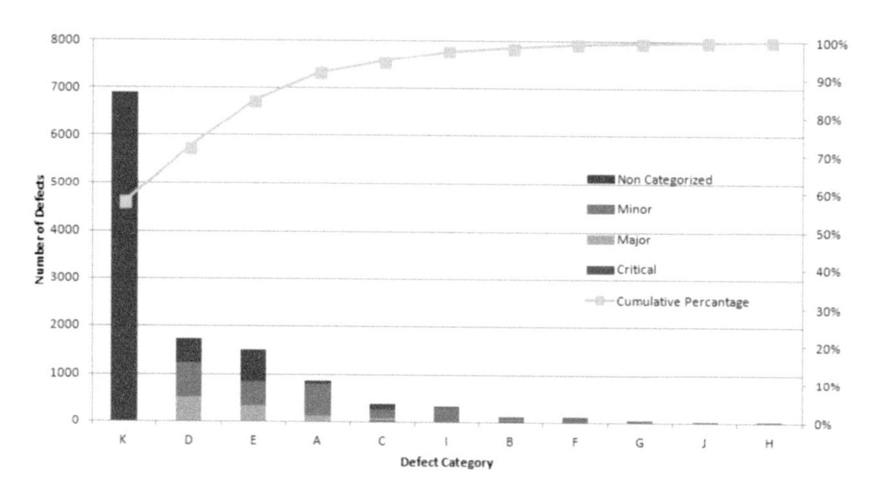

Fig. 2 Defect Categories for Type-2 Projects

in manufacturing and service industries [26], namely People (GP), Methods (GM), Tools (GT), Inputs (GI) and Material (GMa), with the use of expert judgements. However, GT and GMa were merged to a single group during the brainstorming sessions as Tools and Materials (GTM).

3.2 Root Cause Analyses

3.2.1 Root Cause Analysis for Category A Defects

The RCA team gathered to assess the defects of category A (Behavior/ Missing/ Traceability) consisted of 3 software engineers, 1 analysis leader, 1 quality engineer, 1 project manager and 1 process engineer. Each team member was given the defect data prior to RCA meetings and individual evaluations were prepared. The team held 2 meetings, each extending to 6 hours. In these meetings every defect in category A was examined separately, and 17 different root causes grouped under 4 main groups, namely GP, GM, GT and GI were identified with brainstorming sessions. These root causes were emplaced to the corresponding fishbone diagram, given in Fig. 3.

Following the drawing of the fishbone diagram, each team member graded separately the identified 17 root causes at a scale of 0 to 10, 10 denoting the most important item. These grades were then normalized for each team member to allow scale consistency. The normalized grades of each item were summed to identify the main root cause which would be the highest graded item. Each root cause item and the corresponding summed grades are given in Table 4. As a result of the first RCA, the most important reason (main root cause) of Category A defects is "Extenuation of the undertaken task".

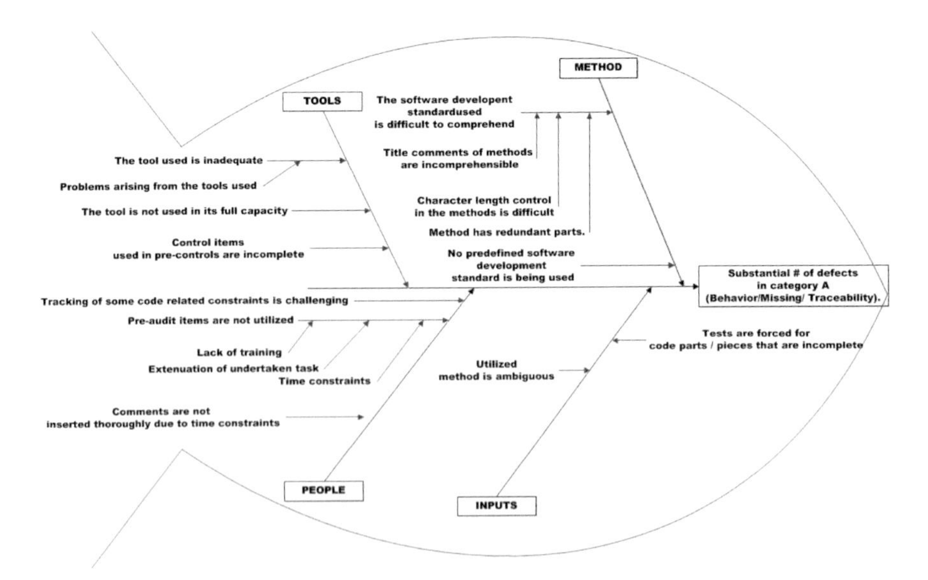

Fig. 3 RCA / Fishbone Diagram for Defects in Category A

Table 4 Root Causes for Defects in Category A

Main Group	Root Cause	Normalized Grade
GP	Extenuation of undertaken task	71.70
GP	Time constraints	61.46
GP	Checklist items are not utilized	60.23
GT	The tool is not used in its full capacity	52.84
GT	Comments are not inserted thoroughly due to time constraints	46.74
GI	Utilized method is ambiguous	45.70
GP	Lack of training	45.34
GI	Tests are forced for code parts / pieces that are incomplete	42.96
GM	The software development standard used is difficult to comprehend	38.76
GT	The tool used is inadequate	35.74
GM	Control of the characters in the method is difficult	35.68
GM	Method has redundant parts	35.27
GM	No predefined software development standard is being used	34.56
GM	Methods title comments are incomprehensible	27.06
GT	Problems arising from the tools used	26.04
GT	Control items which exist in the tool and used in pre-controls are incomplete	23.58
GP	Tracking of some code related constraints is challenging	16.33

RCAs for Category E and D defects were conducted similarly to RCA for defects in Category A, therefore only specific information for these RCAs is given in the following sections.

3.2.2 Root Cause Analysis for Category E Defects

For the assessment and RCA of category E (Documentation/ Understanding) defects, a team consisting of 1 software engineer, 2 test engineers, 2 quality engineers, 1 project manager, 1 process engineer and configuration expert was formed. The team held 3 meetings, each extending to approximately 4 hours. In these meetings every defect in category E was examined separately, and 10 different root causes grouped under 3 main groups, namely GP, GT and GI, were identified with brainstorming sessions. These root causes were emplaced to the corresponding fishbone diagram, given in Fig. 4.

Conducting a grading process similar to category A defects, the main root cause has been identified as "Lack of expertise/ knowledge about the programming language used". Each root cause item and the corresponding summed grades are given in Table 5. As a result of second RCA, the most important cause of Category E defects is "Lack of expertise/ knowledge about the programming language used".

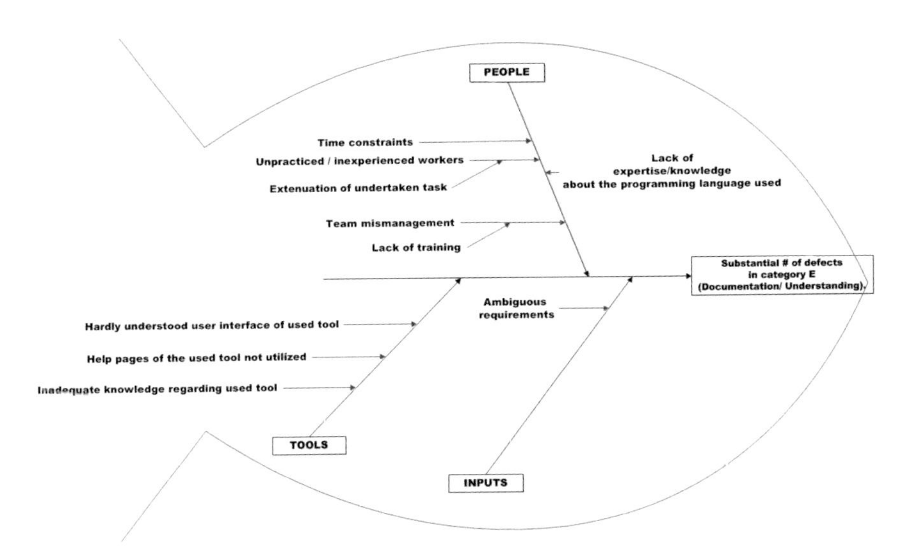

Fig. 4 RCA / Fishbone Diagram for Defects in Category E

Table 5 Root Causes for Defects in Category E

Main Group	Root Cause	Normalized Grade
GP	Lack of expertise/knowledge about the programming language used	87.27
GP	Lack of training	81.82
GP	Extenuation of undertaken task	65.45
GP	Team mismanagement	60.00
GT	Help pages of the used tool not utilized	60.00
GT	Hardly understood user interface of used tool	58.18
GI	Ambiguous requirements	50.91
GP	Time constraints	49.09
GP	Unpracticed / inexperienced workers	43.64
GT	Inadequate knowledge regarding used tool	43.64

3.2.3 Root Cause Analysis for Category D Defects

For the assessment and RCA of category D (Conflict/ Consistency/ Verifiability) defects a team consisting of 5 test engineers, 1 team leader, 1 quality engineer, 1 project manager, and 1 process engineer was formed. The team held 2 meetings, each extending to approximately 3 hours. In these meetings every defect in category E was examined separately, and 28 different root causes grouped under 4 main groups, namely GP, GTM, GM and GI, were identified with brainstorming sessions. These root causes were emplaced to the corresponding fishbone diagram, given in Fig. 5.

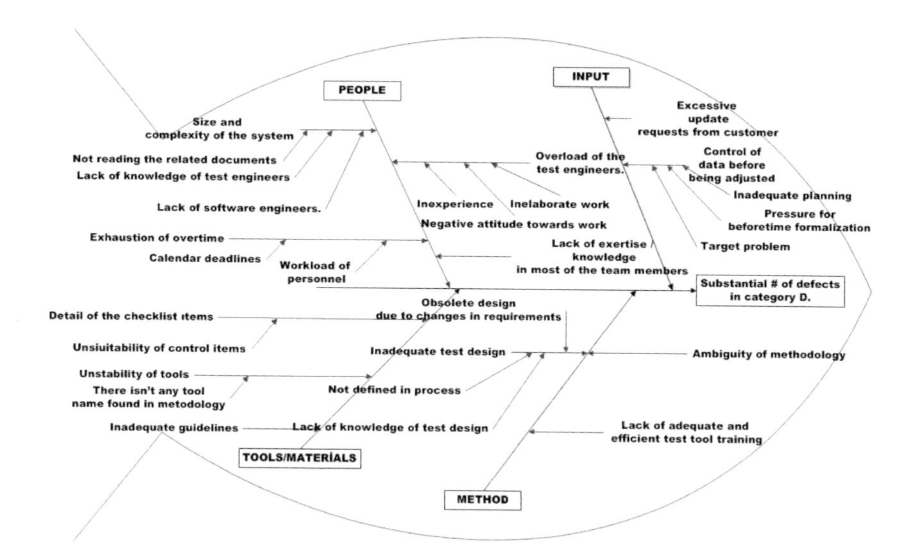

Fig. 5 RCA / Fishbone Diagram for Defects in Category D

Conducting a grading process similar to category A and E defects, the main root cause has been identified as "Lack of knowledge of test engineers". Each root cause item and the corresponding summed grades are given in Table 6. As a result of final RCA, the most important reason of Category D defects is "Lack of knowledge of test engineers".

3.3 *Case Study Effort Analysis*

The effort spent by the team members in the conducted RCAs has been recorded in order to provide an insight for any study to follow. The total effort spent has been calculated as 6,5 person months. In total, 24 people have attended meetings of approximately 30 hours in length. The effort spent for the data preparation prior to RCAs was 302 hours. The presentation and documentation effort of the RCA findings was recorded as 94 hours.

4 Conclusion

Within the conducted analyses a total of 36,424 defect items have been investigated belonging to 13 different software projects conducted in a span of 6 years. Prior to analysis, any inconsistencies or errors in the defect data have been corrected by examining each defect item with respect to accuracy, integrity and consistency considerations. The RCA analyses conducted in this case study is an amalgamation of different best-practice methods and approaches mentioned in the literature, aforementioned in sections 2 and 3. This study has shown that RCA, which is traditionally

Table 6 Root Causes for Defects in Category D

Main Group	Root Cause	Normalized Grade
GP	Lack of knowledge of test engineers	31.15
GM	Ambiguity of methodology	30.18
GM	Lack of adequate and efficient test tool training	29.68
GM	Not defined in process	29.29
GT/M	Detail of the checklist items	29.06
GT/M	No predefined tools in methodology	26.76
GT/M	Inadequate guidelines	24.93
GI	Excessive update requests from customer	24.69
GT/M	Unsuitability of control items	23.36
GT/M	Instability of tools	23.04
GI	Inadequate planning	22.33
GI	Pressure for beforetime formalization	22.12
GP	Calendar deadlines	21.99
GM	Obsolete design due to changes in requirements	20.09
GP	Size and complexity of the system	19.97
GI	Control of data before being adjusted	19.75
GP	Negative attitude towards work	19.13
GP	Lack of expertise / knowledge in most of the team members	18.50
GP	Exhaustion due to overtimes	18.50
GM	Lack of knowledge of test design	17.86
GP	Workload of the resource	17.73
GP	Inexperience	17.57
GP	Lack of work and task details	17.52
GP	Not reading the related documents	17.32
GM	Inadequate test design	16.69
GI	Target problem	16.36
GP	Lack of software engineers	14.38
GP	Work overload of test engineers	11.05

used in the manufacturing industry, can be implemented in software organizations if the defect data is recorded accordingly. Fishbone diagrams are appropriate for displaying all possible root causes together. As well as Pareto charts focus in the majority of defects through these charts required effort decreases. We believe that, this generated practical method and employed approach can be generalized and can be used in any software organization that plans a defect analysis and records the defect data categorically.

The analysis on defect data has proven important insights regarding the organization infrastructure and the encountered defects. It is discovered that prior to product delivery to the customer, most of the defects are detected and resolved. However, the defects that are communicated from the customer in SCRs and that typically are labeled as major or critical, need to be addressed by the organization thoroughly as the solution of these defects lies in far-back phases and the cost of resolving is high. Organization X considers the undertaking of trainings and the assignment of expert members to the phases that these defects originate, as an approach for the reduction of these defects.

When the category types of the defect items are investigated, 74% of the defects fall in the categories of A (Behavior/ Missing/ Traceability), E (Documentation/ Understanding) and D (Conflict/ Consistency/ Verifiability). Therefore, RCA has been conducted only on the defects falling in these three categories. The main root cause for the defects in category A has been identified as "Extenuation of the undertaken task", for category E as "Lack of expertise/knowledge about the programming language used" and for category D as "Lack of knowledge of test engineers". The defect data gathering and adjustments and the conducted RCAs have been accepted as a first step of a long term defect management approach of the organization. The results obtained have been presented to the upper management of Organization X and a group of actions undertaken within the overall approach of SPI are summarized as:

- employee seminars to communicate the work and tasks conducted and its importance,
- courses and trainings for further education regarding the programming language used in each project,
- action plans for the encouragement of expert engineers to emerge within the existing staff,
- and trainings for the sharing of test knowledge and expertise among development teams.

The upper management has accepted a plan of re-conducting the RCAs for a new set of defect data after the completion of the aforementioned SPI activities. Another approach is the conduction of RCAs as an activity within the development lifecycle of software projects in predefined periods, with the aim of minimizing and ideally eliminating all defects from the product before it is delivered to the customer.

Several limitations regarding the undertaken approach exist. First of all, even though Pareto charts reduce the time and effort to conduct RCA, they only focus to a group of defects, resulting in some major items not being analyzed. For organizations that more resources are available, RCA may be conducted for all defect categories. Moreover, grades given to root causes in fishbone diagrams are subjective and may vary according to RCA team members. Therefore, it is crucial to involve different experts to RCA teams and provide for the standardization of the analysis.

A deficiency of the conducted research is the fact that the defect data recorded is not in accordance with any defect classification schemes in literature. A proposed improvement for this shortcoming has been the reconstruction of defect recording approach in accordance to the ODC methodology. The upper management has undertaken a research on the alteration of defect recording and the establishment of an infrastructure that would allow the use of the ODC methodology.

The fact that some of the defect data recorded was incomplete and inconsistent has resulted to extended effort for resolving the existing problems prior to RCAs by conducting interviews with the defect recorders. However, as some of the employees that have recorded the problematic defect items were no longer working in the organization, several defect items were resolved by communicating other team members. This may have resulted to incorrect solutions and assumptions, however, in order to maintain the overall integrity of the defect data these defect items were not excluded from the analysis. As a result of these findings, training sessions have been conducted in Organization X to train employees with respect to correct and consistent defect recording.

As such an RCA has been conducted for the first time since 2006 in the organization, the know-how of the RCA team members regarding the approach was inadequate and the resulting fishbone diagrams have not been in the expected clarification level. However, due to the commitment of the upper management the diagrams are consistent with respect to each other and have provided important insights. We believe that with the lessons learned and acquired knowledge, the organization at present can conduct RCA and develop fishbone diagrams more efficiently.

The results of our study can be compared with other RCAs conducted in software development organizations with respect to several dimensions. When the three RCAs of this study are combined, we can see that 52 unique root causes have been identified, grouped under 4 main groups. Yu [14] discusses an RCA with 33 root causes grouped under 6 main categories, whereas Lehtinen et al. [9] in four different case studies have discovered 52-108 causes in preliminary cause collection and 80-137 in causal analysis workshops, and citing Card they refer to an RCA with 100 target problems. The total effort of the 3 RCAs in our study has been calculated as 6,5 person months. Lehtinen et al. [9] have spent 73-98 person-hours to conduct the cases in their studies, and the authors citing Mayes state that the required effort to conduct an RCA method consists of 47 developers participating in a kickoff and a causal analysis meeting, each lasting 2 h, and 810 action team members using 10% of their time for action team duties. It is evident that the relatively increased effort in our study is mainly due to the high number of defects analyzed.

As a future study to follow this research, a transition plan from the current state to an infrastructure that will allow the recording and analysis of defect data with respect to the ODC model is planned. The transition plan will identify the shortcomings of the current defect recording system, the advantages of using ODC and how the defect management infrastructure needs to be altered in order to accommodate the extra information fields that ODC requires. Another future work for software development organizations that already store the defect data in ODC format would be the analysis of the defects both in simple RCA and RCA following ODC, and the comparison of the results that these two approaches yield. The comparison of these two approaches can further be extended by including guidelines for organizations, stating when and in which case to use which analysis.

References

1. Buglione, L., Abran, A.: Introducing root-cause analysis and orthogonal defect classification at lower CMMI maturity levels. In: Proceedings of MENSURA, Cdiz, Spain (2006)
2. Akman, G., Yilmaz, C.: Innovative capability, innovation strategy and market orientation: an empirical analysis in Turkish software industry. International Journal of Innovation Management 12(01), 69–111 (2008)
3. Gouws, J., Gouws, L.: Fundamentals of software engineering project management. Mlikan Pty Ltd. (2004)
4. Clark, B., Zubrow, D.: How good is the software: a review of defect prediction techniques. In: Proceedings of the Software Engineering Symposium (2001)
5. Mcdonald, M., Musson, R., Smith, R.: The practical guide to defect prevention. Microsoft Press, Washington (2008)
6. Norris, M., Rigby, P.: Software engineering explained. John Wiley and Sons Ltd. (1992)
7. Kumaresh, S., Baskaran, R.: Experimental design on defect analysis in software process improvement. In: Proceedings of the Recent Advances in Computing and Software Systems, RACSS (2012)
8. Raninen, A., Toroi, T., Vainio, H., Ahonen, J.J.: Defect data analysis as input for software process improvement. In: Dieste, O., Jedlitschka, A., Juristo, N. (eds.) PROFES 2012. LNCS, vol. 7343, pp. 3–16. Springer, Heidelberg (2012)
9. Lehtinen, T.O., Mntyl, M.V., Vanhanen, J.: Development and evaluation of a lightweight root cause analysis method (ARCA method)-Field studies at four software companies. Information and Software Technology 53(10), 1045–1061 (2011)
10. Kumaresh, S., Baskaran, R.: Defect analysis and prevention for software process quality improvement. International Journal of Computer Applications 8(7), 42–47 (2010)
11. Reid, I., Smyth-Renshaw, J.: Exploring the fundamentals of root cause analysis: are we asking the right questions in defining the problem? Quality and Reliability Engineering International 28(5), 535–545 (2012)
12. Chulani, S., Boehm, B.: Modeling software defect introduction and removal: COQUALMO (COnstructive QUALity MOdel). Center for Software Engineering, University of Southern California (1999)
13. Card, D.N.: Myths and strategies of defect causal analysis. In: Proceedings of the Pacific Northwest Software Quality Conference (2006)
14. Yu, W.D.: A software fault prevention approach in coding and root cause analysis. Bell Labs Technical Journal 3(2), 3–21 (1998)
15. Leszak, M., Perry, D.E., Stoll, D.: Classification and evaluation of defects in a project retrospective. The Journal of Systems and Software 61(3), 173–187 (2002)
16. Lehtinen, T.O., Mntyl, M.V.: What are problem causes of software projects? Data of root cause analysis at four software companies. In: The Proceedings of the Empirical Software Engineering and Measurement (ESEM) Symposium (2011)
17. Chillarege, R., Bhandari, I.S., Chaar, J.K., Halliday, M.J., Moebus, D.S., Ray, B.K.R., Wong, M.Y.: Orthogonal defect classification - a concept for in-process measurements. IEEE Transactions on Software Engineering 18(11), 943–956 (1992)
18. Soylemez, M., Tarhan, A., Dikici, A.: An analysis of defect root causes by using orthogonal defect classification. In: Proceedings of the 6th National Software Engineering Conference, Ankara, Turkey (2012) (in Turkish)
19. Yin, Z., Caesar, M., Zhou, Y.: Towards understanding bugs in open source router software. ACM SIGCOMM Computer Communication Review 40(3), 34–40 (2010)

20. Schroeder, B., Gibson, G.A.: A large-scale study of failures in high-performance computing systems. IEEE Transactions on Dependable and Secure Computing 7(4), 337–350 (2010)
21. Lethbridge, T.C., Sim, S.E., Singer, J.: Studying software engineers: data collection techniques for software field studies. Empirical Software Engineering 10(3), 311–341 (2005)
22. Runeson, P., Hst, M.: Guidelines for conducting and reporting case study research in software engineering. Empirical Software Engineering 14(2), 131–164 (2009)
23. Shul, F., Basili, V., Boehm, B., Brown, W.A., Costa, P., Lindvall, M., Port, D., Rus, I., Tesoriero, R., Zelkowitz, M.: What we have learned about fighting defects. In: Proceedings of the Eighth IEEE Symposium on Software Metrics (2002)
24. Florac, W.A., Carleton, A.D.: Measuring the software process. Addison Wesley, Indianapolis (2004)
25. Andersen, B., Fagerhaug, T.: Root cause analysis: simplified tools and techniques. ASQ Quality Press (2006)
26. Young, S.: Quality management. MIM Malta Institute of Management (2005)

Extending UML/MARTE-GRM for Integrating Tasks Migrations in Class Diagrams

Amina Magdich, Yessine Hadj Kacem, and Adel Mahfoudhi

Abstract. There is a growing interest in modeling Real-Time Embedded Systems (RTES) using high-level approaches. The recent extension of Unified Modeling Language (UML) profile for Modeling and Analysis of Real-Time Embedded systems (MARTE) is enclosing a lot of stereotypes and sub-profiles providing support for designers to beat the shortcomings of complex systems development. In particular, the MARTE/GRM (Generic Resource Modeling) package offers stereotypes for annotating class diagrams with the needed information which will be extracted to fulfill a scheduling phase. However, GRM does not allow designers to specify data to be used neither in half-partitioned nor in global scheduling approaches; indeed, it does not support the modeling of task migration concept. Thus, we propose through this paper an extension of MARTE/GRM sub-profile to consider the modeling of information needed for the half-partitioned and global scheduling step.

1 Introduction

Model-based development is an adequate approach used when dealing with critical systems since it helps designers to overcome the increasing complexity challenge and to model systems at a high abstraction level. In particular, the UML/MARTE profile fosters an adequate solution to support the whole life cycle development due to its rich set of available annotations. In a model driven development approach, the model transformation is linking across the various models and steps.

In general, a RTES development approach is based substantially on three major phases: modeling, scheduling and implementation [9]. The modeling step is founded on the use of GRM to annotate a static view (class diagram) specifying the system properties. This view represents an entry for the scheduling step. Thus,

Amina Magdich · Yessine Hadj Kacem · Adel Mahfoudhi
CES Laboratory, ENIS Soukra km 3,5, B.P.: 1173-3000 Sfax TUNISIA
e-mail: {amina.magdich,yessine.hadjkacem,
adel.mahfoudhi}@ceslab.org

R. Lee (Ed.): *SERA*, SCI 496, pp. 73–84.
DOI: 10.1007/978-3-319-00948-3_5

the data specified in this model will be extracted and used during the scheduling step. Thereby, the GRM view must match with the entry of the scheduling step (Fig. 1).

In fact, three scheduling approaches are described in the literature : the partitioned, the half-partitioned and the global approaches [4]. MARTE/GRM does not support modeling of information to be used neither in the context of half-partitioned nor in the global scheduling approaches since it does not allow modeling of data describing task migration between the available computing resources. In this context, we propose an extension of MARTE/GRM to improve the existing annotations. We propose in this paper the main changes to be made on the MARTE/GRM subprofile for modeling the needed information for half-partitioned or global scheduling phases.

This paper is structured as follows. Section 2 highlights the various related works. Section 3 introduces the scheduling theory. In Section 4, we highlight the importance of the use of MARTE within the context of RTES development. In Section 5, we emphasise our contribution. In section 6, we validate our proposal using a pedagogic example and we summarize by a conclusion in Section 7.

2 State-of-the-Art and Related Works

Many researchers have benefited from the MARTE profile to develop complex systems at a high-level of abstraction. In [12], the authors have benefited from MARTE to model systems containing both of functional and non functional properties and then they attempted to extract the temporal information included in the built models. Nevertheless, the proposed approach has treated multiprocessor systems according to a partitioned scheduling approach inhibiting task migration. Other works have dealt with development of complex systems while benefiting from the use of MARTE within the context of scheduling analysis such as in [3], [11], [8] and [7].

In [9], a model-based approach, founded on the use of MARTE to cover all the development steps of RTES, was been carried out to develop multiprocessor systems. Yet, the proposed approach does not tackle with task migration during the scheduling analysis step since it used the partitioned scheduling approach (the algorithm used was Rate Monotonic). In fact, originally MARTE did not provide possibilities to model the scheduling of multiprocessor systems allowing task migration. In this context, a research study [10] has been carried out to extend the subprofile MARTE/SAM (Schedulability Analysis Modeling) for modeling scheduling analysis in the context of half-partitioned or global approaches. This contribution is used to model temporal (activity diagrams) scenarios supporting task migration. The corresponding SAM view represents an entry for a schedulability analysis tool (Fig. 1).

Actually, our development approach begins with a system modeling based on MARTE/GRM. The extracted information from GRM view will serve during the scheduling step. For that, and first of all, to use the SAM extension [10] when modeling systems supporting task migration, GRM must be able to support this concept.

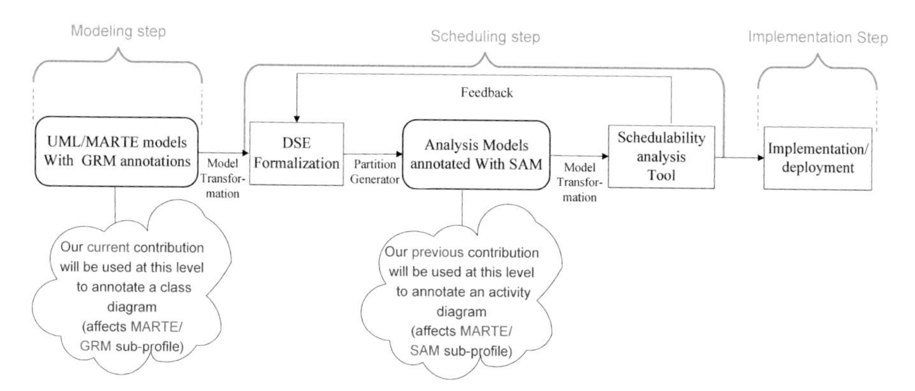

Fig. 1 A model-based development flow with MARTE extensions

This is the contribution to be explained throughout this paper. Our proposal represents a continuation of the contribution outlined in [10].

3 Scheduling Theory

This is a theory [14] which can be applied on-line or off-line in order to assign tasks on the available execution resources while meeting tasks deadlines. Three types of scheduling approaches intended for multiprocessor systems are described in the literature: the partitioned, half-partitioned and global scheduling approaches. In this paper, we focus mainly on the half-partitioned and global approaches.

3.1 Partitioned Approach

It consists of partitioning tasks to assign each one to a single computing resource [2]. This type of scheduling is performed on each processor, so each execution resource must be associated with a scheduler. Such an approach does not allow reaching optimality since it prohibits task migration.

3.2 Global Approach

The global scheduling approach provides a total liberty regarding task migration that achieve optimality [2]. However, this flexibility may have a cost during execution. That is why it is better to limit the number of task migration. This led researchers to limit the number of preemptions by establishing the half-partitioned scheduling approach. Algorithms of the global approach need only one scheduler annotated `main scheduler` to schedule the active tasks.

3.3 *Half-Partitioned Approach*

The half-partitioned scheduling approach allows scheduling with controlled task migration in order to reach optimality [2]. It uses mainly a main scheduler and schedulers associated to processors.

4 Model Driven Engineering (MDE) and RTES Development

The Model Driven Engineering [13] is a software development paradigm aiming to raise the level of abstraction while developing RTES and then overcoming their increasing complexity. It enables the automation of development flows and the reuse of built models. Besides, MDE ensures the entry and validation of constraints through the use of the Object Constraint Language (OCL) [5]. Moreover, it ensures the independence between the different development phases.

4.1 *MARTE Capabilities for RTES Modeling*

MARTE [6] is an extension of UML profile providing support for specification, modeling and verification steps. It is a framework that sets the crucial properties required to develop RTES. This profile provides a unified co-design containing both RTES Hardware and Software components. Thanks to its wealth in stereotypes, MARTE allows annotating models with functional and non-functional properties. In addition, it offers a rich set of annotations for modeling schedulability analysis. MARTE encloses a set of sub-profiles such as: GRM (SRM (Software Resource Modeling), HRM (Hardware Resource Modeling)), GQAM (Generic Quantitative Analysis Modeling (SAM, PAM (Performance analysis Modeling)), etc.

SRM is used to model Software resources such as tasks and HRM is used to model Hardware resources such as computing resources, bus of communication, memories, etc. The sub-profile SAM is used to model systems temporal behaviour for schedulability analysis. Concerning PAM, it is exploited especially to model the properties that are related to performance check. In this paper, we will focus especially on the GRM sub-profile since it is the package affected by our proposal.

4.2 *GRM*

GRM is a sub-profile of MARTE providing the concepts necessary to model a general platform for executing RTES. It provides concepts promoting modeling of Software and Hardware resources at a high-level of abstraction. It also provides mechanisms to manage access to different execution resources. The GRM sub-profile allows only the modeling of systems to be scheduled in the context of partitioned scheduling approaches; originally it does not support modeling of systems to be scheduled according to the half-partitioned or global scheduling approaches.

4.3 DRM (Detailed Resource Modeling)

DRM is a specialization of the sub-profile GRM enclosing SRM and HRM packages which support the modeling of Hardware and Software resources. Since SRM and HRM are two specializations of GRM, the SW/HW allocation model benefits from the unified structure of these models.

4.3.1 SRM

SRM is the sub-profile describing the software resources of the application such as tasks and virtual memories. It encloses a set of stereotypes with a variety of fields leading to an explicit description of the characteristics of the modeled software resources.

4.3.2 HRM

HRM is the sub-profile including most of Hardware concepts under a hierarchical taxonomy with several categories according to their natures, features, technology and forms. Hence, it is used to describe the components of the physical platform (the platform of execution) such as processors, memories, bus, etc.

5 Our Proposal: GRM Extension

The modeling step of RTES uses the GRM sub-profile, it allows specifying the different properties and provides as a result a unified model of Hardware and Software components. The information specified in the GRM view will be needed during the scheduling step. In fact, the GRM meta-model supports the modeling of different systems as it models all the temporal features needed in the scheduling step except those used in the context of scheduling with task migration (the half-partitioned and the global approaches). Consequently, we seek to improve GRM meta-model in order to support modeling of systems allowing task migration. The amendments to be done on the GRM sub-profile can be useful for both of half-partitioned and global approaches. But regardless, we will point out the amendments to be relevant for the two approaches together, the changes to be useful only for the half-partitioned approach and those only for the global one.

5.1 Amendments to Be Used in Both Approaches

The global or half-partitioned scheduling approaches allow task migration, so a task can be allocated across multiple processors for different periods of time. Thus, the multiplicity of the attribute corresponding to the execution resource on which a task is allocated must be [0..*] instead of [0..1]. Hence, we propose to modify in the association linking the two classes `schedulableResource` and `ExecutionHost` of the package GQAM_Resources (Fig. 2 and Fig. 3).

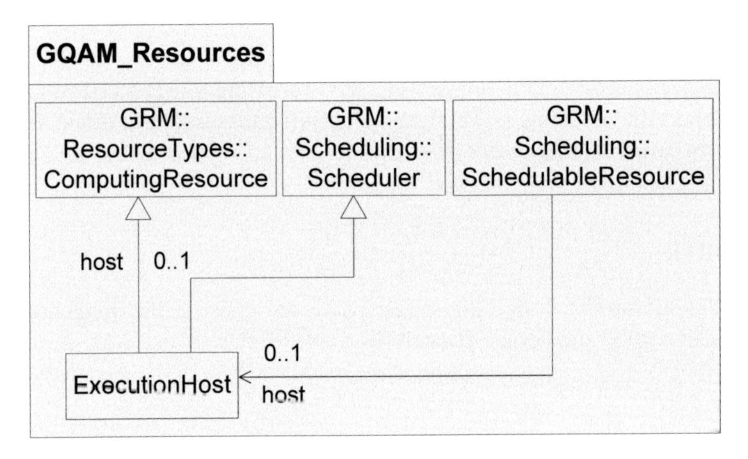

Fig. 2 Meta-model of the GQAM package

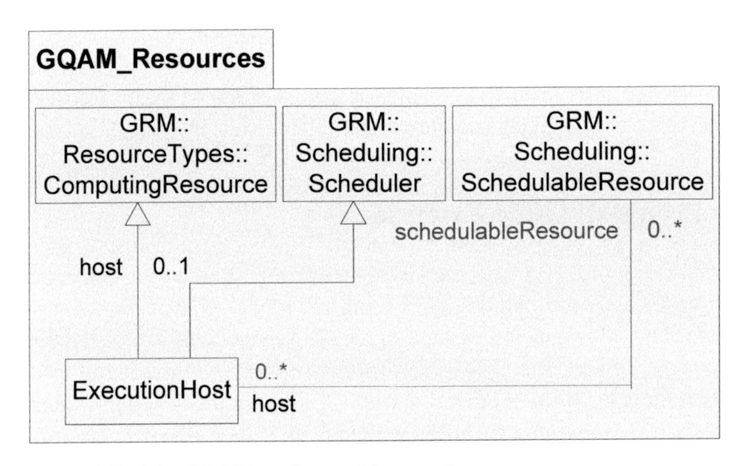

Fig. 3 Meta-model of the GQAM package with amendments

This change affects MARTE/GQAM, but it also affects the use of the GRM sub-profile since the class `ExecutionHost` inherits from `ComputingResource` which is a class of the GRM package. Following this change, we will have in the class `schedulableResource` an attribute `host: GaExecHost [0..*]` instead of `host: GaExecHost[0..1]` (Fig. 5). I.e. `GaExecHost` is the stereotype corresponding to the class `Execution Host` and annotating the execution hosts which are modeled in the GRM view and which will be called during the creation of the SAM view of course with a multiplicity of [0..*].

While migrating from one processor to another, the execution time of a task is not the same, then we try to add an attribute denoting the execution time `execT: NFP_Duration [0..*]` in the class `SchedulableResource` (Fig. 4 and Fig. 5). This will be ordered with the attribute `GaExecHost`, to

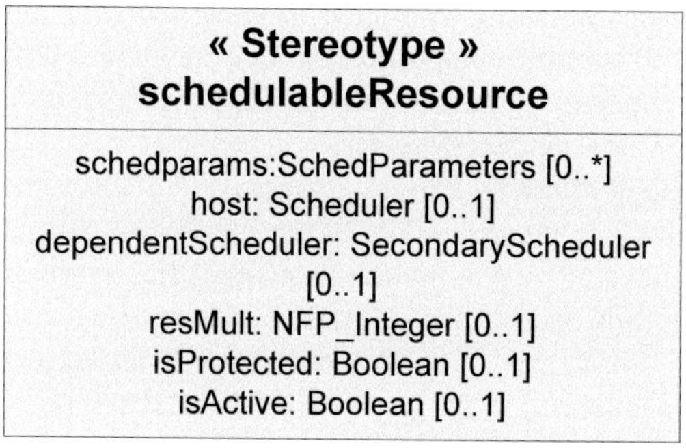

« Stereotype »
schedulableResource

schedparams:SchedParameters [0..*]
host: Scheduler [0..1]
dependentScheduler: SecondaryScheduler [0..1]
resMult: NFP_Integer [0..1]
isProtected: Boolean [0..1]
isActive: Boolean [0..1]

Fig. 4 The old meta-model of the schedulableResource stereotype

« Stereotype »
schedulableResource

schedparams:SchedParameters [0..*]
host: Scheduler [0..*]
dependentScheduler: SecondaryScheduler [0..1]
resMult: NFP_Integer [0..1]
isProtected: Boolean [0..1]
isActive: Boolean [0..1]
Host:GaExecHost [0..*]
execT:NFP_Duration [0..*]
P_Host:ComputingResource [0..*]
P_execT:NFP_Duration [*]

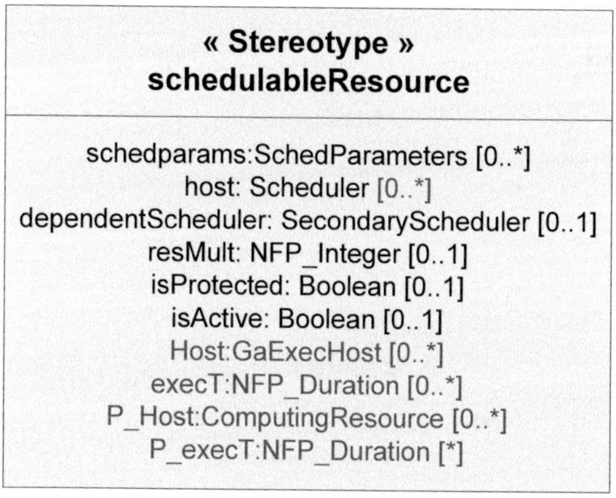

Fig. 5 The new meta-model of the schedulableResource stereotype

indicate the execution time of the $task_i$ when running on the execution $resource_i$. Indeed, adding the attribute `execT` facilitates the representation and the understanding of the GRM view.

5.2 *Changes to Be Used within a Half-Partitioned Approach*

Among a half-partitioned scheduling approach, both a main scheduler and schedulers associated to processors are used. Since this approach allows task migration, a task can be allocated, not simultaneously, on different processors/ schedulers. Accordingly, the multiplicity of `schedulers: NamedElement [0..1]` should be [0..*] (schedulers is an attribute of the stereotype `swSchedulableResource`). Likewise, the multiplicity of `host scheduler [0..1]` must be [0..*]. i.e. `host: scheduler` is an association that connects the two classes `scheduler` and `SchedulableResource` (Fig. 6 and Fig. 7).

A half-partitioned scheduling approach enables restricted task migration, we need then to know the possible computing resource on which a task can be allocated. Hence, we add an association P_Host linking the SchedulableResource and computingResource classes (Fig. 7).

Consequently, the attribute P_Host: ComputingResource [0..*] will be added as an attribute in the stereotype SchedulableResource (Fig. 5).

A $task_i$ beeing allocated on a computing $resource_i$, has a specific execution time. We propose then to add in the class SchedulableResource a new attribute P_execT: NFP_Duration [0..*] ordered with the attribute P_Host (Fig. 5). The information indicated in the attribute P_execT will be of course extracted to guide the scheduling phase and mainly for optimization when assigning tasks to processors.

The difference between the attributes execT: NFP_Duration [0..*] and P_execT: NFP_Duration [0..*] is that execT specifies the spent execution time of a task after its running on a processor (this attribute will take the value given by the scheduling tool after a feed back), but P_execT specifies the estimated execution time.

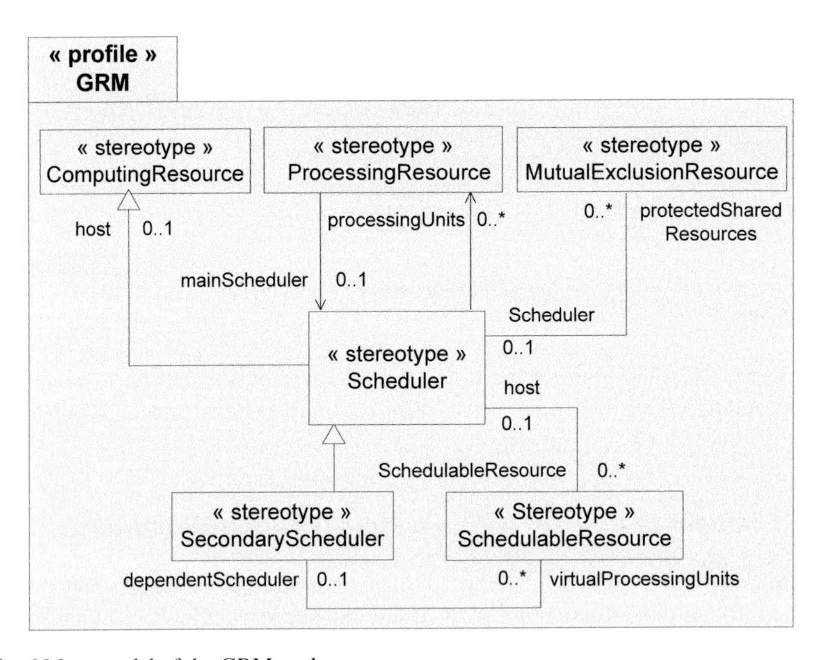

Fig. 6 Meta-model of the GRM package

The value of P_execT may be used for optimization while assigning tasks on available processors. Otherwise, execT is a result specified after the scheduling step, but P_execT is itemized before scheduling and will be extracted to guide the scheduling phase. It concerns only the modeling of systems to be scheduled with the

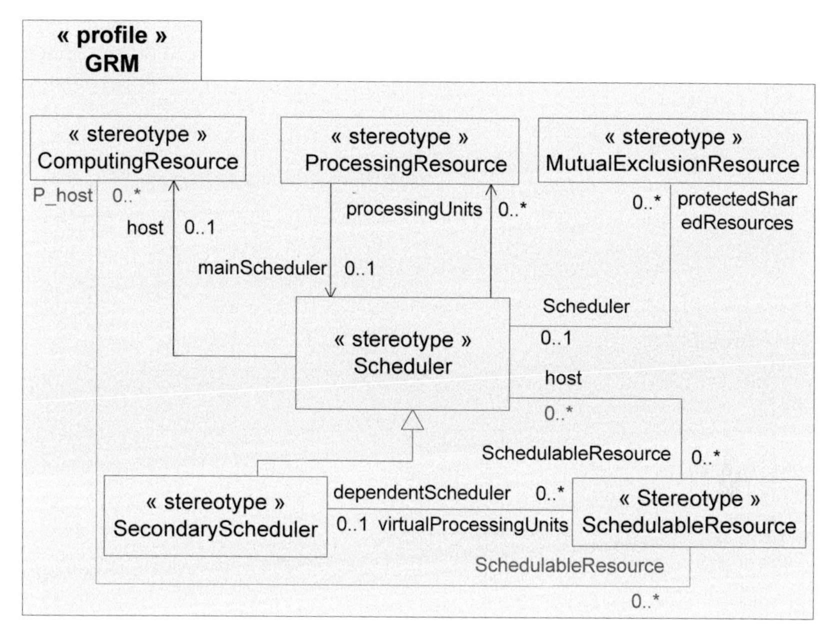

Fig. 7 The new meta-model of the GRM package

half-partitioned approach, however `execT` concerns systems to be scheduled with any approach.

6 Pedagogic Case Study

To better explain our GRM extension, we depend on a pedagogic example modeled using a class diagram. The figure Fig 8 is composed by `a model` which may be reusable for many systems and `a model instance` that concerns only one system (the studied system).

In our case, the model instance models a system composed by an architecture on which is allocated the application. The proposed application is composed by four tasks T1, T2, T3 and T4. Concerning the architecture, it is composed by four execution hosts, memories, a battery and a bus. The allocation of the application on the target architecture is controlled through mutual exclusion resources designed by classes annotated `swMutualExclusionResources`; res_i is controlling the access to $proc_i$.

The different tasks can have dynamic periods, priorities and deadlines due to the concept of task migration, but we use the same corresponding values to facilitate our example. Anyway, we can add the different values and they will be ordered with the attribute `Host: GaExecHost` and `ExecT: NFP_Duration`.

The used scheduling approach is a half-partitioned one, so we can find restricted or free task migration. T1 can be allocated only on P1, P2 and P3 which

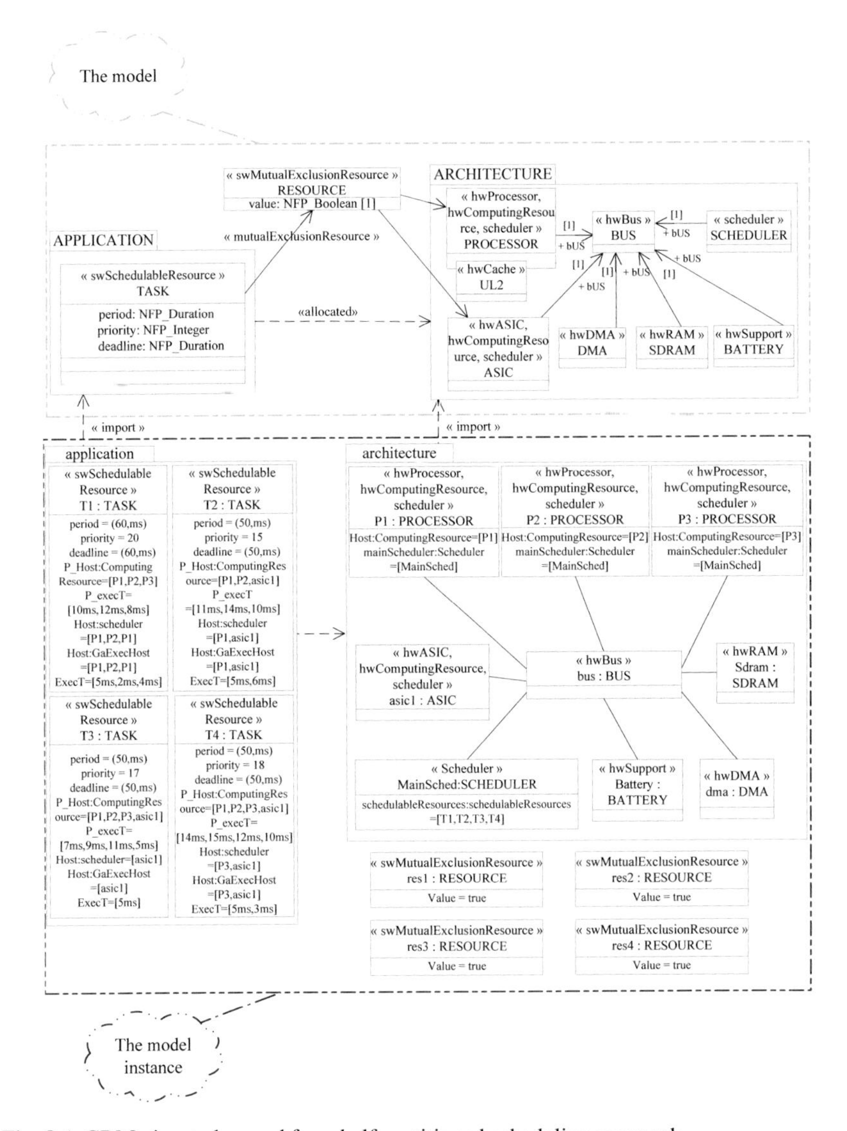

Fig. 8 A GRM view to be used for a half-partitioned scheduling approach

is indicated through `P_Host: ComputingResource=[P1,P2,P3]`. T2 can be allocated on P1, P2 and asic1. However, T3 and T4 can be allocated on all the different computing resources. The estimated execution time corresponding to the possible processors is indicated through `P_execT`: for example for T1, `P_execT: NFP_Duration= [10ms,12ms,8ms]` ordered with the values of `P_Host: ComputingResource`. In fact, T1 and T2 are scheduled firstly via the mainScheduler `MainSched` and then through the possible schedulers associated with the corresponding processors (P1,P2,P3 for T1 and P1,P2,asic1 for T2). T3 and T4 are scheduled only through the main scheduler. Then, we notice that all tasks must be scheduled firstly or only by the main scheduler, this is specified through `schedulableResources: schedulableResources=[T1,T2,T3,T4]` which is an attribute of `scheduler` annotating the main scheduler `MainSched`. After scheduling all tasks, we can specify in our GRM view the used allocations through the attributes `Host: GaExecHost` and the corresponding attribute `ExecT: NFP_Duration`. For example for T1, It was running firstly on P1 for an execT=5ms, and then after migration it was running on P2 for 2ms and finally it has migrated to run on P1 for 4ms.

7 Conclusion

The use of model driven engineering in the context of RTES development increases the level of abstraction and overcomes the growing complexity challenge. In the same context, MARTE facilitates the modeling of critical systems thanks to the big set of stereotypes that it offers. In this paper, we have proposed an extension of MARTE/GRM since it does not support the modeling of concepts in connection with the half-partitioned and global scheduling approaches that allow task migration. The benefit of our approach is the ability to model properties which will be extracted, in the context of a model-based approach, to serve during the scheduling step using algorithms of the half-partitioned approach or the global one. Our approach has been enabled in the papyrus tool [1] which is an editor of MARTE-based modeling.

References

1. `http://www.papyrusuml.org`
2. Carpenter, J., Funk, S., Holman, P., Srinivasan, A., Anderson, J., Baruah, S.: A categorization of real-time multiprocessor scheduling problems and algorithms. In: Handbook on Scheduling Algorithms, Methods, and Models. Chapman Hall/CRC, Boca (2004)
3. Espinoza, H., Medina, J., Dubois, H., Gérard, S., Terrier, F.: Towards a uml-based modelling standard for schedulability analysis of real-time systems. In: MARTES Workshop at MODELS Conference (2006)
4. Goossens, J.: Introduction à l'ordonnancement temps réel multiprocesseur, pp. 157–166 (2007)
5. Object Management Group. UML 2.0 OCL Specification. OMG Adopted Specification ptc/03-10-14. Object Management Group (October 2003)
6. OMG Object Management Group. A UML Profile for MARTE: Modeling and Analysis of Real-Time Embedded systems, Beta 2, ptc/2008-06-09. Object Management Group (June 2008)
7. Hagner, M., Huhn, M.: Tool support for a scheduling analysis view. In: The Workshop "Modeling and Analysis of Real-Time and Embedded Systems with the MARTE UML Profile" at DATE 2008 (Design, Automation & Test in Europe) (2008)

8. Kacem, Y.H., Karamti, W., Mahfoudhi, A., Abid, M.: A petri net extension for schedulability analysis of real time embedded systems. In: The 16th International Conference on Parallel and Distributed Processing Techniques and Applications, PDPTA 2010, pp. 304–314 (2010)
9. Kacem, Y.H., Mahfoudhi, A., Karamti, W., Abid, M.: Using model driven engineering and uml/marte for hw/sw partitioning. International Journal of Discrete Event Control Systems (IJDECS) 1(1), 57–67 (2011)
10. Magdich, A., Kacem, Y.H., Mahfoudhi, A., Abid, M.: A MARTE extension for global scheduling analysis of multiprocessor systems. In: The 23th IEEE International Symposium on Software Reliability Engineering (ISSRE) (November 2012) (to appear)
11. Medina, J., Cuesta, A.G.: Experiencing the uml profile for marte in the generation of schedulability analysis models for mast. In: The 2nd Workshop on Model Based Engineering for Embedded Systems Design, M-BED (2011)
12. Peraldi-Frati, M.-A., Sorel, Y.: From high-level modelling of time in MARTE to real-time scheduling analysis. In: MoDELS 2008 Workshop. on Model Based Architecting and Construction of Embedded Systems on ACES-MB, Toulouse, France, pp. 129–143 (September 2008)
13. Schmidt, D.C.: Model-driven engineering. IEEE Computer 39(2) (February 2006)
14. Sha, L., Abdelzaher, T., Årzén, K.-E., Cervin, A., Baker, T., Burns, A., Buttazzo, G., Caccamo, M., Lehoczky, J., Mok, A.K.: Real time scheduling theory: A historical perspective. Real-Time Syst. 28(2-3), 101–155 (2004)

Towards a Protocol Algebra Based on Algebraic Specifications

Iakovos Ouranos and Petros Stefaneas

Abstract. We sketch some first steps towards the definition of a protocol algebra based on the framework of behavioural algebraic specification. Following the tradition of representing protocols as state machines, we use the notion of Observational Transition System to express them in an executable algebraic specification language such as CafeOBJ. This abstract approach allows defining several useful operators for protocol reasoning and proving properties of them using theorem proving techniques and CafeOBJ term rewriting machine. The proposed protocol algebra is inspired by the module algebra and the hierarchical object composition technique.

Keywords: formal methods, algebraic specification, behavioural specification, protcol algebra, CafeOBJ.

1 Introduction

Distributed protocols typically involve several modules interacting with one another in a complex manner. The design and verification of them are two of the most difficult and critical tasks in the current computing systems development. Many new protocols are designed in the current software industry, and the tradition is to develop them from scratch, i.e. specify them informally and implement them directly into software using a programming language. This makes protocol development even more painful and risky. A solution to these problems seems to be formal algebraic specification techniques. In this paper we propose the *behavioral*

Iakovos Ouranos
Hellenic Civil Aviation Authority, Heraklion Airport, Greece
e-mail: iouranos@central.ntua.gr

Petros Stefaneas
National Technical University of Athens, School of Applied Math. & Phys. Sciences, Greece
e-mail: petros@math.ntua.gr

R. Lee (Ed.): *SERA*, SCI 496, pp. 85–98.
DOI: 10.1007/978-3-319-00948-3_6

specification paradigm [1-3] as a suitable basis for formal protocol reasoning. We model protocols as *Observational Transition Systems* [21-22] a special kind of *behavioral objects* [4] with a flexible modular structure. This formal object oriented approach makes it possible to *reuse* protocol specifications and verifications, *combine/compose* them hierarchically to form more complex protocols, *prove* invariant properties of them and *define operations* between protocols formally.

The main advantages of our approach are a) based on behavioral specification, which follows an equational style of specification, it is easier to read, understand and learn, b) object orientation provides a more flexible way for handling protocol development, c) it is compatible with semi automatic verification techniques such as structural induction and coinduction, d) specifications can be executed by systems that support behavioural specification such as CafeOBJ [5] and finally, e) it improves reliability since it permits the verification to be carried out at the level of design.

This work was inspired by the hierarchical object composition technique presented in [4], the module algebra [6-7] and of course the Observational Transition Systems

(OTSs) [21-22]. We have also adapted some results and definitions of [15] and [10] to the behavioral specification framework.

Since every protocol specification consists of several interacting modules and we can compose protocol specifications to form more complex with the aid of hidden algebra, we can build protocol specifications hierarchically and define operations between protocols in the framework of *protocol algebra*. In our previous works [8-9] we have proposed an algebraic framework for modeling of mobile systems and protocols based on behavioral specification and the hierarchical object composition technique. Additionally, several authors have indicated the need for applying formal techniques and modularization to protocol modeling and design. In [10] the author proposes a technique for parallel composition of protocols that allows protocols to share messages and variables. The work reported in [11] uses category theory to define interfaces for basic modules and composition. The concept of protocol object is defined in [12] together with BAST protocol class library, which follows a layered architecture. In [13] a compositional technique to design multifunction protocols has been proposed. In [14] and [15] two different protocol algebras are proposed which are oriented to security and commitment protocol design, respectively. Finally, there are many approaches which focus on implementation aspects of the composition of a protocol from micro-protocols developed for specific services, such as Appia [18], Cactus [19] and Samoa [20].

The rest of the paper is organized as follows: section 2 introduces behavioral specification and OTSs. Section 3 defines some key notions related to protocol reasoning in terms of behavioral specification. In section 4 we define some basic operators of our *protocol algebra* and give some basic properties of them, while section 5 discusses protocol verification. Finally, section 6 concludes the paper.

2 Behavioural Specification and OTSs

Hidden algebra is the logical formalism underlying behavioral specification. It gives a general semantics for distributed concurrent systems extending ordinary general algebra with sorts representing states of abstract machines, rather than data elements, and also introduces a new satisfaction between algebras and sentences, called *behavioral satisfaction*. The goal of hidden algebra is to significantly decrease the difficulty of proving properties of distributed concurrent systems. In the following we review the basic concepts of hidden algebra. For more details and proofs someone can consult [3-4].

Definition 1. A ***hidden algebraic signature (H, V, F, F^b)*** consists of disjoint sets *H* of *hidden sorts*, *V* of *visible sorts*, a set *F* of *$(H \cup V)$*-sorted operation symbols, and a distinguished subset $F^b \subseteq F$ of *behavioral operations*. Behavioral operations are required to have at least one hidden sort in their arity. An operation symbol which has visible arity and sort is called *data operation*. The hidden sorts denote sets of states of objects, the visible sorts denote data types, the operations $\sigma \in F^b_{w \to s}$ can be thought as methods whenever *s* is hidden, and as attributes whenever *s* is visible.

Definition 2. An ***(H, V, F, F^b)-algebra*** is an *$(H \cup V, F)$*-algebra. Given an (H, V, F, F^b)-algebra *A*, a hidden congruence ~ on *A* is just an F^b-congruence which is identity on the visible sorts. The largest hidden F-congruence $\sim_A$ on A is called behavioral equivalence.

Theorem 1. ***Behavioral equivalence always exists.***

Definition 3. A ***behavioral theory (Σ, E)*** consists of a hidden algebraic signature Σ and a set *E* of Σ-sentences.

Definition 4. A ***behavioral object B*** is a pair consisting of a behavioural theory $((H_B, V_B, F_B, F_B{}^b), E_B)$ and a hidden sort $h_B \in H_B$ such that each behavioral operation in $F_B{}^b$ is monadic, i.e. it has only one hidden sort in its domain.

The hidden sort h_B denotes the space of the states of *B*. The visible sorted behavioral operations on h_B are called *B-observations* and the h_B-sorted behavioral operations on h_B are called B-*actions*. The h_B-sorted operations with a visible sorted domain are called *constant states*.

Definition 5. For any behavioral object B, a ***B-algebra*** is just an algebra for the signature of B satisfying the sentences E_B of the theory of the object *B*. The class of B-algebras is denoted by ***Alg(B)***.

Definition 6. Given an object *B*, ***two B-algebras*** *A* and *A'* are ***equivalent***, denoted $A \equiv A'$, when

- they coincide on the data
- $A_{h_B} = A_{h_{B'}}$ and $\sim_A = \sim_{A'}$ (on the sort h_B), and
- $A_\sigma = A_{\sigma'}$ for each B-action σ.

Definition 7. Two ***behavioral objects*** *B* and *B'* are ***equivalent,*** denoted $B \equiv B'$, when there exists a pair of mappings Φ : Alg(B) $\rightarrow$ Alg(B') and Ψ : Alg(B') $\rightarrow$ Alg(B) which are inverse to each other modulo algebra equivalence, i.e. A $\equiv$ $\Psi(\Phi(A))$ for each B-algebra A and A' $\equiv$ $\Phi(\Psi(A))$ for each B' -algebra A'.

Observational Transition Systems or OTSs can be regarded as a proper sub-class of behavioral specifications, corresponding to conventional transition systems. The main difference between OTSs and behavioral specifications is that in the case of the latter, even if every observation function returns the same value for two states, the two states may be different, while in the case of OTSs, behavioral equivalence coincides with observational equivalence, i.e. when every observation function returns the same value for two states, these states are equal with respect to the OTS.

3 Reasoning About Protocols

Since distributed protocols are mainly distributed systems of interacting objects, the object oriented behavioural specification paradigm is a suitable framework for reasoning about protocols. A protocol is specified as a set of modules each of which models either a data part or a state machine. This approach follows the tradition of specifying protocols as state transition systems.

Definition 8. A ***protocol specification*** is an OTS ***P*** where

- the hidden sort $h_p \in H_P$ denotes the set of the states *S* of the protocol,
- the P-actions denote the set of transitions that under conditions, change or not change the state of the protocol,
- the initial state $s_0 \in S$ is specified as a constant state,
- the set $F \subseteq S$ of final states are all possible reachable states (def. 12) of the OTS,
- the P-observations "observe" quantities of the protocol at possible states, and
- the data types used by the protocol are specified as submodules of the specification.

Definition 9. Two protocols are ***equivalent***, whenever their specifications are equivalent (def. 7).

Definition 10. Two protocol states are equivalent, if and only if they are *observationally equivalent*.

Definition 11. A ***protocol run R*** is an infinite or finite sequence of states $\langle s_0, s_j, \ldots, s_k, \ldots \rangle$ that the protocol allows.

Definition 12. A state of a protocol P is called ***reachable*** if it appears in a run of the P.

Definition 13. A ***protocol run*** R_i **subsumes** another ***protocol run*** R_j if and only if, for every state s_j that occurs in R_j, there exists a state s_i in R_i that is observationally equivalent and has the same temporal order relative to other states in R_i as s_j does with states in R_j.

Definition 14. A protocol run R_i is similar to protocol run R_j if and only if the two runs are of equal length and every i^{th} state of R_i is equivalent to the ith state of R_j.

Hierarchical object composition technique based on behavioral specification has been proposed in [4] and is a suitable framework for composing systems by components. Here we argue that the hierarchical model for composition is suitable not only for composing protocols by participants, but also for composing *protocols* by *sub-protocols* (or *systems* by *sub-systems*). The main advantage of this method is not only the reuse of specifications but also the reuse of proofs. If a property of a *component protocol* holds and has been verified, then we can reuse the verification to prove properties of the *compound protocol*, under certain conditions.

Definition 15. The component protocols $\mathbf{P_i}$, $i=1,\ldots,n$ with n the number of the components, that form a compound protocol $\mathbf{P_j}$ are called ***sub-protocols*** of it.

Definition 16. A protocol that has not sub-protocols, (i.e. is not composed by other protocols) is called a ***base-level protocol***. A base level protocol is composed at most by its agents.

Additionally, agents (or component objects) of a base level protocol are either *base level objects* or *compound objects* (i.e. objects composed by components).

The hierarchical object composition method uses *projections operations* to compose objects. Informally, projection operations "project" the state of components at the level of compound object.

Definition 17. Two actions of a compound protocol are in the same ***action group*** when they change the state of the same component protocol via a projection operation.

Parallel protocol composition (i.e. without synchronization) is the most fundamental form of behavioural protocol composition. From a methodological perspective, to compose protocol components in parallel:

- involving the corresponding projection which expresses the relationship between them;
- we write equations expressing the fact that each compound protocol action affects only the states of its corresponding component;
- each compound protocol observation is specified as an abbreviation of a component protocol observation;
- each compound protocol state constant is projected to a state constant on each component.
- we add a new hidden sort for the states of the compound protocol;
- for each component protocol action, we add a corresponding compound action and an equation

Example 1. Let us consider the following simple data transfer protocol: A sender **S** wants to deliver natural numbers 0, 1, 2, … to a receiver **R** from a list via a cell. The sender puts a number in the cell, and receiver gets a number from the cell, if the cell is not empty. Sender and receiver share a boolean variable called flag, which is initially false. While flag is false, sender repeatedly puts a number in the cell. The receiver upon receiving a number stores it to a list and sets the flag to true. If flag is true, sender picks the next number and sets flag to false. This simple protocol's specification consists of abstract data type specifications for data types such as lists and cell, and a behavioural object. We assume that we have two specifications of the protocol P_1 and P_2. The first models the communication between sender **S** and receiver $\mathbf{R_1}$, while the second the communication between sender **S** and receiver $\mathbf{R_2}$. The parallel composition of P_1 and P_2 results to a protocol of a sender **S** that sends data to receivers $\mathbf{R_1}$ and $\mathbf{R_2}$ independently. This means that the behaviour of each component protocol does not affect the behaviour of the other component protocol (which is projected at the level of compound protocol with projection operations).

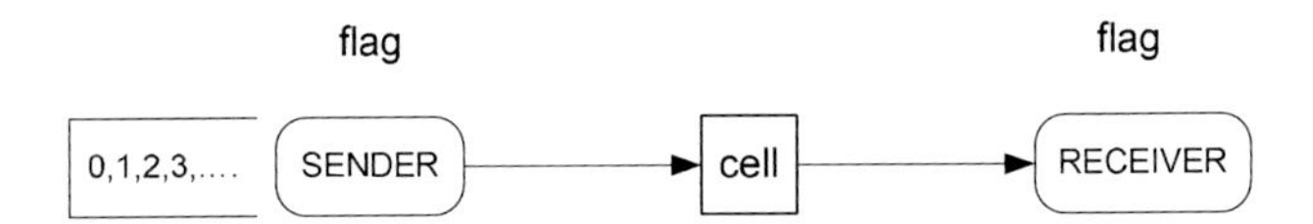

In the case of *synchronized parallel composition*, we may also allow on the compound protocol actions other than those of the components. These actions may change the states either of a number of component protocols (broadcasting) or of one (client-server computing).

Example 2. We consider the component protocols P_1 and P_2 of example 1. A synchronized composition of them may result in a protocol with a sender sending the same sequence of data to both receivers. To send the next number, the flags shared by sender with both receivers must have been set to true. In this case the behaviour and states of the compound protocol depends on the component protocols.

We note here that we omit the formal definition of behavioural protocol composition because it can be easily derived from the parallel composition of behavioural objects defined in [4]. Additionally, parallel protocol composition inherits several semantic properties of object composition such as *associativity* and *commutativity*.

Apart from composition, another way of combining protocols follows a layered style of structuring (protocol stack). According to this style, protocols utilize services rendered by other protocols, and extend their services to be used in conjunction with other protocols to achieve the overall desired objective. To capture this kind of protocol interactions, the specification of a protocol can import another protocol specification. It is permitted by the underlying behavioural specification paradigm, and we distinguish three kinds of importation: *protecting*, *extending* and *using*. For complex protocols, a module may import several other modules.

Example 3. Let us consider the TCP/IP protocol stack. It consists of four layers each of which consists of a number of protocols. A protocol of the top layer (application layer) such as File Transfer Protocol uses the protocols of the lower layer (host-to-host transport layer) to deliver data. So, the behavioural specification of FTP imports the specification or the *sum* of specifications of the corresponding protocols to provide the service.

4 Protocol Algebra

Here we introduce the operators of *protocol algebra*. These are *sum/combination* (+), *import* ($\lhd$), *composition* (|| for parallel / $\otimes$ for synchronized), *subsumption* ([), *renaming* (*) and *subprotocol relation* ([|). In addition the temporal ordering of states in a protocol run r is denoted by $<_r$, the behavioural equivalence between two states of a protocol by ~, and the equivalence between two protocols by $\equiv$.

A. Protocol Sum

Since we specify protocols as behavioural objects, the protocol sum corresponds to the sum of the corresponding behavioural theories. Therefore, the properties are:

Proposition 1: Given protocol specifications P1, P2 and P3, then:

1. P1 + P2 = P2 + P1.
2. P1 + (P2 + P3) = (P1 + P2) + P3.
3. P1 + P1 = P1.

Proof: Each protocol specification is a behavioural object say P1 = $((H_{P1}, V_{P1}, F_{P1}, F_{P1}^{b}), E_{P1})$ with hidden sort $h_{P1} \in H_{P1}$ and P2 = $((H_{P2}, V_{P2}, F_{P2}, F_{P2}^{b}), E_{P2})$ with hidden sort $h_{P2} \in H_{P2}$. Since the sum of them is defined to be the union of the sets that a behavioral object consists of, the proof of the above uses that union has the three corresponding properties.

The protocol sum operator is used when importing/reusing more than one protocol specifications.

B. Protocol Import

Definition 18. We denote the partial order of protocol imports by $\lhd_{mode}$, where mode may be *protecting*, *extending* or *using* [16-17]. When we specify a protocol P1 that reuses the specification of protocol P2 in protecting mode we write P2 $\lhd_{pr}$ P1.

A protocol specification may reuse more than one protocol specifications. In this case we use the sum operator, e.g. for protocols P1, P2, P3, (P1+P2) $\lhd_{pr}$ P3, i.e. protocol P3 imports the specification of protocols P1 and P2 in protecting mode.

Proposition 2: Given protocol specifications P1, P2, and P3, then if P2 $\lhd_{mode}$ P1 and P3 $\lhd_{mode}$ P2 implies P3 $\lhd_{mode}$ P1.

C. Protocol Composition

Since protocols are described as behavioural objects, hierarchical object composition technique can be applied to the composition of protocols. We distinguish two types of composition, the parallel composition, that is the most fundamental, and the synchronized parallel composition, where communication between component objects occurs. Below, we adapt the results of [4], to protocol composition.

- Parallel protocol composition

Definition 19. Let P1 and P2 protocol specifications. By P1 || P2 is denoted the class of protocols P which are parallel compositions of protocols P1 and P2.

Proposition 3: For any protocol specifications P1, P2, for each parallel composition P $\in$ P1 || P2, we have that $\alpha \sim_A \alpha'$ if and only if $A_{\pi 1}(\alpha) \sim_{A1} A_{\pi 1}(\alpha')$ and $A_{\pi 2}(\alpha) \sim_{A2} A_{\pi 2}(\alpha')$, for each P-algebra A, elements $\alpha, \alpha' \in A_{hB}$, and A_i is the reduct of A to B_i for each $i \in \{1,2\}$.

For all protocols P1 and P2, all P, P' $\in$ P1 || P2 are equivalent protocols, i.e. P$\equiv$P'. Additionally, parallel composition has several expected semantic properties such as associativity and commutativity that are applied to protocol composition:

For all protocols P1, P2, P3

- P1 || P2 = P2 || P1, and
- P(12)3 $\equiv$P1(23) for all P(12)3 $\in$ P12 || P3 and P1 || P23, where PIJ is any composition in PI || PJ.

- Synchronized protocol composition

This is the most general form of protocol composition. It supports dynamic compositions and synchronization, while a kind of communication between components (either protocols or objects) is allowed. The class of protocols P which are synchronized compositions of protocols P1 and P2 is denoted by P1 $\otimes$ P2.

Theorem 2. For any protocols P1 and P2, for each composition with synchronization $P \in P1 \otimes P2$, we have that $\alpha \sim_A \alpha'$ if and only if $(\forall W_i)\ A_{\pi i}(\alpha, W_i) \sim_{Ai} A_{\pi i}(\alpha', W_i)$ for $i \in \{1,2\}$ for each P-algebra A, elements $\alpha, \alpha' \in A_{hB}$, and A_i is the reduct of A to B_i for each $i \in \{1,2\}$.

We note that in case a protocol P is composed by protocols P1 and P2, then also $(P1+P2) \lhd P$.

D. Protocol Subsumption

In definition 13 we introduce the notion of protocol run subsumption. Here we define it using operators from our protocol algebra and introduce the notion of *protocol subsumption.*

Let r_i, r_j be protocol runs and relation $r_j [\ r_i$ denoting subsumption of r_i by r_j. Then:

$r_j [\ r_i \Leftrightarrow \forall s_i \in r_i, \exists s_j \in r_j : s_j \sim s_i$ and $\forall s_i' \in r_i, \exists s_j' \in r_j : s_j' \sim s_i' \Rightarrow (s_i <_{ri} s_i' \Rightarrow s_j <_{ri} s_j')$. It is obvious that longer runs subsume shorter ones provided they have behaviourally equivalent states in the same order. Subsumption of runs has also the two following properties:

- Reflexivity: Every protocol run subsumes itself.
- Transitivity: If $r_j [\ r_i$ and $r_k [\ r_j$ then $r_k [\ r_i$.

Definition 20. A protocol P_j subsumes a protocol P_i if and only if, every run of P_j subsumes a protocol run of P_i. The set of protocol runs that a protocol allows are denoted by $\{P_i\}$ and $\{P_j\}$ correspondingly. We write $\forall r_j \in \{P_j\}\ \exists r_i \in \{P_i\}$: $r_j [\ r_i$.

The protocol runs that a protocol allows are defined at the equation part of its specification.

Definition 21. Two protocols P1, P2 are ***similar*** if and only if for every run in one protocol there exists at least one similar run in the other protocol and vice versa.

Definition 22. The composition of two protocols P1 and P2, creates a protocol whose runs may subsume some run from {P1} and some run from {P2}.

E. Subprotocol relation

Definition 23. A protocol P is a subprotocol of Q (P [| Q) if and only if Q is a compound protocol and P is a component of it.

As it is obvious, the operator [| is related to composition operators || and $\otimes$. Given protocols P1, P2 and P, if P = P1 || P2 or P = P1 $\otimes$ P2 then P1 [| P and P2 [| P.

Theorem 3. A compound protocol subsumes its subprotocols.

Proof. A compound protocol has actions that correspond to the actions of its components (action groups). Since every protocol run is sequence of states and each state depends on the application of actions, every component protocol run corresponds to a protocol run of the compound protocol. From def. 20 the compound protocol subsumes its components. Also, from def. 21 a compound and a component protocol are similar.

F. Protocol renaming

Protocol renaming is necessary when reusing protocol specifications, since in most cases we wish to avoid having shared variables between two protocol specifications that are going to be combined /composed. We use the operator * for renaming a module. A convention we adopt is that if we have a protocol specification **P** and we want to rename it to **P1** we write φ***P**, where φ is a signature morphism.

F. Applying protocol algebra

Here we present how we can apply our algebra to the examples 1 and 2. The base level protocol DTP (Data Transfer Protocol) imports other modules specifying data types, so we write `(NAT + CELL + LIST)` $\lhd_{pr}$ `DTP`. The renaming of `DTP` results to two identical protocols `DTP1`, `DTP2` that inherit all properties of `DTP`. We have φ_1 = {`Protocol -> Protocol1, init -> init1`} and φ_2 = {`Protocol -> Protocol2, init -> init2`} and so, `DTP1` = φ_1* `DTP`, `DTP2` = φ_2* `DTP`. Additionally, the parallel composition of `DTP1` and `DTP2`, which is denoted as `DTP1` || `DTP2`, results to `2DTP`. Since `DTP1` = φ_1* `DTP` and `(NAT + CELL + LIST)` $\lhd_{pr}$ `DTP` $\Rightarrow$ `(NAT + CELL + LIST)` $\lhd_{pr}$ `DTP1`. The same occurs for `DTP2`. Additionally, due to def. 23, `DTP1` and `DTP2` are subprotocols of `DTP`, i.e. `DTP1` [| `DTP`, `DTP2` [| `DTP`. The protocol runs of `DTP` is denoted by `{DTP}`, and the temporal ordering of the states that constitute it is: $s_0 <_{\{DTP\}} s_1 <_{\{DTP\}} s_2 <_{\{DTP\}} s_3 <_{\{DTP\}} s_4 <_{\{DTP\}} s_5 <_{\{DTP\}} s_6 <_{\{DTP\}} s_7$. Here we denote the transition relation $<s_i, a, s_{i+1}>$ which means that each protocol state (target state) s_{i+1} emerges by another state s_i (source state) after applying an action `a`. Similarly to parallel composition, the synchronized parallel composition, with state space `2Protocol`, is denoted by `DTP1` $\otimes$ `DTP2`. We also can show the subsumption of runs that the protocol allows: Let $R1_{DTP} = \langle s_0\ s_1\ s_2 \rangle$ and $R2_{DTP} = \langle s_0\ s_1\ s_2\ s_3\ s_4 \rangle$. Then it is obvious that $R2_{DTP}$ [$R1_{DTP}$. In order to examine the correctness of a possible run we can simulate it with CafeOBJ system. For example, a run that begins from a state other than s_0 = `init` is not valid. The subsumption of runs can be very useful especially when dealing with security protocols. For example, we can adapt the protocol specification so as the protocol support additional services/options to customers of an e-commerce site.

Reusing protocol specifications can be very useful for protocol development. This is because, we can specify and verify a protocol design once and then reuse it. We can create libraries of base level protocols that will speed up the process of designing a new protocol. It will be also easier to handle the compound protocols.

5 Verifying Protocol Specifications

The two main proof methods that are supported by behavioral specification are structural induction and coinduction. They are used to verify safety (invariant) and behavioral properties of systems. These two kinds of properties are of major importance for protocol specification. In [4] many important results have been presented for the compositionality of verifications. The approach supports reusability of proofs which simplifies the verification process.

Definition 24. An ***invariant property*** of a protocol is a state predicate which holds in all reachable states of the protocol.

Theorem 4. Let $\mathbf{P_1}$ and $\mathbf{P_2}$ be protocols and $\mathbf{P_{12}}$ the parallel composition of them. If **I** is an invariant of $\mathbf{P_1}$ ($\mathbf{P_2}$) which refers only to protocol observations of $\mathbf{P_1}$ ($\mathbf{P_2}$), then **I** is an invariant of $\mathbf{P_{12}}$.

Proof. Since $\mathbf{P_{12}}$ is the parallel composition of $\mathbf{P_1}$ and $\mathbf{P_2}$, for each protocol run of $\mathbf{P_{12}}$ there exists a corresponding protocol run of $\mathbf{P_1}$ ($\mathbf{P_2}$). Additionally, since **I** refers only to observations of $\mathbf{P_1}$ ($\mathbf{P_2}$), the protocol run affects the state of each protocol independently. So, if the invariant holds for the component protocol it holds also for the compound protocol.

Behavioral properties of protocols may include the behavioral equivalence of two protocol states. To prove such properties, coinduction method is used. In general, to prove $(\forall X)t = t'$ the steps which are followed are:

- define a hidden equivalence relation _R_
- prove that _R_ is a congruence,
- prove that t R t'.

The correctness of the coinduction method follows from the fundamental result characterizing the behavioral equivalence as the largest hidden congruence. Notice that _R_ does not refer to a particular relation on a particular model but it is rather interpreted in all models as any other operation symbol.

The behavioral equivalence proof is very important for protocol reasoning, since many relations between protocols are based on it.

By iterative application of Theorem 2, in the case of protocol composition, the behavioral equivalence for the compound protocol is just the conjunction of the behavioral equivalences of the base-level protocols, which are generally simpler and may checked automatically by systems that support behavioral proofs such as CafeOBJ. This means that many times the behavioral proofs are almost automatic without having to use the usual coinduction method.

CafeOBJ also provides a methodology to prove the correctness of composition of protocols. It is based on the idea that a composition is correct when the composed protocol is the refinement of its components and for the concurrent part the commutativity equations corresponding to the concurrency of methods/attributes belonging to different components holds.

6 Conclusions and Future Work

We have sketched some first steps towards the definition of a protocol algebra based on the behavioral algebraic specification framework. Hierarchical object composition technique based on it also can be utilized to compose protocols by sub-protocols. The proposed protocol algebra is based on the basic operators of module algebra to handle protocol reasoning and design and it seems to be a promising approach. Our framework can be implemented by CafeOBJ, an algebraic specification language that supports behavioural specification and verification techniques such as the OTS/CafeOBJ method and coinduction.

We have already applied a combination of behavioural specification and composition to the modeling of mobile systems in [8]. In the future there is much work to be done. We aim to conduct case studies to show the applicability of our framework in the area of specifying and verifying business protocols based on the commitments approach [9]. The definition of the hierarchical composition of OTSs is another work to be done. Finally, a library of verified base level protocol specifications can be developed for highly reusable protocol development.

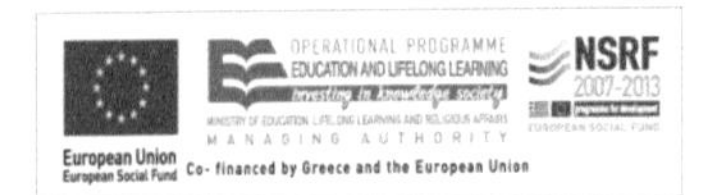

Aknowledgements. This research has been co-financed by the European Union (European Social Fund – ESF) and Greek national funds through the Operational Program "Education and Lifelong Learning" of the National Strategic Reference Framework (NSRF) - Research Funding Program: THALIS.

References

1. Diaconescu, R.: Behavioural Coherence in Object-Oriented Algebraic Specification. Universal Computer Science 6(1), 74–96 (2000)
2. Goguen, J.A., Diaconescu, R.: Towards an algebraic semantics for the object paradigm. In: Ehrig, H., Orejas, F. (eds.) ADT 1992 and COMPASS 1992. LNCS, vol. 785, pp. 1–34. Springer, Heidelberg (1994)
3. Goguen, J., Malcolm, G.: A hidden agenda. Technical Report CS97-538, University of California at San Diego (1997)
4. Diaconescu, R.: Behavioral Specification for Hierarchical Object Composition. Theoretical Computer Science 343, 305–331 (2005)
5. Diaconescu, R., Futatsugi, K.: CafeOBJ report. World Scientific (1998)
6. Bergstra, J.A., Heering, J., Klint, P.: Module Algebra. Journal of ACM 37(2), 335–372 (1990)
7. Diaconescu, R., Goguen, J.A., Stefaneas, P.: Logical Support for Modularisation. In: Huet, G., Plotkin, G. (eds.) Logical Environment, pp. 83–130 (1993)
8. Ouranos, I., Stefaneas, P., Frangos, P.: An Algebraic Framework for Modeling of Mobile Systems. IEICE Trans. Fund. E90-A(9), 1986–1999 (2007)
9. Mallya, A.P., Singh, M.P.: An algebra of commitment protocols. Journal of Autonomous Agents and Multi - Agent Systems 14(2), 143–163 (2007)
10. Singh, G.: A compositional approach for designing protocols. In: Proceedings of the International Conference on Network Protocols, San Francisco, CA, USA, pp. 98–107 (1993)
11. Sinha, P., Suri, N.: Modular Composition of Redundancy Management Protocols in Distributed Systems: An Outlook on Simplifying Protocol Level Formal Specification and Verification. In: Proceedings of ICDCS-21, pp. 255–263 (2001)
12. Garbinato, B., Felber, P., Guerraoui, R.: Modeling Protocols as Objects for Structuring Reliable Distributed Systems. In: Proceedings of the Communication Networks and Distributed Systems Modeling and Simulation Conference (CNDS 1997), Phoenix, Arizona, pp. 165–171 (1997)
13. Singh, G., Mao, Z.: Structured design of communication protocols. In: Proceedings of the IEEE International Conference on Distributed Computing Systems (1986)
14. Hagalisletto, A.M.: Protocol Algebra. In: Proceedings of the 11th IEEE Symposium on Computers and Communications (2006)
15. Mallya, A.U., Singh, M.: An algebra for commitment protocols. Autonomous Agents and Multiagent Systems 4(2), 143–163 (2007)
16. Futatsugi, K., Goguen, J., Jouannaud, J.-P., Meseguer, J.: Principles of OBJ2. In: Proceedings of the 12th ACM Symposium on Principles of Programming Languages, pp. 52–66. ACM (1985)
17. Goguen, J., Winkler, T., Meseguer, J., Futatsugi, K., Jouannaud, J.-P.: Introducing OBJ. Technical report, SRI International, Computer Science Laboratory (1993)
18. Miranda, H., Pinto, A., Rodrigues, L.: Appia: A flexible protocol kernel supporting multiple coordinated channels. In: Proceedings of the 21st Int. Conf. on Distributed Computing Systems (ICDCS 2001), Washington - Brussels - Tokyo, pp. 707–710 (2001)
19. Wong, G., Hiltunen, M., Schlichting, R.: CTP: A configurable and extensible transport protocol. In: Proceedings of the 20th Annual Conference of IEEE Communications and Computer Societies (INFOCOM 2001), Anchorage, Alaska (2001)

20. Wojciechowski, P., Rütti, O., Schiper, A.: SAMOA: Framework for synchronization augmented microprotocol approach. In: Proceedings of Int. Parallel and Distributed Processing Symposium (IPDPS 2004), Santa Fe, US (2004)
21. Ogata, K., Futatsugi, K.: Some Tips on Writing Proof Scores in the OTS/CafeOBJ Method. In: Futatsugi, K., Jouannaud, J.-P., Meseguer, J. (eds.) Algebra, Meaning, and Computation. LNCS, vol. 4060, pp. 596–615. Springer, Heidelberg (2006)
22. Ogata, K., Futatsugi, K.: Proof scores in the oTS/CafeOBJ method. In: Najm, E., Nestmann, U., Stevens, P. (eds.) FMOODS 2003. LNCS, vol. 2884, pp. 170–184. Springer, Heidelberg (2003)

A Model-Based Testing Approach Combining Passive Conformance Testing and Runtime Verification: Application to Web Service Compositions Deployed in Clouds

Sébastien Salva and Tien-Dung Cao

Abstract. This paper proposes a model-based testing approach which combines two monitoring methods, runtime verification and passive testing. Starting from ioSTS (input output Symbolic Transition System) models, this approach generates monitors to check whether an implementation is conforming to its specification and meets safety properties. This paper also tackles the trace extraction problem by reusing the notion of proxy to collect traces from environments whose access rights are restricted. Instead of using a classical proxy to collect traces, we propose to generate a formal model from the specification, called Proxy-monitor, which acts as a proxy and which can directly detect implementation errors. We apply and specialise this approach on Web service compositions deployed in PaaS environments.

Keywords: Passive Testing, Runtime Verification, Proxy, ioco, Web services, Clouds.

1 Introduction

Software testing is a large process, more and more considered by IT (Information technologies) companies, used to check the correctness or quality of software, that are notions required by end customers. In particular, Model-based Testing, which is the topic of this paper, is an approach where the system to test is formally described with specification models which express its functional behaviours. Beyond the use of formal techniques, these models offer the advantage to automate some (and eventually all) steps of the testing process. Usually, the latter is performed with active approaches: basically, test cases are constructed from the specification and are

Sébastien Salva
LIMOS CNRS UMR 6158, University of Auvergne, France
e-mail: sebastien.salva@udamail.fr

Tien-Dung Cao
School of Engineering, Tan Tao University, Vietnam
e-mail: dung.cao@ttu.edu.vn

R. Lee (Ed.): *SERA*, SCI 496, pp. 99–116.
DOI: 10.1007/978-3-319-00948-3_7

experimented on its implementation to check whether the implementation meets desirable behaviours w.r.t. a test relation which defines the confidence level of the test between the specification and implementations. Active testing may give rise to some inconvenient though, e.g., the repeated or abnormal disturbing the implementation.

Two other complementary approaches are employed to cover implementations over a longer period of time without disturbing them: passive testing and runtime verification. The former relies upon a monitor which passively observes the implementation reactions, without requiring pervasive testing environments. The sequences of observed events, called *traces*, are analysed to check whether they meet the specification. Runtime verification, originating from the Verification area, addresses the monitoring and analysis of system executions to check that strictly specified properties hold in every system states.

Both approaches share some important research directions, such as methodologies for checking test relations and properties, or trace extraction techniques. This paper explores these directions and describes a testing technique which combines the two previous approaches. The main contributions can be summarised threefold:

1. Combination of runtime verification and ioco passive testing: we propose to monitor an implementation against a set of safety properties which express that "nothing bad ever happens". These ones are known to be monitorable and can be used to express a very large set of properties, e.g., security vulnerabilities. We combine this monitoring approach with a previous work dealing with ioco passive testing [13]. Ioco [15] is a well-known conformance test relation which defines the conforming implementations by means of suspension traces (sequences of actions and quiescence). So, starting from an ioSTS (input output Symbolic Transition System) model, our method generates monitors to check whether an implementation is ioco-conforming to its specification and meets safety properties,
2. Trace extraction: to collect traces on a system in production, it is required to have the sufficient access rights on the implementation environment to install testing tools. More and more frequently over recent years, these environment access rights are restricted. For instance, Web server access rights are often strictly limited for security reasons. Another example concerns Clouds. Clouds, and typically PaaS (Platform as a service) layers are virtualised environments where Web services and applications are deployed. This virtualisation of resources combined with access restriction make difficult the trace extraction. We address this issue by using the notion of transparent proxy and by assuming that the implementation can be configured to pass through a proxy (usually the case for Web applications). But, instead of using a classical proxy to collect traces, we propose to generate a formal model from the specification, called Proxy-monitor, which acts as a proxy and which can directly detect implementation errors,
3. Analysis overhead: The proposed algorithms also offer the advantage of performing synchronous (receipt of an event, error detection, forward of the event to its recipient) or asynchronous analyses (receipt and forward of an event, error detection) whereas the use of a basic proxy allows asynchronous analysis only. The overhead, with both synchronous and asynchronous analyses, is measured and discussed in the experiment part.

The paper is structured as follows: initial notations and definitions are given in Section 2. Section 3 gives some definitions about runtime verification and ioco passive testing. The combination of both approaches is defined in Section 4. We apply, in Section 5, the concept of Proxy-monitor on Web service compositions deployed in Windows Azure which is the Cloud platform of Microsoft[1]. Finally, we review some related works in Section 6 and Section 7 concludes the paper.

2 Model Definition and Notations

In this paper, we focus on models called input/output Symbolic Transition Systems (ioSTS). An ioSTS is a kind of automata model, extended with two sets of variables, and with guards and assignments on transitions, which give the possibilities to model the system state and constraints on actions.

Below, we give the definition of an ioSTS extension, called ioSTS suspension which also expresses quiescence i.e., the authorised deadlocks observed from a location. Quiescence is modelled by a new symbol $!\delta$ and an augmented ioSTS denoted $\Delta(ioSTS)$. For an ioSTS $\mathcal{S}$, $\Delta(\mathcal{S})$ is obtained by adding a self-loop labelled by $!\delta$ for each location where no output action may be observed. The guard of this new transition must return true for each value which does not allow firing a transition labelled by an output. More details about ioSTSs can be found in [9].

Definition 1 (ioSTS suspension). An ioSTS suspension is a tuple $< L, l0, V, V0, I, \Lambda, \rightarrow >$, where:

- L is the finite set of locations, with $l0$ the initial one,
- V is the finite set of internal variables, while I is the finite set of parameters. We denote D_v the domain in which a variable v takes values. The internal variables are initialised with the assignment V_0 on V, which is assumed to be unique,
- Λ is the finite set of symbols, partitioned by $\Lambda = \Lambda^I \cup \Lambda^O \cup \{!\delta\}$: Λ^I represents the set of input symbols, (Λ^O) the set of output symbols,
- $\rightarrow$ is the deterministic finite transition set. A transition $(l_i, l_j, a(p), G, A)$, from the location $l_i \in L$ to $l_j \in L$, denoted $l_i \xrightarrow{a(p),G,A} l_j$ is labelled by an action $a(p) \in \Lambda \times \mathcal{P}(I)$, with $a \in \Lambda$ and $p \subseteq I$ a finite set of interaction variables. G is a guard over $(p \cup V \cup T(p \cup V))$ which restricts the firing of the transition. $T(p \cup V)$ are boolean terms over $p \cup V$. Internal variables are updated with the assignment function A of the form $(x := A_x)_{x \in V}$ A_x is an expression over $V \cup p \cup T(p \cup V)$.

Web service compositions exhibit special properties relative to the service-oriented architecture (operations, partners, etc.). This is why we adapt the ioSTS action modelling. To represent the communication behaviours of Web service compositions with ioSTSs, we firstly assume that an action $a(p)$ expresses either the call of a Web service operation op with $a(p) = opReq(p)$, or the receipt of an operation response with $a(p) = opResp(p)$, or quiescence. The set of parameters p must gather also some specific variables:

[1] `http://www.windowsazure.com`

- the variable *from* is equal to the calling partner and the variable *to* is equal to the called partner,
- Web services may engage in several concurrent interactions by means of several stateful instances called sessions, each one having its own state. For delivering incoming messages to the correct running session when several sessions are running concurrently, the usual technical solution is to add, in messages, a correlation value set which matches a part of the session state [1]. A correlation set is modelled with a parameter denoted *coor* in *p*.

The use of correlation sets also requires the following hypotheses which result from the correlation sets functioning. Particularly, the last one is required to correlate some successive operation calls with the same composite service instance:

Session identification: the specification is well-defined. When a message is received, it always correlates with at most one session.
Message correlation: except for the first operation call which starts a new composition instance, a message $opReq(p)$, expressing an operation call, must contain a correlation set $coor \subseteq p$ such that a non-empty subset $c \subseteq coor$ is composed of parameter values given in previous messages.

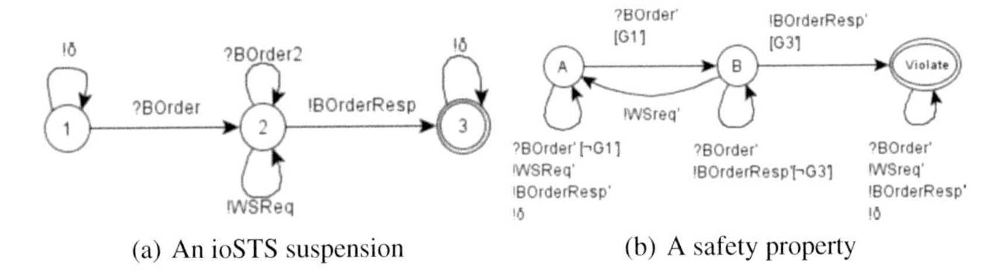

(a) An ioSTS suspension (b) A safety property

Fig. 1 ioSTS specifications

These notation are expressed in the straightforward example of Figure 1(a). For each symbol of Figure 1(a), Table 1 gives the corresponding action, guard and assignment. This specification describes the functioning of a *BookSeller* service. A client places an order composed of a list of books with *BookSeller* by supplying an ISBN list and the quantity of books ordered. *BookSeller* calls a service *Wholesaler* with *WholeSalerReq* to buy each book one by one. For one composition instance, we have two sessions of Web services connected together with correlations sets. Each session is identified with its own correlation set e.g., *BookSeller* with $c1 = \{account = "custid"\}$, and *Wholesaler* with $c2 = \{account = "custid", isbn = "2070541274"\}$. As these two correlation sets respect the Message correlation assumption, we can correlate the call of *Wholesaler* with one previous call of *BookSeller* even though several sessions are running in parallel.

An ioSTS is also associated to an ioLTS (Input/Output Labelled Transition System) to formulate its semantics. Intuitively, the ioLTS semantics is a valued automaton often infinite: the ioLTS states are labelled by internal variable values while

Table 1 Symbol table

Symbol	Action	Guard	Update
?BOrder	?BookOrderReq(List Books, quantity, account,from,to,corr)	$G1$=[from="Client" ∧to="BR"∧ $corr = \{account\}$]	q:=quantity, b:=ListBooks, c1:=corr
?BOrder2	?BookOrderReq(List Books, quantity, account,from,to,corr)	$[\neg G1]$	
!BOrderResp	!BookOrderResp(resp, from, to, corr)	$G3$=[from="BR"∧ to= "Client"∧ resp="Order done"∧corr=c1]	
?R1	?BookOrderResp ?WholeSalerReq		
?R2	?BookOrderResp ?WholeSalerReq ?δ	[≠G3] [≠G2]	
!WSReq	!WholeSalerReq(isbn, from, to, corr)	$G2$=[isbn=b[q]∧ $q \geq 1$∧from= "BR"∧to= "WS"∧ $corr = \{a, isbn\}$]	$q := q - 1$
?BOrderReq'	?BookOrderReq(List Books, quantity,account)	G1'=[quantity≥ 1]	
!WSReq'	!WholeSalerReq(isbn)		
!BOrderResp'	!BookOrderResp(resp)	G3'=[end(resp)="done"]	
!BOrder [G1']	?BookOrderReq(List Books, quantity,)	[G1']	q:= quantity, b:=ListBooks, c1:= corr
?BOrderResp[$G3 \wedge G3'$]	!BookOrderResp(resp)	$[G3 \wedge G3']$	
?BOrderResp[$\neg G3 \wedge G3'$]	!BookOrderResp(resp, from, to, corr)	$[\neg G3 \wedge G3']$	
?R3	?BookOrderResp ?WholeSalerReq ?δ	$[\neg G3 \wedge \neg G3']$ [≠G2]	

transitions are labelled by actions and parameter values. The semantics of an ioSTS $\mathcal{S} =< L, l0, V, V0, I, \Lambda, \rightarrow >$ is an ioLTS $||\mathcal{S}|| =< Q, q_0, \Sigma, \rightarrow >$ composed of valued states in $Q = L \times D_V$. $q_0 = (l0, V0)$ is the initial one, Σ is the set of valued symbols and $\rightarrow$ is the transition relation. The complete definition of ioLTS semantics can be found in [9].

Runs and traces of ioSTS can be defined from their semantics:

Definition 2 (Runs and traces). For an ioSTS $\mathcal{S}$, interpreted by its ioLTS semantics $||\mathcal{S}|| =< Q, q_0, \Sigma, \rightarrow >$, a run $q_0\alpha_0...\alpha_{n-1}q_n$ is an alternate sequence of states and valued actions. $Run_F(\mathcal{S}) = Run_F(||\mathcal{S}||)$ is the set of runs of $\mathcal{S}$ finished by a state in $F \times D_V \subseteq Q$ with F a set of locations of $\mathcal{S}$.

It follows that a trace of a run r is defined as the projection $proj_{\Sigma}(r)$ on actions. $Traces_F(\mathcal{S}) = Traces_F(||\mathcal{S}||)$ is the set of traces of runs finished by states in $F \times D_V$.

The parallel product is a classical state-machine operation used to produce a model representing the shared behaviours of two original automata. For ioSTSs, these ones are to be compatible:

Definition 3 (Compatible ioSTSs). An ioSTS $\mathcal{S}_1 = < L_1, l0_1, V_1, V0_1, I_1, \Lambda_1, \rightarrow_1>$ is compatible with $\mathcal{S}_2 = < L_2, l0_2, V_2, V0_2, I_2, \Lambda_2, \rightarrow_2>$ iff $V_1 \cap V_2 = \emptyset$, $\Lambda_1^I = \Lambda_2^I$, $\Lambda_1^O = \Lambda_2^O$ and $I_1 = I_2$.

Definition 4 (Parallel product $||$). The parallel product of two compatible ioSTSs $\mathcal{S}_1 =< L_1, l0_1, V_1,$
$V0_1, I_1, \Lambda_1, \rightarrow_1>$ and $\mathcal{S}_2 = < L_2, l0_2, V_2, V0_2, I_2, \Lambda_2, \rightarrow_2>$, denoted $\mathcal{S}_1||\mathcal{S}_2$, is the ioSTS $\mathcal{P} =< L_{\mathcal{P}}, l0_{\mathcal{P}}, V_{\mathcal{P}}, V0_{\mathcal{P}}, I_{\mathcal{P}}, \Lambda_{\mathcal{P}}, \rightarrow_{\mathcal{P}}>$ such that $V_{\mathcal{P}} = V_1 \cup V_2$, $V0_{\mathcal{P}} = V0_1 \wedge V0_2$, $I_{\mathcal{P}} = I_1 = I_2$, $L_{\mathcal{P}} = L_1 \times L_2$, $l0_{\mathcal{P}} = (l0_1, l0_2)$, $\Lambda_{\mathcal{P}} = \Lambda_1 = \Lambda_2$. The transition set $\rightarrow_{\mathcal{P}}$ is the smallest set satisfying the following inference rule:

$$\frac{l_1 \xrightarrow{a(p),G_1,A_1}_{\mathcal{S}_1} l_2, l_1' \xrightarrow{a(p),G_2,A_2}_{\mathcal{S}_2} l2'}{(l_1, l_1') \xrightarrow{a(p),G_1 \wedge G_2, A_1 \cup A_2}_{\mathcal{P}} (l_2, L_2')}$$

We end this Section with the definition of the ioSTS operation $refl$ which exchanges input and output actions of an ioSTS.

Definition 5 (Mirrored ioSTS and traces). Let $\mathcal{S}$ be an ioSTS. $refl(\mathcal{S}) =_{def}< L_{\mathcal{S}}, l0_{\mathcal{S}}, V_{\mathcal{S}}, V0_{\mathcal{S}}, I_{\mathcal{S}}, \Lambda_{refl(\mathcal{S})}, \rightarrow_{\mathcal{S}}>$ where $\Lambda^I_{refl(\mathcal{S})} = \Lambda^O_{\mathcal{S}}$, $\Lambda^O_{refl(\mathcal{S})} = \Lambda^I_{\mathcal{S}}$.

We extend the $refl$ notation on trace sets. $refl : (\Sigma^*)^* \rightarrow (\Sigma^*)^*$ is the function which constructs a mirrored trace set from an initial one (for each trace, input symbols are exchanged with output ones and vice-versa).

3 Passive Testing with Proxy-Testers and Runtime Verification

To reason about conformance and property satisfiability, one assume that an implementation can be modelled with an ioLTS I. I is also assumed to have the same interface as the specification (actions with their parameters) and is input-enabled to accept any action.

For readability, the proofs of the propositions given below can be found in [14].

3.1 Verification of Safety Properties

The primary goal of runtime verification is to check whether an implementation I, from which traces can be observed, meets a set of properties expressed in trace predicate formalisms such as regular expressions, temporal logics or state machines. Given that we wish to merge the verification of safety properties with an ioSTS-based conformance, it sounds natural to also model them with a specific state machine model. We propose to take back the notion of observers [6] which capture the negation of a safety property by means of final "bad" locations. Runs which lead to these locations represent behaviours which violate the property.

Definition 6 (Observer). An Observer is a deterministic ioSTS $\mathcal{O}$ composed of a non empty set of violation locations $Violate_{\mathcal{O}} \subset L_{\mathcal{O}}$. $\mathcal{O}$ must be both input and output-enabled, i.e. for each state $(l,v) \in L_{\mathcal{O}} \times D_{\mathcal{O}}$, and for each valued action $(a(p),\theta) \in \Lambda_{\mathcal{O}} \times D_p$, there exists $(l,v) \xrightarrow{a(p),\theta} (l',v') \in \rightarrow_{||\mathcal{O}||}$. Given an ioSTS $\mathcal{S}$, $Comp(\mathcal{S})$ stands for the set of compatible Observers of $\mathcal{S}$.

For the specification $\mathcal{S}$, an Observer $\mathcal{O}$ has to be input and output-enabled and compatible with $\mathcal{S}$. These assumptions are required to model a safety property which is violated by all the traces in $Traces_{Violate_{\mathcal{O}}}(\mathcal{O})$ and which is satisfied by all the traces in $(\Sigma_{||\mathcal{S}||})^* \setminus Traces_{Violate_{\mathcal{O}}}(\mathcal{O})$. Consequently, given an implementation I, one can say that I satisfies the Observer $\mathcal{O}$ if I does not yield any trace which also violates $\mathcal{O}$:

Definition 7 (Implementation satisfies Observer). Let $\mathcal{S}$ be an ioSTS and I an implementation. I satisfies the Observer $\mathcal{O} \in Comp(\Delta(\mathcal{S}))$, denoted $I \models \mathcal{O}$, if $Traces(\Delta(I)) \cap Traces_{Violate_{\mathcal{O}}}(\mathcal{O}) = \varnothing$.

Figures 1(b) and Table 1 illustrate an example of Observer for the specification of Figure 1(a). It means that "the receipt of an order confirmation ending with "done", without requesting *WholeSaler*, must never occurs".

Two Observers $\mathcal{O}_1$ and $\mathcal{O}_2$, describing two different safety properties, can be interpreted by the Observer $\mathcal{O}_1||\mathcal{O}_2$. In the remainder of the paper, we shall consider only one Observer, assuming that it may represent one or more safety properties.

3.2 *Ioco Testing with Proxy-Testers*

In the paper, conformance is expressed with the relation *ioco* [15], which intuitively means that I is ioco-conforming to its specification $\mathcal{S}$ if, after each trace of the ioSTS suspension $\Delta(\mathcal{S})$, I only produces outputs (and quiescence) allowed by $\Delta(\mathcal{S})$. For ioSTSs, ioco is defined as:

Definition 8. Let I be an implementation modelled by an ioLTS, and $\mathcal{S}$ be an ioSTS. I is ioco-conforming to $\mathcal{S}$, denoted I *ioco* $\mathcal{S}$ iff $Traces(\Delta(\mathcal{S})).(\Sigma^O \cup \{!\delta\}) \cap Traces(\Delta(I)) \subseteq Traces(\Delta(\mathcal{S}))$.

We have shown in our previous work [13] that *ioco* can be checked on implementations by means of a passive testing technique relying upon the concept of Proxy-tester. A Proxy-tester formally expresses the functioning of a transparent proxy, able to collect traces and to detect non-conformance without requiring to be set up in the same environment as the implementation one. We recall here some notions about Proxy-testers.

The Proxy-tester of a deterministic ioSTS $\mathcal{S}$ is derived from its Canonical tester $Can(\mathcal{S})$. This model is composed of the transitions of $\rightarrow_{\Delta(refl(\mathcal{S}))}$, i.e. the specification transitions labelled by mirrored actions (inputs become outputs and vice-versa).

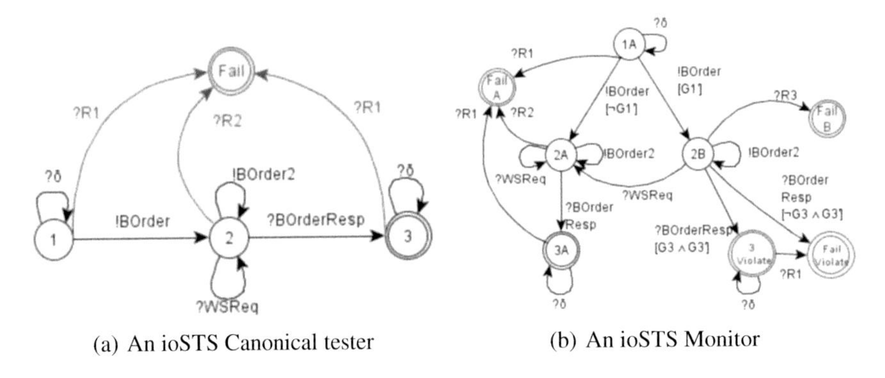

(a) An ioSTS Canonical tester (b) An ioSTS Monitor

Fig. 2 Canonical tester and monitor examples

It is also enriched with transitions leading to a new location *Fail*, exhibiting the receipt of unspecified actions (expressing incorrect behaviours).

Instead of giving the definition of the Canonical tester, which can be found in [14], we illustrate in Figure 2(a) and Table 1 the Canonical tester of the ioSTS depicted in Figure 1(a). The specification actions are mirrored and, for instance, if we consider the location 2, new transitions to *Fail* are added to model the receipt of unspecified events (messages or quiescence).

The Proxy-tester of an ioSTS $\mathcal{S}$ corresponds to an augmented Canonical tester where all the transitions, except those leading to Fail, are *doubled* to express the receipt of an event and the forwarding to its addressee.

Definition 9 (Proxy-tester). The Proxy-tester of the ioSTS $\mathcal{S} =< L_{\mathcal{S}}, l0_{\mathcal{S}}, V_{\mathcal{S}}, V0_{\mathcal{S}}, I_{\mathcal{S}}, \Lambda_{\mathcal{S}}, \rightarrow_{\mathcal{S}}>$ is the ioSTS $Pr(Can(\mathcal{S}))$ where Pr is an ioSTS operation such that $Pr(Can(\mathcal{S})) =_{def}< L_{\mathcal{P}} \cup LF_{\mathcal{P}}, l0_{Can(\mathcal{S})}, V_{Can(\mathcal{S})} \cup \{side, pt\}, V0_{Can(\mathcal{S})} \cup \{side := "", pt := ""\}, I_{Can(\mathcal{S})}, \Lambda_{\mathcal{P}}, \rightarrow_{\mathcal{P}}>$. $LF_{\mathcal{P}} = LF_{Can(\mathcal{S})} = \{Fail\}$ is the Fail location set. $L_{\mathcal{P}}$, $\Lambda_{\mathcal{P}}$ and $\rightarrow_{\mathcal{P}}$ are constructed with the following rules:

$$\frac{l_1 \xrightarrow{!a(p),G,A}_{Can(\mathcal{S})} l_2, l_2 \notin LF_{Can(\mathcal{S})}}{l_1 \xrightarrow{?a(p),G,A\cup\{p_t:=p,side:=""\}}_{\mathcal{P}} (l_1,l_2,a(p),G) \xrightarrow{!a(p),[p=p_t],\{(x:=x)_{x\in V_{Can(\mathcal{S})}},side:="Can",pt:=pt\}}_{\mathcal{P}} l_2}$$

$$\frac{l_1 \xrightarrow{?a(p),G,A}_{Can(\mathcal{S})} l_2, l_2 \notin LF_{Can(\mathcal{S})}}{l_1 \xrightarrow{?a(p),G,A\cup\{p_t:=p,side:="Can"\}}_{\mathcal{P}} (l_1,l_2,a(p),G) \xrightarrow{!a(p),[p=p_t],\{(x:=x)_{x\in V_{Can(\mathcal{S})}},side:="",pt:=pt\}}_{\mathcal{P}} l_2}$$

$$\frac{l_1 \xrightarrow{a(p),G,A}_{Can(\mathcal{S})} l_2, l_2 \in LF_{Can(\mathcal{S})}}{l_1 \xrightarrow{a(p),G,A\cup\{side:="Can",pt:=pt\}}_{\mathcal{P}} l_2}$$

Intuitively, the two first rules double the transitions whose terminal locations are not in the Fail location set LF to express the functioning of a transparent proxy. The first rule means that, for an event (action or quiescence) initially sent to the implementation, the Proxy-tester waits for this event and then forwards it. The two transitions are separated by a unique location composed of the tuple $(l_1, l_2 a(p) G)$ to ensure that these two transitions, and only them, are successively fired. The last rule enriches the resulting ioSTS with transitions leading to Fail. A new internal variable, denoted *side*, is also added to keep track of the transitions provided by the Canonical tester (with the assignment side:="Can"). This distinction will be useful to define partial traces of Proxy-testers and to express conformance with them.

Previously, we have also intentionally enriched Proxy-tester transitions with an assignment on the variable *side*. The assignments $side = "Can"$ mark the transitions carrying actions provided by the Canonical-tester. These assignments help to extract partial runs and traces in Proxy-testers:

Definition 10 (Partial runs and traces). Let $\mathcal{P}$ be a Proxy-tester and $||\mathcal{P}|| = P =< Q_P, q0_P, \sum_P, \rightarrow_P >$ be its ioLTS semantics. We define $Side : Q_P \rightarrow D_{V_{\mathcal{P}}}$ the mapping which returns the valuation of the *side* variable of a state in Q_P. $Side^E(Q_P) \subseteq Q_P$ is the set of states $q \in Q_P$ such that $Side(q) = E$.

Let $Run(\mathcal{P})$ be the set of runs of $\mathcal{P}$. We denote $Run^E(\mathcal{P})$ the set of partial runs derived from the projection $proj_{Q_P \sum_P Side^E(Q_P)}(Run(\mathcal{P}))$. It follows that $Traces^E(\mathcal{P})$ is the set of partial traces of (partial) runs in $Run^E(\mathcal{P})$.

With these notations, we have showed that *ioco* can be rephrased with Proxy-tester traces by [14]:

Proposition 1
$I \; ioco \; \mathcal{S} \Leftrightarrow Traces(\Delta(I)) \cap refl(Traces^{Can}_{Fail}(Pr(\, Can(\mathcal{S})))) = \varnothing$

4 Combining Runtime Verification and Proxy-Testing

4.1 Proxy-Tester and Observer Composition

Canonical testers are enough for detecting all the implementations that are not ioco-conforming to a given specification since they reflect incorrect behaviours in Fail states. Observers offer at least one similarity with Canonical testers since they describe undesired behaviours. This similarity tends to combine them to produce a model which could be used to detect both property violations and non-conformance. This product is called *Monitor*. It refines the original Canonical tester behaviours by separating the traces which violate safety properties among all the traces which may be observed from the implementation under test. A monitor is defined as:

Definition 11 (Monitor). Let $\Delta(\mathcal{S})$ be an ioSTS suspension and $\mathcal{O} \in Comp(\Delta(\mathcal{S}))$ be an Observer. The Monitor of the Canonical tester $Can(\mathcal{S})$ and of the Observer $\mathcal{O}$ is the ioSTS $\mathcal{M} = Can(\mathcal{S})||(refl(\mathcal{O})$.

As an example, the Monitor constructed from the previous Canonical tester (Figure 2(a)) and the Observer of Figure 1(b) is depicted in Figure 2(b) and Table 1. It contains different verdict locations: Fail received from the Canonical tester, Violate received from the Observer and a combination of both Fail/Violate which denotes non-conformance and the violation of the safety property. For example, the trace "$?BookOrder(,1,"custid")$ $!BookOrderResp("done")$" violates the Observer of Figure 1(b) because *WholeSaler* is not called. This trace reflects also an incorrect behaviour because the response received "done" is incorrect. We should have received "Order done".

The combination of Canonical tester locations with Observer ones leads to new locations labelled by local verdicts. We define these locations exhibiting verdicts by verdict location sets:

Definition 12 (Verdict location sets). Let $Can(\mathcal{S})$ be a Canonical tester and $\mathcal{O} \in Comp(\Delta(\mathcal{S}))$ be a compatible Observer with $\Delta(\mathcal{S})$. The parallel product $\mathcal{M} = Can(\mathcal{S})||refl(\mathcal{O})$ produces several sets of verdict locations defined as follows:

1. **VIOLATE** = $(L_{Can(\mathcal{S})} \setminus \{Fail\}) \times Violate_{\mathcal{O}}$,
2. **FAIL** = $\{Fail\} \times (L_{\mathcal{O}} \setminus Violate_{\mathcal{O}})$,
3. **FAIL/VIOLATE** = $\{(Fail, Violate_{\mathcal{O}})\}$.

In particular, we denote $LF_{\mathcal{M}} = FAIL \cup FAIL/VIOLATE$, the Fail location set of $\mathcal{M}$.

Monitors share many similarities with Canonical testers: they have a mirrored alphabet and a verdict location set LF. Typically, they are specialised Canonical testers recognising also property violations. To passively monitor an implementation, it sounds natural to apply the concept of Proxy-tester on Monitors. This gives a final model called Proxy-monitor:

Definition 13 (Proxy-monitor). Let $\mathcal{M}$ be a Monitor resulting from the parallel product $Can(\mathcal{S})||refl(\mathcal{O})$ with $\mathcal{S}$ an ioSTS and $\mathcal{O} \in Comp(\Delta(\mathcal{S}))$ an Observer compatible with the suspension of $\mathcal{S}$.

We call $Pr(\mathcal{M})$, the Proxy-monitor of $\mathcal{M}$.

Proxy-monitors are constructed as Proxy-testers except that the Fail location sets are different. For a Proxy-tester, there is only one Fail location, whereas a Proxy-monitor has a Fail location set $LF_{\mathcal{M}}$ equals to $FAIL \cup FAIL/VIOLATE$ since it stems from a composition between an Observer and a Canonical tester. Except this difference, transitions of the Monitor are still doubled in its Proxy-monitor.

Before focusing on test verdicts which can be obtained from Proxy-monitors, it remains to define formally the notion of passive monitoring of an implementation I by means of a Proxy-monitor. This product cannot be defined without modelling the external environment, e.g., the client side, which interacts with the implementation. We assume that this external environment can be also modelled with an ioLTS Env which can interact with I (hence $refl(Env)$ is compatible with I and $Traces(\Delta(Env))$ is composed of sequences in $refl((\Sigma_{\Delta(I)})^*)$).

Definition 14 (Monitoring of an implementation). Let $PM = < Q_{PM}, q0_{PM}, \Sigma_{PM}, \rightarrow_{PM} >$ be the ioLTS semantics of a Proxy-monitor $Pr(\mathcal{M})$ derived from an ioSTS $\mathcal{S}$ and an Observer $\mathcal{O} \in Comp(\Delta(\mathcal{S}))$. $QF_{PM} \subseteq Q_{PM} = LF_{Pr(\mathcal{M})} \times D_{V_{Pr(\mathcal{M})}}$ is its Fail state set. $I = < Q_I, q0_I, \Sigma_I \subseteq \Sigma_M, \rightarrow_I >$ is the implementation model, assumed compatible with $\mathcal{S}$ and $Env = < Q_{Env}, q0_{Env}, \Sigma_{Env} \subseteq \Sigma_P, \rightarrow_{Env} >$ is the ioSTS modelling the external environment, compatible with $refl(I)$.

The monitoring of I by $Pr(\mathcal{M})$ is expressed with the product $||_p(Env, PM, I) = < Q_{Env} \times Q_{PM} \times Q_I, q0_{Env} \times q0_{PM} \times q0_I, \Sigma_{PM}, \rightarrow_{||_p(Env,PM,I)} >$ where the transition relation $\rightarrow_{||_p(Env,PM,I)}$ is defined by the smallest set satisfying the following rules. For readability reason, we denote an ioLTS transition $q_1 \xrightarrow[\text{"}E\text{"}]{?a} q_2$ if $Side(q_2) = E$ (the variable *side* is valued to E in q_2).

$$\frac{q_1 \xrightarrow{!a}_{\Delta(Env)} q_2, q_2'' \xrightarrow{?a}_{\Delta(I)} q_3'', q_1' \xrightarrow[\text{"''"}]{?a}_{PM} q_2' \xrightarrow[\text{"}Can\text{"}]{!a}_{PM} q_3'}{q_1 q_1' q_2'' \xrightarrow[\text{"''"}]{?a}_{||_p(Env,PM,I)} q_2 q_2' q_2'' \xrightarrow[\text{"}Can\text{"}]{!a}_{||_p(Env,PM,I)} q_2 q_3' q_3''}$$

$$\frac{q_2 \xrightarrow{?a}_{\Delta(Env)} q_3, q_1'' \xrightarrow{!a}_{\Delta(I)} q_2'', q_1' \xrightarrow[\text{"}Can\text{"}]{?a}_{PM} q_2' \xrightarrow[\text{"''"}]{!a}_{PM} q_3'}{q_2 q_1' q_1'' \xrightarrow[\text{"}Can\text{"}]{?a}_{||_p(Env,PM,I)} q_2 q_2' q_2'' \xrightarrow[\text{"''"}]{!a}_{||_p(Env,PM,I)} q_3 q_3' q_2''}$$

$$\frac{q_2 \xrightarrow{?\delta}_{\Delta(Env)} q_3, q_1'' \xrightarrow{!a}_{\Delta(I)} q_2'', q_1' \xrightarrow[\text{"}Can\text{"}]{?a}_{PM} q_2', q_2' \in QF_{PM}}{q_2 q_1' q_1'' \xrightarrow[\text{"}Can\text{"}]{?a}_{||_p(Env,PM,I)} q_2'}$$

The verdict list can now be drawn up from Definition 12. Concretely, the observed traces lead to a set of verdicts, extracted from the verdict location sets which indicate specification and/or safety property fulfilments or violations:

Proposition 2 (Test verdicts). *Consider an external environment Env, an implementation I monitored with a Proxy-monitor $Pr(\mathcal{M})$, itself derived from an ioSTS $\mathcal{S}$ and an Observer $\mathcal{O} \in Comp(\Delta(\mathcal{S}))$. Let $OT \subseteq Traces(||_p(Env, PM, I))$ be the observed trace set. If there exists $\sigma \in OT$ such that:*

1. *σ belongs to $Traces_{FAIL/VIOLATE}(||_p(Env, PM, I))$, then I does not satisfy the safety property and I is not ioco-conforming to $\mathcal{S}$,*
2. *σ belongs to $Traces_{FAIL}(||_p(Env, PM, I))$, then I is not ioco-conforming to $\mathcal{S}$,*
3. *σ belongs to $Traces_{VIOLATE}(||_p(Env, PM, I))$, then I does not satisfy the safety property.*

Intuitively, the sketch of proof of the above Proposition is based on some successive Trace set replacements. For example with 1), we have $Traces_{FAIL/VIOLATE}(||_p(Env, PM, I)) \neq \varnothing$. By considering successively Definition 11, Definition 13 and Definition 14, $Traces_{FAIL/VIOLATE}$ $(||_p$ $(Env, PM, I))$ can be replaced by $refl(Traces(\Delta(I))) \cap (Traces_{Fail}(Can(\mathcal{S})) \cap Traces_{Violate_{\mathcal{O}}}(refl(\mathcal{O}))) \neq \varnothing$. We deduce that $Traces^{Can}_{FAIL/VIOLATE}(||_p(Env, PM, I)) \neq \varnothing$ iff $refl(Traces(\Delta(I))) \cap Traces_{Fail}(Can(\mathcal{S})) \neq \varnothing (a)$ and iff $refl(Traces(\Delta(I))) \cap refl(Traces_{Violate_{\mathcal{O}}}(\mathcal{O})) \neq \varnothing (b)$. From (a), we have $\neg I$ ioco $\mathcal{S}$, from (b), we have $I \nvDash \mathcal{O}$. The complete proof is given in [14].

5 Application to Web Service Composition Deployed in Clouds

We consider having a Web service composition deployed in a PaaS environment and we assume that each partner participating to the composition (Web services and clients) are configured to pass through a passive tester. The latter, whose architecture is depicted in Figure 3, is mainly based upon Proxy-monitors and aims to collect all the traces of Web service composition instances. To consider these instances and to detect non-conformance or violations of safety properties, several analyser instances, based upon a Proxy-monitor model, are executed in parallel. Any incoming message received from the same composition instance must be delivered to the same analyser instance: this step is performed by a module called *entry-point* which routes messages to the correct analyser instance by means of correlation sets.

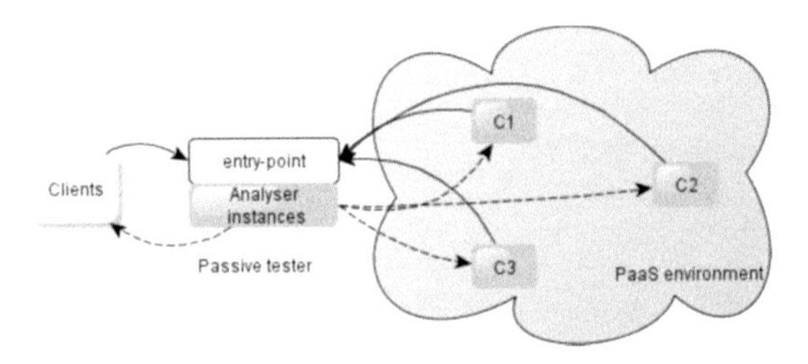

Fig. 3 The passive tester architecture

The entry-point functioning is given in Algorithm 1. The latter handles a set L of pairs (p_i, PV) with p_i an analyser instance identifier and PV the set of parameter values received in previous messages. For each received message, this set is used to correlate it with an existing composition instance in reference to the *Message correlation* hypothesis. Whenever a message $(e(p), \theta)$ is received, its correlation set c is extracted to check if an exiting analyser instance is running to accept it. This instance exists if L contains a pair (p_i, PV) such that a non-empty subset $c' \subseteq c$ is composed of values of PV (correlation hypothesis). In this case, the correlation set has been constructed from parameter values of messages received previously. If an instance is already running, the message is forwarded to it. Otherwise, (line 7), a new one is started. If an analyser instance p_i has returned a trace set (line 11), then the latter is stored in $Traces(Pr(\mathcal{M}))$ and the corresponding pair (p_i, PV) is removed from L.

Algorithm 2 describes the functioning of an analyser. Basically, it waits for an event (message or quiescence), covers Proxy-monitor transitions, and constructs traces to detect non conformance or property violations when a verdict location is reached. Algorithm 2 is based upon a forward checking approach: it starts from the initial state i.e., $(l0_{Pr(\mathcal{M})}, V0_{Pr(\mathcal{M})})$ and constructs a run denoted *Run*. Whenever an event $(e(p), \theta)$ is received (valued action or quiescence), with eventually θ a valuation over p (line 2), it looks for the next transitions which can be fired (line 5). Each transition must have the same start location as the one found in the final state (l, v) of the run *Run*, the same action as the received event $e(p)$ and its guard must be

Algorithm 1. Entry-point

input : Proxy-monitor $Pr(\mathcal{M})$
output: $Traces(Pr(\mathcal{M}))$

1 $L = \varnothing$;
2 **while** *message* $(e(p), \theta)$ **do**
3   extract the correlation set c in θ;
4   **if** $\exists (p_i, PV) \in L$ *such that* $c' \subseteq c$ *and* $c' \subseteq PV$ **then**
5   forward $(e(p), \theta)$ to p_i; $PV = PV \cup \theta$;
6   **else**
7   create a new $Pr(\mathcal{M})$ instance p_i;
8         $L = L \cup (p_i, \{\theta\})$; send $(e(p), \theta)$ to p_i;
9   **if** $\exists (p_i, PV) \in L$ *such that* p_i *has returned the trace set* T **then**
10         $Traces(Pr(\mathcal{M})) = Traces(Pr(\mathcal{M})) \cup T$;
11         $L = L \setminus \{(p_i, PV)\}$;

satisfied over the valuation $v \cup \theta$. If this transition reaches a verdict location (Definition 12) then the algorithm constructs a new *Run* (lines 8-11) and ends. Otherwise, the event $(e(p), \theta)$ is forwarded to the called partner with the next transition t_2 (lines 12 to 14). *Run* is completed with r' followed by the sent event and the reached state $q_{next2} = (l_{next2}, v'')$. Then, the algorithm waits for the next event. It ends when Fail and/or Violate is detected or when no new event is observed after a delay sufficient to detect several times quiescent states (set to ten times in the algorithm with qt). It returns the trace T derived from *Run*.

Algorithm 2 reflects exactly the definition of the monitoring of an implementation (Definition 14). It collects valued events and constructs traces of $||_p(Env, PM, I)$ by supposing that both I and Env are ioLTS suspensions. Lines (5-15) implement the rules of Definition 14. In particular, when a verdict location lv is reached (line 8 or 11), the analyser has constructed a run, from its initial state which belongs to $Run_V(||_p(Env, PM, I))$ with V a verdict location set. From this run, we obtain a trace of $Traces_V(||_p(Env, PM, I))$.

So, with Proposition 2, we can state the correctness of the algorithm with:

Proposition 3. *The algorithm has reached a location verdict in:*

- *FAIL/VIOLATE* $\Rightarrow Traces_{FAIL/VIOLATE}(||_p(Env, PM, I)) \neq \varnothing \Rightarrow I \nvDash (\mathcal{O}, Violate_{\mathcal{O}})$ *and* $\neg(I$ *ioco* $\mathcal{S})$,
- *FAIL* $\Rightarrow Traces_{FAIL}(||_p(Env, PM, I)) \neq \varnothing \Rightarrow \neg(I$ *ioco* $\mathcal{S})$,
- *VIOLATE* $\Rightarrow Traces_{VIOLATE}(||_p(Env, PM, I)) \neq \varnothing \Rightarrow I \nvDash (\mathcal{O}, Violate_{\mathcal{O}})$.

Both the previous algorithms perform a synchronous analysis. Algorithm 1 receives a message, transfers it to Algorithm 2, which constructs a run from Proxy-monitor transitions before eventually forwarding the message to its addressee. However, this analysis can be done asynchronously to reduce the checking overhead with slight modifications: as soon as Algorithm 1 receives a message, it can forward it directly. Then, the message can be also given to Algorithm 2 which constructs its run only.

Algorithm 2. Proxy-Monitor-based analyser algorithm

input : A Proxy-monitor $Pr(\mathcal{M})$
output: Trace

1 $Run := \{(q_0 = (l0_{Pr(\mathcal{M})}, V0_{Pr(\mathcal{M})}))\}$; $qt = 0$;
2 **while** $Event(e(p), \theta) \wedge$ *Fail is not detected* $\wedge qt < 10$ **do**
3 　**if** $e(p) = !\delta$ **then**
4 　　$qt := qt + 1$;
5 　**foreach** $t = l \xrightarrow{e(p),G,A} l_{next} \in \rightarrow_{Pr(\mathcal{M})}$ *such that Run ends with* (l, v) *and* $\theta \cup v \models G$ **do**
6 　　$q_{next} = (l_{next}, v' = A(v \cup \theta))$;
7 　　$r' = Run.(e(p), \theta).q_{next}$;
8 　　**if** $l_{next} \in VIOLATE \cup FAIL/VIOLATE$ **then**
9 　　　Violation is detected; $Run := r'$;
10 　　**if** $l_{next} \in FAIL \cup FAIL/VIOLATE$ **then**
11 　　　Fail is detected; $Run := r'$;
12 　　**if** $l_{next} \notin VIOLATE \cup FAIL/VIOLATE \cup FAIL$ **then**
13 　　　Execute($t_2 = l_{next} \xrightarrow{!e(p),G_2,A_2}_{Pr(\mathcal{M})} l_{next2}$) ; `// forward` $(!e(p), \theta)$
14 　　　$q_{next2} := (l_{next2}, A_2(\theta \cup v'))$; $Run := r'.(!e(p), \theta).q_{next2}$;
15 return the trace $T = \{proj_{\Sigma_{||Pr(\mathcal{M})||}}(Run)\}$;

5.1 Experimentation

We have implemented this approach in a tool called *CloudPaste* (Cloud PASsive TEsting [2]) to assess the feasibility of the approach. We experimented it with the Web service composition of Figure 1(a), developed with SOAP Web services in C# and deployed in Windows Azure. The Azure PaaS layer supports proxy configuration, i.e. services can be configured to pass through proxies that can be hosted inside or outside of the Cloud. The guard solving in Algorithm 2 is performed by the SMT (Satisfiability Modulo Theories) solver Z3 [3] that we have chosen since it offers good performance, takes several variable types and allows a direct use of arithmetic formulae. However, it does not support String variables. So, we extended the Z3 expression language with terms, which refer to the ioSTS definition, and in particular with String-based terms. A term stands for a function over internal variables and parameters which returns a Boolean. Basically, our tool takes Z3 expressions enriched with terms, terms are evaluated and replaced with Boolean values. Then, a Z3 script, composed of the internal variables, the parameters and the guard, is dynamically written before calling Z3. If the guard is satisfiable (not satisfiable), Z3 returns *sat* (*unsat* respectively). Z3 returns *unknown* when it cannot determine whether a formula is satisfiable or not.

[2] `http://sebastien.salva.free.fr/cloudpaste/cloudpaste.html`
[3] `http://z3.codeplex.com/`

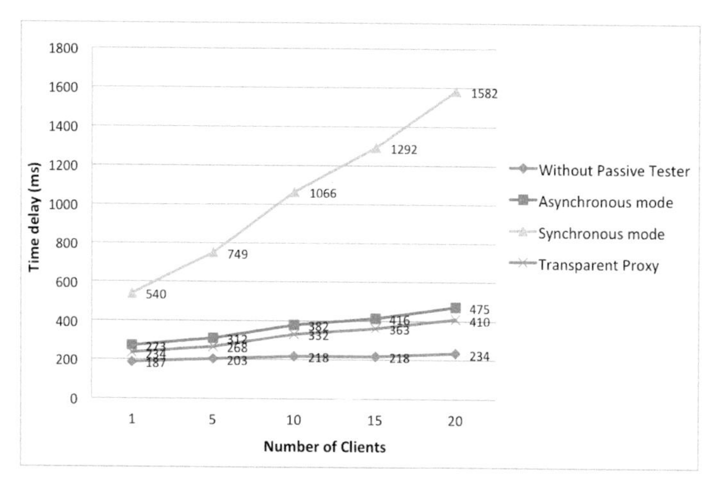

Fig. 4 Time processing measurements in Window Azure

We generated the Proxy-monitor from the ioSTS of Figure 1(a) combined with five safety properties with a tool generating Canonical testers and Proxy-monitors. The first property is the one described in Section 3.1. The other properties are based on security vulnerabilities. Client applications were simulated with at most 20 instances of Java applications performing one request to the *BookSeller* Web service with correct lists of two ISBNs. The passive tester was installed in a Windows server hosted in Azure. The detection of quiescence was implemented with a timeout set to 10s with respect to the HTTP timeout (usually set between 3 and 100 seconds).

Figure 4 depicts the average time processing of one client (milliseconds) when one up to twenty clients are running. The curves represent respectively the average time, without passive-tester, with the use of the transparent proxy *Charles* [4], with CloudPaste in asynchronous mode and in synchronous mode. In asynchronous mode, CloudPaste processes messages with a slightly higher time delay than Charles (with 20 clients, 102ms per message with Charles, 118ms per message with CloudPaste). This time delay is far lower than the quiescence timeout (and than the HTTP timeout as well). In synchronous mode, the checking overhead is higher with an average time of 135ms per message for 1 client and 395ms per message for 20 clients. This big difference results from the constraint solver calls and from the lack of optimisation of our code (Z3 is not yet called in parallel in CloudPaste). Nevertheless, in synchronous mode, the time processing is still lower than the timeout set to observe quiescence (the testing process can be done) and than the HTTP timeout (messages can be forwarded correctly). This mode is also particularly interesting since it offers the advantage to eventually implement recovery action calls, e.g., error compensation or implementation reset, when an error is detected. Error recovery is not possible with classical proxies or in asynchronous mode. These results tend

[4] `http://www.charlesproxy.com`

to show that our approach represents a good solution for testing and that it can be done in real-time.

6 Related Works

The works proposed in the literature either dealing with runtime verification, e.g., [3, 4, 8] or with passive testing, e.g., [11, 5, 2, 12] rely on three main methods for trace observation. Monitors or passive testers can be encapsulated within the implementation environment [4, 5], i.e. it is modified or completed with new test modules e.g., workflow engines. Traces can be also observed with probes, e.g., sniffer-based tools, deployed in the implementation environment [3, 11, 8, 12]. With these two methods, it is required to assume that the implementation environment access rights are granted and that it may be modified. This prerequisite condition cannot be always satisfied with any implementation environment. Installing a sniffer-based tool in a PaaS platform is not possible since services are geographically deployed in a dynamic manner and since the access and the modification of PaaS and IaaS layers are not authorised. The same issue is usually raised with Web servers: Web applications are tested by means of active methods with a testing server and are then deployed into another production server whose access rights are restricted for security reasons. Another possibility consists in adding directly probes into the system code [7] but this is occasionally considered only since it has the disadvantage of modifying the implementation behaviours for testing. Our work focuses on these issues, by proposing the use of the proxy concept for testing. A first naive solution would be to collect traces with a proxy e.g. SOAPUI[5], to eventually prune/modify them to obtain usual traces (those that would be collected directly from the implementation) and to analyse them with a specification to detect errors. Our proposal consists in generating automatically another model called *Proxy-tester* from a specification and to use a passive tester performing an analysis directly with Proxy-testers.

Few works have also focused on the combination of runtime verification with conformance testing [3, 6]. The latter consider active testing and therefore a combination of properties with classical test cases which are later actively executed on the system: in [3], test cases are derived from a model describing system inputs and properties on these inputs. Once test cases are executed, the resulting traces are analysed to ensure that the properties hold. Runtime verification and active testing have been also combined to check whether a system meets a desirable behaviour and conformance w.r.t. ioco [6]. In these previous works, the combination of active testing with runtime verification helps to choose, in the set of all possible test cases, only those expressing behaviours satisfying the given specification and safety properties. The other behaviours (those satisfying the specification but not the safety property and vice-versa) are not considered. Our proposal solves this issue by defining differently specifications and safety properties so that the resulting monitors could cover any behaviours passively over a long period of time.

[5] www.soapui.org/

7 Conclusion

We have proposed a testing approach combining ioco passive testing with runtime verification of safety properties. A monitor, called Proxy-monitor, is automatically generated from safety properties and specifications modelled with ioSTSs. Proxy-monitors are then used to detect whether the implementation is not ioco-conforming to its specification or if the former violates properties. Proxy-monitors are also based upon the notion of transparent proxy to ease the extraction of traces from environments in which testing tools cannot be deployed. Our approach can be applied on different types of communication software, e.g., Web service compositions, in condition that they could be configured to send messages through a proxy. In the experimentation part, we have also showed that the overhead obtained by the use of our approach remains reasonable and is much lower than the HTTP timeout.

In this paper, we have dealt with deterministic ioSTS specifications to rephrase *ioco*, like many testing approaches proposed in the literature. However, nondeterministic ioSTSs can be considered as well by apply determinization techniques [10]. In a future work, we could also consider nondeterministic ioSTSs with a weaker test relation than ioco to generate nondeterministic Proxy-testers. Another immediate line of future work concerns the enrichment of the experimentation with larger Web service compositions deployed in different Clouds, each having its own possibilities and restrictions.

References

1. Ws-bpel, Oasis Consortium (2007), `http://docs.oasis-open.org/wsbpel/2.0/OS/wsbpel-v2.0-OS.html`
2. Andrés, C., Cambronero, M.E., Núñez, M.: Passive testing of web services. In: Bravetti, M. (ed.) WS-FM 2010. LNCS, vol. 6551, pp. 56–70. Springer, Heidelberg (2011)
3. Arthoa, C., Barringerb, H., Goldbergc, A., Havelundc, K., Khurshidd, S., Lowrye, M., Pasareanuf, C., Rosug, G., Seng, K., Visserh, W., Washingtonh, R.: Combining test case generation and runtime verification. Theoretical Computer Science 336(2-3), 209–234 (2005)
4. Barringer, H., Gabbay, D., Rydeheard, D.: From runtime verification to evolvable systems. In: Sokolsky, O., Taşıran, S. (eds.) RV 2007. LNCS, vol. 4839, pp. 97–110. Springer, Heidelberg (2007)
5. Cavalli, A., Benameur, A., Mallouli, W., Li, K.: A Passive Testing Approach for Security Checking and its Practical Usage for Web Services Monitoring. In: NOTERE 2009 (2009)
6. Constant, C., Jéron, T., Marchand, H., Rusu, V.: Integrating formal verification and conformance testing for reactive systems. IEEE Trans. Softw. Eng. 33(8), 558–574 (2007), doi:10.1109/TSE.2007.70707
7. d'Amorim, M., Havelund, K.: Event-based runtime verification of java programs. In: Proceedings of the Third International Workshop on Dynamic Analysis, WODA 2005, pp. 1–7. ACM, New York (2005), doi:10.1145/1082983.1083249
8. Falcone, Y., Jaber, M., Nguyen, T.-H., Bozga, M., Bensalem, S.: Runtime verification of component-based systems. In: Barthe, G., Pardo, A., Schneider, G. (eds.) SEFM 2011. LNCS, vol. 7041, pp. 204–220. Springer, Heidelberg (2011)

9. Frantzen, L., Tretmans, J., Willemse, T.A.C.: Test Generation Based on Symbolic Specifications. In: Grabowski, J., Nielsen, B. (eds.) FATES 2004. LNCS, vol. 3395, pp. 1–15. Springer, Heidelberg (2005)
10. Jéron, T., Marchand, H., Rusu, V.: Symbolic determinisation of extended automata. In: Navarro, G., Bertossi, L., Kohayakawa, Y. (eds.) TCS 2006. IFIP, vol. 209, pp. 197–212. Springer, Boston (2006)
11. Lee, D., Chen, D., Hao, R., Miller, R.E., Wu, J., Yin, X.: Network protocol system monitoring: a formal approach with passive testing. IEEE/ACM Trans. Netw. 14, 424–437 (2006)
12. Nguyen, H.N., Poizat, P., Zaidi, F.: Online verification of value-passing choreographies through property-oriented passive testing. In: Ninth IEEE International Symposium on High-Assurance Systems Engineering, pp. 106–113 (2012)
13. Salva, S.: Passive testing with proxy-testers. International Journal of Software Engineering and Its Applications (IJSEIA). Science & Engineering Research Support Society (SERSC) 5 (2011)
14. Salva, S.: A model-based testing approach combining passive testing and runtime verification. Tech. rep., LIMOS, LIMOS Research report RR13-04 (2013), http://sebastien.salva.free.fr/useruploads/files/RR-13-04.pdf
15. Tretmans, J.: Test generation with inputs, outputs and repetitive quiescence. Software - Concepts and Tools 17(3), 103–120 (1996)

An Empirical Study on the Relationship between User Characteristics and Quality Factors for Effective Shopping Mall Websites Implementation

Donghwoon Kwon, Young Jik Kwon, Yeong-Tae Song, and Roger Lee

Abstract. In This paper, we have investigated how user characteristics affect quality factors for an effective Shopping mall websites implementation. User characteristics consist of gender, age, school year, department, experience, and purchasing experience during a specified period. We also selected a total of 14 quality factors from the literature review such as design, customer satisfaction, etc. As a proof of our hypothesis to investigate how those user characteristics and quality factors are interrelated, we have used 6 hypotheses. To verify them, the results have analyzed the SAS 9.2 statistic package tool and we have asked 519 participants to fill out a questionnaire for 5 Chinese and 8 Korean websites.

Keywords: Shopping Mall Websites Implementation, User Characteristics, Quality Factors.

1 Introduction

In the recent years, the size of the online shopping market has been growing rapidly. For example, the size of the online market in China was 265 billion yuan

---

Donghwoon Kwon · Yeong-Tae Song
Department of Computer & Information Science, Towson University, USA
e-mail: dkwon3@students.towson.edu, ysong@towson.edu

Young Jik Kwon
Department of Information & Communication, Daegu University, Korea
e-mail: yjkwon@daegu.ac.kr

Roger Lee
Software Engineering & Information Technology Institute,
Central Michigan University, USA
e-mail: lee1ry@cmich.edu

R. Lee (Ed.): *SERA*, SCI 496, pp. 117–127.
DOI: 10.1007/978-3-319-00948-3_8

in 2009 which was approximately 42.1 billion U.S. dollars, but it would reach 1.3 trillion yuan (206.4 billion U.S. dollars) in 2013 which is five times bigger than the size of online shopping market in 2009 [1]. According to the Korean National Statistical Office, the size of cyber shopping in Korea was 7,277 billion Korean Won in the third quarter of 2011 (6.52 billion U.S. dollars) which 16.3% was increased compared to 2010 [2]. However, while the size of the online shopping market is getting bigger, quality of most shopping mall websites is not improved accordingly. Many shopping mall websites users usually prefer sophisticated design, smooth communication, a variety of contents, better technology and reliability, etc. of the web site. In reality, however, most shopping mall websites do not fulfill such quality criteria due to the fact that it is required to spend more money, time, and resources to implement the high quality online shopping web site.

For this reason, this paper proposes the ways how to implement an effective shopping mall websites through root cause analysis and hypothesis verification.

2 Related Work

Conte et al. (2007) conducted website usability evaluation based on four web design perspectives: conceptual, presentation, navigation, and structural [3]. Cindy et al. (2005) conducted research that designing a shopping mall website using 3D technology is feasible and much more effective than 2D [5]. Albuquerque and Belchior (2002) conducted an E-commerce website quality evaluation using the communication factor, which is a part of various quality factors, and the fuzzy model stage [4]. Hai and Tu (2010) conducted research regarding a P2P E-commerce model based on users' interest community [6]. Shim and Suh defined the Customer Relationship Management (CRM) strategies for a small-sized shopping mall website [7]. Zhou and Zhang (2009) analyzed how E-commerce website quality affects user satisfaction based on TAM and information system success model [8]. Fengtao and Dengbai (2011) examined that information and information system satisfaction are directly related to customer satisfaction [9]. Xu and Liu (2010) defined that online stickiness, online satisfaction, online trust, and repurchase intention are interrelated [10]. Peng (2011) demonstrated the relationship between website quality and customer behavior of the transaction [11]. Sun and Wu (2010) conducted research about customer loyalty in online shopping [12]. Zhou and Niu (2010) defined how to design the online shopping system based on a software bus and components [13]. Yoo et al. (2008) demonstrated that 3 types of trust such as calculus-based, knowledge-based, and identification-based strongly affects customer satisfaction and loyalty [14]. Zubing and Guohe (2009) defined that the product brand, category, and security of the transaction directly have influence on customer behavior [15].

Based on literature research above, totaled 14 quality factors were drawn for our study: design, communication, community, security, user satisfaction, repurchase intention, transaction, customer loyalty, product, technology, trust, contents, interaction, and size. The specific quality factors of literature research are shown in Table 1.

Table 1 Quality Factors Used in the Literature

Authors	Quality Factors
Conte et al. (2007)	Design
Albuquerque and Belchior (2002)	Security, trustworthiness, content adequacy, technology suitability, etc.
Cindy et al. (2005)	Design and 3D technology
Hai and Tu (2010)	User's interest community
Shim and Suh (2010)	Transaction, comments, bulletin, etc.
Zhou and Zhang (2009)	Trust, satisfaction, etc.
Fengtao and Dengbai (2009)	Customer trust, customer loyalty, etc.
Xu and Liu (2010)	Online stickiness, online satisfaction, online trust, perceived website value, repurchase intention
Peng (2011)	Purchase tendency, web content, etc.
Sun and Wu (2010)	Product quality perception, customer satisfaction, customer trust, customer loyalty, etc.
Zhou and Niu (2010)	Component-based design
Yoo et al. (2008)	Trust in transaction, customer satisfaction, customer loyalty
Zubing and Guohe (2009)	Trust, security, brand, satisfaction, transaction, design, etc.

3 An Empirical Study

3.1 Outline

The research model is shown in Figure 1.

As mentioned earlier, all independent and dependent variables are same, and we selected 8 Korean shopping mall websites and 5 Chinese shopping mall websites. This is because we aim to conduct research as objectively as possible and respondents are from Korea and China.

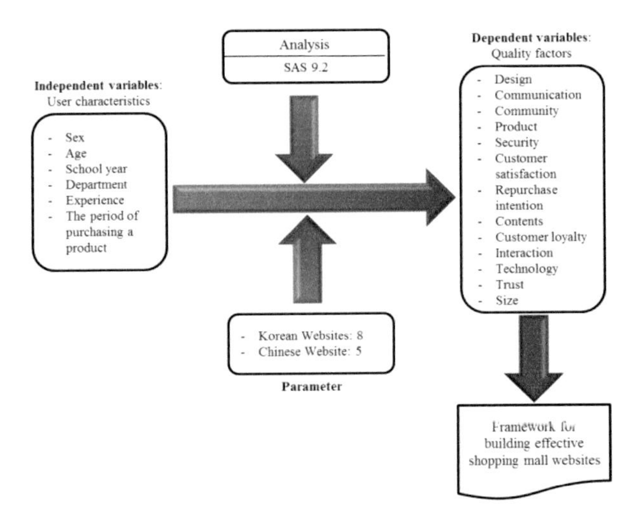

Fig. 1 The Overall Research Model

3.2 *Methodology*

3.2.1 Questionnaire Design

Each user characteristic and quality factor contains sub factors as shown in the Table 2. We have converted them into questionnaire items. However, due to the volume of sub factors, only the significant sub factors are shown below.

3.2.2 Questionnaire Performing Survey

The selected sub factors in the previous section were used for the questions in the questionnaire. To measure the degree of consensus for each questionnaire from the participants, 5 point Likert scale was used. The websites used for the survey are:

Korea: www.lotteshopping.com, www.ehyundai.com, www.auction.co.kr, www.ebay.com, www.gmarket.com, www.interpark.com, www.11st.co.kr, personal shopping malls, etc.

China: www.taobao.com, www.amazon.cn, www.dangdang.com, www.paipai.com, www.china.alibaba.com, etc.

Table 2 Detailed Sub Factors

Categories	Main Factors	Sub Factors
User characteristics	Sex	Male or female
	Age	Under 20, 21~25, 26~30, 31~35, and over 35
	School year	Freshmen, sophomore, junior, senior, and graduate students
	Dept.	Engineering, Information & communication, etc.
	Experience	Yes or no
	The period of purchasing a product (Month)	1~2, 3~4, 5~6, over 6, etc.
Quality factors	Design	Images, graphics, font size, etc.
	Communication	Bulletin board, users' review and comments, etc.
	Community	A sense of belonging, influence on users' life, etc.
	Product	Product categories, product quality, etc.
	Security	Privacy agreement, security system, secure payment, etc.
	User satisfaction	Transaction satisfaction, delivery satisfaction, etc.
	Repurchase intention	Intention of repurchasing a same product, intention of recommendation, etc.
	Contents	Display, shopping mall symbols, icon clarity, etc.
	Transaction	Convenience of account registration and login, efficiency of shipping tracking, etc.
	Customer loyalty	Affinity, sense of favor, etc.
	Interaction	Online chatting, Q&A, customer representative kindness, etc.
	Technology	Error frequency, shortcut, etc.
	Trust	Overall reliability of the shopping mall, etc.
	Size	The size of suppliers, a number of members and users, etc.

3.3 The Demographic of the Participants

The occurrence of each user characteristic from the questionnaire is analyzed and shown in the Table 3.

Table 3 Frequency Analysis of User Characteristics

User Characteristics		Frequency (Persons)	Missing Value
Sex	Male	316	6
	Female	197	
Age	Under 20	253	2
	21~25	237	
	26~30	27	
School Year	Freshmen	289	12
	Sophomore	85	
	Junior	65	
	Senior	56	
	Graduate Students	12	
Department	Eng.	53	2
	Info.&Comm.	88	
	Economics& Business	152	
	College of administrator	73	
	Rehabilitation sciences	33	
	Other Depts.	118	
Exp.	Yes	485	5
	No	29	
The period of purchasing a product	1~2 mo.	215	28
	3~4 mo.	95	
	5~6 mo.	28	
	Over 6 mo.	99	
	etc.	54	

3.4 Reliability Coefficient

According to Cronbach, a question is considered reliable because the alpha value of the question is greater than 0.70. Cronbach's alpha is used to measure the reliability of the questionnaire. The Cronbach's alpha value of each quality factor is shown in Table 4.

Table 4 Reliability Coefficient using Cronbach's Alpha

Quality Factors	Cronbach's alpha	Reliability
Design	0.882198	Since the Cronbach's value of each quality factor is higher than 0.70, the questionnaire of this study is reliable.
Communication	0.761989	
Community	0.801693	
Product	0.809899	
Security	0.887740	
Customer satisfaction	0.906327	
Repurchase intention	0.781777	
Contents	0.792084	
Transaction	0.785264	
Customer loyalty	0.900060	
Interaction	0.877644	
Technology	0.858624	
Trust	0.878629	
Size	0.847242	

4 Hypotheses Formulation

We decomposed the main hypothesis into several specific hypotheses based on a mathematical set which is shown in Figure 2; in other words, it is to figure out how the elements of the set "User Characteristics" affect each element of the set "Quality Factors".

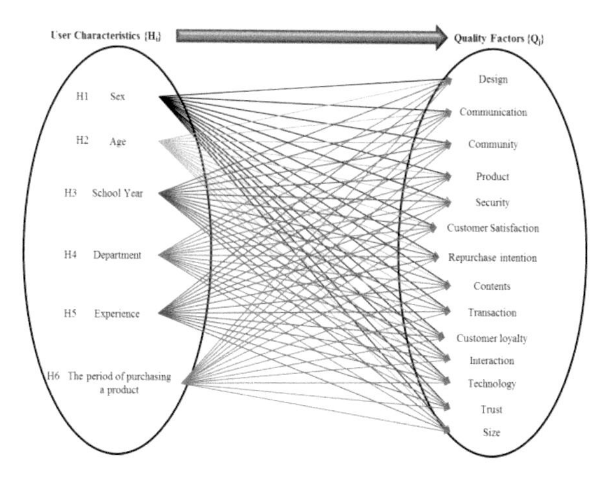

Fig. 2 Hypothesis Model

5 Hypotheses Verification

As mentioned earlier, the SAS 9.2 statistic tool was used to verify the hypotheses. If there were 2 groups, we conducted T-test and if there were 3 groups, we conducted Analysis of Variable (ANOVA). Furthermore, all hypotheses were verified at the 0.95 confidence interval. The detailed statistical analysis results are summarized in Figure 3.

5.1 H1 Verification Analysis

The results show that the user characteristic, sex, does not affect design, community, product, security, customer satisfaction, contents, transaction, interaction, trust, and size while it affects communication, repurchase intention, customer loyalty, and technology. For this reason, to implement the effective online shopping website, design, community, product, security, user satisfaction, contents, transaction, interaction, trust, and size must be considered with distinction of sex. However, the factor of communication, repurchase intention, customer loyalty, and technology must be reflected with consideration of sex.

5.2 H2 & H3 Verification Analysis

The results show that the user characteristics, age and school year, affect every quality factor. Thus, all quality factors must be reflected with consideration of age and school year to implement websites.

5.3 H4 Verification Analysis

The results show that quality factors such as design, communication, product, security, transaction, and size are not affected by the user characteristic, the department. It is not necessary to consider the department for those quality factors. However, community, user satisfaction, repurchase intention, contents, customer loyalty, interaction, technology, and trust are affected by the department so that those user characteristics must be reflected for those quality factors.

5.4 H5 Verification Analysis

The results show that experience does not affect community, product, trust, and size while it affects design, communication, security, user satisfaction, repurchase intention, contents, transaction, customer loyalty, interaction, and technology. Therefore, experience must be considered for quality factors such as design, communication, security, user satisfaction, repurchase intention, contents, transaction, customer loyalty, interaction, and technology.

User Characteristics / Hypotheses / Quality Factors	Sex		Age		School Year	
	Pr	Ha or Hr	Pr	Ha or Hr	Pr	Ha or Hr
Design	0.4207	Hr	0.0004	Ha	0.0001	Ha
Commmunication	0.0209	Ha	0.0005	Ha	0.0001	Ha
Community	0.7273	Hr	0.0001	Ha	0.0001	Ha
Product	0.0855	Hr	0.0001	Ha	0.0001	Ha
Security	0.9070	Hr	0.0001	Ha	0.0001	Ha
Customer Satisfaction	0.7222	Hr	0.0001	Ha	0.0001	Ha
Repurchase Intention	0.0078	Ha	0.0001	Ha	0.0001	Ha
Contents	0.1130	Hr	0.0001	Ha	0.0001	Ha
Transaction	0.5875	Hr	0.0001	Ha	0.0001	Ha
Customer Loyalty	0.0028	Ha	0.0001	Ha	0.0001	Ha
Interaction	0.3936	Hr	0.0001	Ha	0.0001	Ha
Technology	0.0259	Ha	0.0001	Ha	0.0001	Ha
Trust	0.1842	Hr	0.0001	Ha	0.0001	Ha
Size	0.4510	Hr	0.0001	Ha	0.0001	Ha

User Characteristics / Hypotheses / Quality Factors	Dept.		Experience		The period of purchasing a product	
	Pr	Ha or Hr	Pr	Ha or Hr	Pr	Ha or Hr
Design	0.4058	Hr	0.0009	Ha	0.0039	Ha
Communication	0.0562	Hr	0.0003	Ha	0.7580	Hr
Community	0.0006	Ha	0.1655	Hr	0.0038	Ha
Product	0.3033	Hr	0.1286	Hr	0.0079	Ha
Security	0.2926	Hr	0.0001	Ha	0.6022	Ha
Customer Satisfaction	0.0202	Ha	0.0001	Ha	0.0181	Ha
Repurchase Intention	0.0441	Ha	0.0001	Ha	0.0004	Ha
Contents	0.0321	Ha	0.0467	Ha	0.0011	Ha
Transaction	0.1098	Hr	0.0051	Ha	0.0446	Ha
Customer Loyalty	0.0226	Ha	0.0030	Ha	0.0001	Ha
Interaction	0.0319	Ha	0.0132	Ha	0.2819	Hr
Technology	0.6247	Ha	0.0309	Ha	0.1303	Hr
Trust	0.0092	Ha	0.0728	Hr	0.1164	Hr
Size	0.0712	Hr	0.0520	Hr	0.0199	Ha
Ha: Hypothesis Accepted			**Hr: Hypothesis Rejected**			

Fig. 3 Statistical Analysis Results

5.5 *H6 Verification Analysis*

The results show that the period of purchasing a product has an impact on design, community, product, security, user satisfaction, repurchase intention, contents, transaction, customer loyalty, and size. However, it does not have an impact on communication, interaction, technology, and trust. For this reason, it is not required to consider the period of purchasing a product for communication, interaction, technology, and trust.

6 Conclusion

We demonstrated which of the user characteristics has an impact on quality factors for implementing effective online shopping websites based on the statistical analysis and hypotheses verification. Our research shows that some user characteristics such as age and school year affect every quality factor, whereas most user characteristics do not have impact on all quality factors. Although this paper demonstrates a correlation between user characteristics and quality factors, we will also conduct research about a correlation between website quality factors and quality improvement in the future.

References

1. Nakajima, H., Kuzushima, T., Huang, X.: Strategic Use of Online Sales Aimed at China's Rapidly Growing Consumer Market. Nomura Research Institute Paper, pp. 8–9 (May 1, 2012)
2. Korea National Statistical Office, Report of E-Commerce and Cyber Shopping Survey in the Third Quarter 2011, `http://kosis.kr/ups/ups_01List01.jsp?pubcode=OE` (retrieved on March 2, 2012)
3. Conte, T., Massollar, J., Mendes, E., Travassos, G.H.: "Usability Evaluation Based on Web Design Perspectives. In: First International Symposium on Empirical Software Engineering and Measurement, pp. 146–155. IEEE (2007)
4. Albuquerque, A.B., Belchior, A.D.: E-Commerce Website Quality Evaluation. In: Proceedings of the 28th Euromicro Conference. IEEE (2002)
5. Cindy, H.B.B., Chaudhari, N.S., Patra, J.C.: Design of a Virtual Shopping Mall: Some Observations. In: Proceedings of the 2005 International Conference on Cyberworlds (CW 2005). IEEE (2005)
6. Hai, M., Tu, Y.: A P2P E-Commerce Model Based on Interest Community. In: 2010 International Conference on Management of E-Commerce and E-Government, pp. 362–365. IEEE (2010)
7. Shim, B.-S., Suh, Y.-M.: Crm Strategies for a Small-Sized Online Shopping Mall Based on Association Rules and Sequential Patterns. In: 14th Pacific Asia Conference on Information Systems, PACIS, pp. 355–366 (2010)
8. Zhou, T., Zhang, S.: Examining the Effect of E-Commerce Website Quality on User Satisfaction. In: 2009 Second International Symposium on Electronic Commerce and Security, pp. 418–421. IEEE (2009)

9. Fengtao, L., Dengbai, W.: The Impact of Information and Information System Satisfaction on Customer Satisfaction under E-commerce. In: 2011 International Conference on Information Management, Innovation Management and Industrial Engineering, pp. 190–193. IEEE (2011)
10. Xu, J., Liu, Z.: Study of Online Stickiness: its Antecedents and Effect on Repurchase Intention. In: 2010 International Conference on e-Education, e-Business, e-Management and e-Learning, pp. 116–120. IEEE (2010)
11. Peng, H.: Positive Analysis of Chinese Online Customer Behavior During the Transaction. In: 2011 International Conference on Management of e-Commerce and e-Government, pp. 200–203. IEEE (2011)
12. Sun, H., Wu, H.: The Customer Loyalty Research Based on B2C Ecommerce Sites. In: 2010 International Conference on E-Business and E-Government, pp. 3156–3159. IEEE (2010)
13. Zhou, C.-S., Niu, L.-H.: Research on Component Based Online Shopping System Design. In: 2010 International Conference on Computational Aspects of Social Networks, pp. 133–136. IEEE (2010)
14. Yoo, J.-S., Lee, J.-N., Hoffmann, J.: Trust in Online Shopping: The Korean Student Experience. In: Proceedings of the 41st Hawaii International Conference on System Sciences, pp. 1–10. IEEE (2008)
15. Hou, Z., Yu, G.: An Empirical Research on Influence Factors of Online Shopping. In: The 1st International Conference on Information Science and Engineering (ICISE 2009), pp. 2846–2849. IEEE (2009)

Improving Code Generation for Associations: Enforcing Multiplicity Constraints and Ensuring Referential Integrity

Omar Badreddin, Andrew Forward, and Timothy C. Lethbridge

Abstract. UML classes involve three key elements: attributes, associations, and methods. Current object-oriented languages, like Java, do not provide a distinction between attributes and associations. Tools that generate code from associations currently provide little support for the rich semantics available to modellers such as enforcing multiplicity constraints or maintaining referential integrity. In this paper, we introduce a syntax for describing associations using a model-oriented language called Umple. We show source code from existing code-generation tools and highlight how the issues above are not adequately addressed. We outline code generation patterns currently available in Umple that resolve these difficulties and address the issues of multiplicity constraints and referential integrity.

Keywords: Associations, Model Driven Development, UML, Code Generation, Umple, Reverse Engineering.

1 Introduction

A UML association is a relationship among classes within an object-oriented system. It specifies the connections, called links, which may exist among instances of classes at run time. A notation called multiplicity appears at each end of an association to describe the number of links other objects may have to the object in question. In this paper we focus on binary associations, which have just two ends. Examples of UML diagrams with associations can be found in Figures 1 and 2.

Omar Badreddin · Andrew Forward · Timothy C. Lethbridge
School of Electrical Engineering and Computer Science,
University of Ottawa, Canada K1N 6N5
e-mail: {obadr024,aforward,tcl}@eecs.uottawa.ca

R. Lee (Ed.): *SERA*, SCI 496, pp. 129–149.
DOI: 10.1007/978-3-319-00948-3_9

Current programming languages do not directly support associations, which are coded by hand or by using a UML code generator tool. However, current code generators have weaknesses such as not dealing with referential integrity [1].

This paper introduces a syntax called Umple for defining associations textually, as well as a series of patterns for generating high-level language code from associations. We analyse all possible combinations of multiplicity that may appear on association ends, highlighting the underlying complexity in properly implementing associations in a target executable language. We analyse existing UML code generation tools to investigate the completeness of their implementation, or lack thereof. We provide code-generation patterns covering all multiplicity combinations that ensure referential integrity as well as adherence to multiplicity constraints.

Umple enables modelling concepts to be described textually with a similar syntax to Java. We present our code generator for Java, but Umple also supports PHP and Ruby and can conceptually support any object-oriented language.

In translating UML associations into an executable language, it is more efficient (but currently uncommon) to include access methods (get, set, add, remove) to manage links of associations. These methods would maintain the multiplicity constraints of the association and preserve referential integrity – ensuring that both ends of an association are properly updated when adding or removing links.

2 Associations in Practice

In a separate paper in this conference [3] we discussed an empirical study that analyzed the use of attributes in open source systems. We used the same systems as a starting point for the research presented here.

The seven projects selected for analysis include fizzbuzz, ExcelLibrary, ndependencyinjection, Java Bug Reporting Tool, jEdit, Freemaker; and Java Financial Library. We documented all member variables and recorded the project, namespace, object type, and variable name for each, as well as related characteristics such as constructors, and set/get methods.

To find variables representing associations, we used a two-step manual process. The first step recursively eliminated attributes. An attribute is considered to have as its type either: a) a simple data type including String, Integer, Double, Date, Time and Boolean, or b) a *complex attribute* type, i.e. a class that itself only contains instance variables meeting conditions a and b, with the proviso that in this recursive search process, if a cycle is found, then the variable is deemed a candidate association end. This process resulted in 350 candidate association ends.

In the second step, we filtered the set down to 235 association ends by removing *internal* variables. Internal variables represent data that is not an intrinsic part of the permanent state of the object, but is used for some algorithm or temporary process. They are neither set in the constructor nor are available via set/get methods. Examples of the types of such variables include Readers, Streams, and Maps.

Table 1 highlights distribution statistics of the 235 association ends. The results are not mutually exclusive so the column sum will not be 100%.

Table 1 Distribution of set / get and availability in constructor

Category	Freq	%	Description
Set/Get Methods	67	29%	All variables that had both a set and get method.
Set Method	89	38%	All variables that at least had a set method.
Get Method	120	51%	All variables that at least had a get method.
No Set Method	54	23%	All immutable variables, as there is no set method
Only Get Method	39	17%	Internally managed (no set method, not in constructor).

In addition to tracking the distribution of set and get methods, we also noted that some implementations provided direct access to the list structure and others provided methods like add and remove. Of the 235 association ends, 42 (17.9%) were defined using Map, Set, Hash, or List classes and hence likely represented associations with an upper bound greater than one.

When analyzing the open-source systems, it was difficult to match association-end variables to one another because many associations linked to external resources and most likely represented one-way associations. There was little evidence of referential integrity between association ends, implying that the application developer *using* the object model would have to write code to maintain the correct multiplicities and inverse pointers. This difficulty in analyzing how associations are used in practice provides motivation for our work. It would appear to be beneficial for developers to be able to define associations in one location, and have links created, accessed and modified in a consistent manner.

To better understand the types of associations used in practice, we analyzed 1400 associations in UML diagrams of real systems (not association ends as discussed above). These were found in two UML specifications (v1.5 / v2.1.2) and some UML profiles (MARTE, Flow Composition, ECA, Java, Patterns, rCOS).

Table 2 Industry Usage of Association Multiplicities in UML

Industry Usage			Rank in the various sets of examples		
Multiplicity	Frequency	Ratio	Industry	Example	Repository
1--*	273	19.0%	1	1	1
0..1--*	270	18.8%	2	4	2
->	179	12.4%	3	9	4
--	162	11.3%	4	2	3
0..1--1	126	8.8%	5	N/A	7
*->1	86	6.0%	6	N/A	6
*->0..1	73	5.1%	7	N/A	5
0..1--0..1	58	4.0%	8	6	N/A
--n..	54	3.8%	9	N/A	N/A
Other	157	10.9%	N/A	N/A	N/A
Total	1438	100.0%			

We also analyzed UML models in a book by one of the authors [4] (*example models*), and in our repository of UML modeled systems [5] (*repository models)* built using Umple.

The top nine associations patterns by multiplicity, ordered according to the industry examples, are in Table 2. The multiplicities include the following constraints: optional (0), one (1), lower-bound (n), upper-bound (m), and many without bounds (*). For example, a one-to-many multiplicity would be 1 -- *. The rank of actual usage is based on the UML specification. For comparison the rank of the particular multiplicity within the example and repository collection is also shown.

We performed similar analysis based on example usage from [4] and present the results in Table 3.

Table 3 Example Usage [4] of Association Multiplicities

Example Usage			Rank in the various sets of examples		
Multiplicity	**Frequency**	**Ratio**	**Industry**	**Example**	**Repository**
1--*	39	39.8%	1	1	1
--	15	15.3%	4	2	3
1--1	13	13.3%	N/A	3	N/A
0..1--*	11	11.2%	2	4	2
*->1	4	4.1%	N/A	5	6
0..1--0..1	4	4.1%	8	6	N/A
1--0..n	3	3.1%	N/A	7	N/A
*--n	2	2.0%	N/A	8	N/A
->	2	2.0%	3	9	4
Other	5	5.1%	N/A	N/A	N/A
Total	98	100.0%			

Table 4 is ordered based on our UML model repository examples [5], which has over 28 systems in domains like airlines, elevators, traffic lights and the Umple metamodel itself.

Table 4 Usage of Association Multiplicities from Model Repository [5]

Model Repository Usage			Rank in the various sets of examples		
Multiplicity	**Frequency**	**Ratio**	**Industry**	**Example**	**Repository**
1--*	108	43.4%	1	1	1
0..1--*	34	13.7%	2	4	2
--	27	10.8%	4	2	3
->	22	8.8%	3	9	4
*->0..1	21	8.4%	7	N/A	5
*->1	12	4.8%	6	5	6
1--0..1	4	1.6%	5	N/A	7
1--1..*	3	1.2%	N/A	N/A	8
1..*--*	3	1.2%	N/A	N/A	9
Other	15	6.0%	N/A	N/A	N/A
Total	249	100.0%			

The industry and example UML models share five of the top nine multiplicity usage patterns; one-to-many, optional-one-to-many, many-to-many, optional-one-to-one, and optional-one-to-optional-one.

After analyzing over 1,800 different modeled associations, approaches like eUML [6] (where only a subset of the UML multiplicities can be modeled) provide sufficient coverage for most applications. The same is true of the applications analyzed in Section 5, where most code generators provide little capability for association multiplicities beyond differentiating a *one-end* from *many-end*. As shown above, about 5% of the UML specifications fall outside of the simple cases currently supported and it should be of both academic and practical relevance to explore all types of association relationships.

3 Textual Associations in Umple

Umple is a set of extensions to object-oriented languages that provides a concrete textual syntax for UML abstractions like attributes, associations and state machines. Below, we describe how associations are represented in Umple. Please see our separate paper [3] for a discussion of attributes. For more details, and for the motivation regarding why we created Umple, the reader should refer to [7] and [8]. Umple can also be edited directly within a browser [9].

Figure 1 shows two associations. To distinguish between Umple and Java, the Umple examples use dashed borders in light-grey shading, and Java examples use solid-line borders with no shading. The UML class diagram one the right has three classes and two one-to-many associations. The code on the left is the equivalent in Umple . The '--' means that the association is bi-directionally navigable (more on this later). It is also possible to use '->' or '<-' to indicate that navigation is possible in only one direction. The full set of UML multiplicity symbols may be used.

```
class Student {}
class CourseSection {}
class Registration
{
  String grade;
  * -- 1 Student;
  * -- 1 CourseSection;
}
```

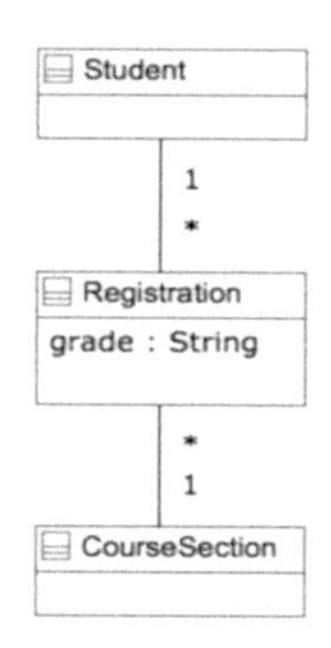

Fig. 1 Umple code/model for part of the registration system

In addition to showing an association embedded in one of the two associated classes, it is also possible to show an association 'on its own'. The association name, *Enrollment* and role names *course* and *attendee* are optional.

```
association Enrollment
{* Registration course -- 1 Student atendee;}
```

Besides providing improved abstraction, explicitly coding associations may speed development and reduce bugs, since the compiler can enforce various design constraints and less code needs to be written. The current implicit nature of associations in standard languages most likely results in code that is bug-prone since there is no general mechanism to enforce things like referential integrity.

4 Analyzing Association Multiplicity

Consider the following association between a Mentor and a Student.

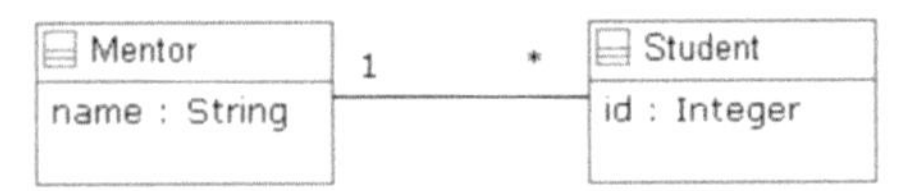

Fig. 2 An example binary association

An association has a multiplicity at each end describing how many instances of one class can be linked with the other class. In Figure 2, a Mentor can link to any number of Students, but a Student must be assigned to one and only one Mentor.

In general, there are nine distinct categories of multiplicity, shown in Table 5. These largely require different treatment in the code, except that the 1..n case can be coded as a special case of m..n and 1..* can be coded as special case of m..*.

Table 5 Multiplicity Possibilities for Associations

Multiplicity Notation	Lower Bound	Upper Bound	Description
0..1	0	1	Optional-One – Item is either present or not
0..n	0	n > 1	At Most n – At most n items, or none at all
0..* (*)	0	undef. > 1	Many – Any number of items present. or none at all
1..1 (1)	1	1	One – The item is mandatory
1..n	1	n > 1	Mandatory at Most n – At least one item is mandatory up to a maximum of n items
1..*	1	undef. > 1	Mandatory Many – One item is mandatory (no max)
n..n (n)	n > 1	n	Exactly n – Exactly m items are mandatory
m..n	m > 1	n > m	From m to n – At least m items are mandatory up to a maximum of n items
m..*	m > 1	undef. > m	At least m – At least m items mandatory (no max)

4.1 Bidirectional Associations Between Two Different Classes

We analyzed associations between two classes that are navigable in both directions. Both linked objects are conceptually *aware* of the relationship. There are 28 different patterns of binary associations, as listed in Table 6.

Note that x -- y is equivalent to y -- x, so for example, 0..1 -- n is equivalent to n -- 0..1; for this reason, the upper diagonal in Table 6 is left blank. Also note that the variables ***m*** and ***n*** are not assumed to be the same on both ends of the association. For example, 0..6 -- 3..4 association would be in the 0..n -- m..n category.

Our examples model a Mentor and Student, allowing us to vary the multiplicity in a sensible way. Figure 3 shows a Student always has one Mentor, but a Mentor can have zero or more students.

Our code written in Java assumes a Mentor has a *student* variable when the multiplicity upper bound (ub) is 1, or *students* if ub > 1. A Student, has a *mentor* (ub =1), or *mentors* (ub > 1). The Umple notation for Figure 3 is:

```
association {1 Mentor -- * Student;}
```

Table 6 All 28 Bi-Directional Non-Reflexive Association Patterns

0..1	**0..n**	*****	**1**	**n**	**m..n**	**m...***
0..1 -- 0..1						
0..1 -- 0..n	0..n – 0..n					
0..1 -- *	0..n -- *	* -- *				
0..1 -- 1	0..n -- 1	* -- 1	1 -- 1			
0..1 -- n	0..n -- n	* -- n	1 -- n	n -- n		
0..1 -- m..n	0..n -- m..n	* -- m..n	1 -- m..n	n -- m..n	m..n -- m..n	
0..1 -- m..*	0..n -- m..*	* -- m..*	1 -- m..*	n -- m..*	m..n -- m..*	m..* -- m..*

Shaded cells show cases where both sides are *mandatory*. Associations with thick borders indicate the common cases as observed in the previous section. Note that 1..* is a subset of the more generic m..* and 1..n is a subset of m..n.

4.2 Unidirectional (Directed) Associations

A directed association is navigable in one direction only. Only one object of the pair is aware of and can manage the relationship. For example, one could write in Umple:

```
association {* Mentor -> * Student;}
```

In this, a Mentor is aware of the associated Students, but a Student is unaware of any Mentor's to which he/she might be associated. The end that is unaware of the link can be unknowingly linked to multiple objects (i.e. a * relationship is generally implied); resulting in seven possible combinations for code generation.

```
* -> 0..1, * -> 1, * -> *, * -> m..n, * -> n, * -> m..*, * -> 0..n
```

Without injecting additional complex code, the system will be unable to manage the association when changes occur to the unaware side, such as a Student deleting itself. If such situations must be managed, then a bi-directional association must be used. In practice, we find that the vast majority of associations would benefit from being bidirectional. Doing so enables functionality that tends to be required anyway, and where the functionality is not immediately required, the code is better suited to meet unanticipated future needs. However, bi-directional association can increase coupling unnecessarily.

4.3 Reflexivity and Symmetry

Where both ends of an association are the same class, we must consider several special cases. A reflexive association allows an object to be linked to other objects of the same class including itself. An example of such an association might be "lives-at-same-address".

It is also common to have *irreflexive* associations with both ends being the same class. The added constraint is that a given object cannot be linked to itself. For example, a *mother* relationship among Person objects is irreflexive as you cannot be your own mother.

A *symmetric* association describes a mapping that reads the same as its inverse; for example, a *spouse* association. An asymmetric association is not reversible; for example a *child* relationship. Finally, an anti-symmetric association is asymmetric except that it allows a relationship to self. For example the relationship *being-present-at-birth* is anti-symmetric.

One could encode the mentor-student example using a single class, where all objects are Persons, and some persons can mentor others. This is an irreflexive asymmetric association, since one cannot mentor oneself and the meaning of the association is different in each direction. The Umple notation would be:

```
association {0..1 Person mentor -- * Person student;}
```

The Umple language natively supports symmetric and asymmetric associations. Anti-symmetric associations are currently not supported as we find them to be quite rare (they may be supported in the future). The distinction between reflexivity and irreflexivity is currently not managed, but applications can relatively easily be coded to prevent (or allow) an object to be linked to itself.

Asymmetric associations require code that is effectively identical to associations between two different classes (see Table 6) - except that the lower bound of both ends must be zero. The top seven associations from Table 6 are therefore possible as asymmetric associations from one class to itself. The reason the lower bound must be zero is to prevent infinite regress. For example, in the case of class Person in the asymmetric association previously discussed, if every mentor Person

must have a student Person, and since every student is also a mentor (by virtue of being a Person) there would be an infinite chain of persons requiring a mentor.

A symmetric association specifies links between different instances of the same class, and must have the same multiplicity on each end. The diagonal of Table 6 gives the cases to consider.

As a result of all the above analysis, a total of 42 different possible association types have been identified (28 for bidirectional associations; 7 for unidirectional associations, and 7 for symmetric associations). In the following section, we highlight certain implications that these association types will have on code generation. This overview serves as a guide when comparing existing code generation tools and also as a template for building code generator for systems programmed in the Umple modeling language.

4.4 Implications for Code Generation

The implementation of associations in a language like Java impacts the following aspects of a class. First, the class will have an additional member variable to reflect the other end of an association. *One* and *optional-one* multiplicities can be declared as a member variable of the other type, while many multiplicities are declared as lists (implemented as a collection class) of objects of the other type.

Second, the constructor may need an additional parameter to ensure *mandatory* association ends like 1 or 1..*.

Finally, the class requires methods to set, get, add and remove links between objects. To be consistent with the model of the associations, the implementation of those methods should enforce the referential integrity between pairs of objects, as well as ensuring that multiplicity constraints are upheld.

In the following section, we analyze how existing code generators deal with the various combinations of multiplicities and to what extent they behave according to the structure outlined above. We then discuss the code generation available from the Umple language.

5 Existing Code Generators

In this section we look at existing tools to see how well they translate the semantics of associations into a programming language.

The UML modeling tools considered were identified from a Gartner report [10] and an online list [11]. We selected four open-source and one closed source application to analyze, as listed in Table 7. Each tool was configured to generate Java code for a simple 1 -- * relationship shown in Figure 2.

The generated code provided in the following sections has only been modified to provide a consistent layout/format, and for space considerations comments have also been removed.

Table 7 UML code generation tools

Tool	Version	Source
ArgoUML	0.26.2	argouml.tigris.org
StarUML	5.0.2.1570	staruml.sourceforge.net
BOUML	4.11	bouml.free.fr
Green	3.1.0	green.sourceforge.net
RSA	7.5	ibm.com/software/awdtools/architect/

5.1 Code Generation Patterns

In general, all tools analyzed provided two basic code generation templates; one for 0..1 and 1 (referred to as one) multiplicitics and a second for m..n multiplicities (referred to as many where $m \geq 0$, $n > m$, $n > 1$).

The template pattern for *one* would generate a member variable to refer to the other association end. The template pattern for *many* would generate a reference to a List or Set structure that could contain multiple references to the other association end. Both examples are shown below.

```
private <ClassName> <assocEndName>;   // ub = 1
private <ListStructure> <assocEndName>; // ub > 1
```

Some tools provide explicit code generation patterns for *n* relationships (where $n > 1$), as well as m..* relationships (where $m \geq 0$). Some tools provided explicit get/set methods in addition to creating the necessary member variables. A discussion of each code generation pattern will be provided based on the tools analyzed.

5.2 ArgoUML

ArgoUML is an open source modeling platform that provides code generation for Java, C++, C#, PHP4 and PHP5. Below is the generated code from Figure 2.

```
import java.util.Vector;
public class Mentor {
  public Integer id;
  public Vector myStudent; }
public class Student {
  public String name;
  public Mentor myMentor; }
```

The generated code provides a mechanism to access "each end" of the relationship. The generator provides little validation or constraint checking to ensure the relationship is maintained, and the variables are made directly available without the inclusion of accessor (get and set) methods.

In general, all 0..1 and 1 multiplicities generate similar structures as seen in the Student class above, and all m..n multiplicities (where $m \geq 0$ and $n > m$ and $n > 1$) generate similar structures to the Mentor class.

5.3 *StarUML*

StarUML is an open source modeling tool. StarUML's generated code does not account for the *many* multiplicity, resulting in unusable generated code. Below is the generated code for the Mentor and Student example:

```
public class Mentor {
  public String name;
  public Student student; }
public class Student {
  public Integer id;
  public Mentor mentor; }
```

5.4 *Bouml*

Bouml is a free tool based on UML 2 that provides source code generation for C++, Java, Idl, Php and Python.

The source code generated below is very similar to that of ArgoUML. This code does not provide any mechanism to test or ensure the constraints outlined in the model; this code must be written by hand after code generation. In addition, the source code is incomplete as no reference the *java.util.List* class is provided, which means that the generated code must be maintained by hand to ensure proper compilation into byte code.

```
class Mentor {
  private List<Student> student;
  private String name; }
class Student {
  private Mentor mentor;
  private int id; }
```

5.5 *Green Code Generator*

Green UML is another UML editor that can generate source code from a class diagram. Below is the generated code for the Mentor and Student example.

```
import java.util.List;
public class Mentor {
  private List<Student> student;
  java.lang.String name;
  public Mentor(List<Student> student2)
   { student = student2; }
}
public class Student {
  private Mentor mentor;
  int id;
  public Student(Mentor mentor2)
   { mentor = mentor2; }
}
```

Green does provide additional code generation support by creating custom constructors based on the association. Green supports the following types of multiplicities: 1, n, m..*, and * (where n > 1 and m >= 0).

Green provides some enforcement of constraints; although the implementation is awkward and not scalable. For example, the implementation of the constraint of a *mandatory* relationship where a Mentor must have *n* Students (e.g. n = 3) is shown below.

```
public class Mentor {
  private Student student3;
  private Student student2;
  private Student student;
  java.lang.String name;
  public Mentor(Student student4,
                Student student5,
                Student student6) {
    student3 = student4;
    student2 = student5;
    student = student6;    } }
```

This implementation provides little opportunity to access or manage the collection of students, and instead each must be accessed explicitly by name. It also does a poor job of maintaining the constraint; as the variables could be set to *null*, violating the model's intention.

Green also provides an enforcement of m..* relationships. Below is an example implementation of a 2..* relationship.

```
import java.util.List;
public class Mentor {
  private List<Student> student;
  java.lang.String name;
  public Mentor(List<Student> student2) {
    student = student2;
    student.add(new Student());
    student.add(new Student());   }  }
```

The implementation above presents two issues. First, the potentially unwanted side effect of creating and inserting additional entities into the list argument (i.e. students). Second, the code generator assumes that a default (and empty) constructor exists for the Student object; an assumption that might not always be valid and could result in a generated system that does not compile.

Although Green UML does attempt to provide some additional source code generation to manage the various types of association multiplicities available; the results provide little, if any, added benefit in representing the model's intentions.

5.6 Rational Software Architect (RSA)

IBM's Rational Software Architect (RSA and RSA Real-Time) are full-fledged development environments that support model-driven development including source code generation from UML diagrams.

```
import java.util.Set;
public class Mentor {
  public Set<Student> students;
  public Set<Student> getStudents()
   { return students; }
  public void setStudents(Set<Student>students)
   { this.students = students; }
}
public class Student {
  public Mentor mentor;
  public Mentor getMentor()
   { return mentor; }
  public void setMentor(Mentor mentor)
   { this.mentor = mentor; }
}
```

RSA's model transformation into Java code provides some flexibility regarding the template patterns including (a) which Java collection to use, and (b) whether or not to include get/set methods for the attributes and association ends. As with all other source code generators, no distinction between the various possible one or many relationships are present in the generated code; leaving the implementation of the modeling constraints up to manually-written code. In addition to providing simple set and get methods, RSA's member variables representing the association ends was also public; presenting an encapsulation issue (especially considering the code already provides set and get methods).

6 Association Code Generation in Umple

The existing UML code generation tools analyzed in the previous section fall short of providing robust code to implement associations. The generated code provided little implementation support either to manage referential integrity or to ensure multiplicity constraints (beyond the 'one' vs. 'many' distinction).

In this section, we present our approach to code generation and identify implementation patterns that go beyond the capabilities of current tools. This approach is instantly accessible from Umple online [5].

6.1 Defining Association Variables

The first pattern to emerge is the distinction between having one object in the association and having many (i.e. upper bound equal to one versus greater than one). For convenience, we will use UB for upper bound and LB for lower bound.

Table 8 Member Variable Patterns

Mult. Constraint	Pattern	Example
UB = 1	ObjectType associationEnd;	Student student;
UB > 1	List<ObjectType> associationEnd;	List<Student> students;

6.2 Constructor Parameters for Associations

The next patterns relate to a class' constructor signature. The constructor defines how objects should be created and indirectly affects the order in which objects can be instantiated. Three signatures emerge from the various multiplicities:

- The association end is not required (LB=0) and not be part of the constructor
- Exactly one, the upper and lower bounds are exactly one
- Mandatory Many, (LB > 0 and UB > LB)

The patterns in Table 9 work well when the multiplicity of at least one end of the association is zero; allowing the creation of one object before the other. Below are example implementations of the constructors above.

Table 9 Constructor Signature Patterns

Multiplicity Constraint	Pattern	Example
LB = 0	Empty	N/A
LB=UB=1	ObjectType anAssociationEnd	Student aStudent;
LB > 0 && UB > 1	List<Student> someAssociationEnds	List<Student> allStudnets;

By using the *setStudent* method (which we discuss in the interface patterns section), we are able to encapsulate *how* students are set; including the verification that the set operation is indeed valid (i.e. association multiplicity constraints are not violated). If we are unable to assign the student, then an exception is thrown. The exact error message is not shown for simplicity.

```
public Mentor(Student aStudent) {
  boolean didAddStudent = setStudent(aStudent);
  if (!didAddStudent) {
    throw new RuntimeException("***"); } }
```

When the upper bound is greater than one (and the lower bound is not zero), we must initialize a list of associated members. We can delegate the action and verification using the *setStudents* (instead of setStudent like above) method.

```
public Mentor(List<Student> allStudents) {
  students = new ArrayList<Student>();
  boolean didAddStudents =
              setStudents(allStudents);
  if (!didAddStudents) {
    throw new RuntimeException("***");  } }
```

A chicken-and-egg issue manifests itself when neither end has a lower bound of zero; meaning that each end requires the other, resulting in deadlock as neither constructor can be called before the other. This issue has been divided into three domains: One to One, One to Mandatory Many and Mandatory Many to

Mandatory Many. To highlight the implementation of each situation above, we added a *name* attribute to the Mentor, and a *number* attribute to the Student class.

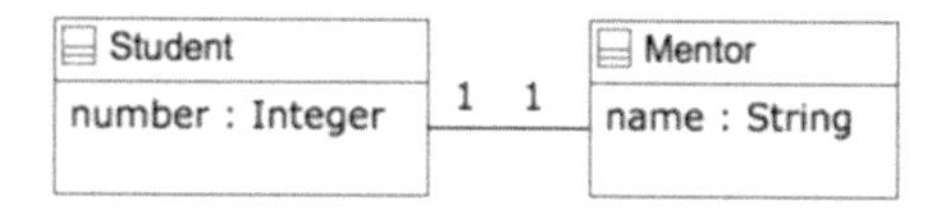

Fig. 3 One-to-one Student and Mentor association

In 3, a Mentor must have exactly one Student, and vice versa.

```
public Mentor(String aName, Student aStudent) {
  name = aName;
  if (aStudent == null || aStudent.getMentor() != null)
    { throw new RuntimeException("***"); }
  student = aStudent; }
public Mentor(String aName, int aNumberForStudent) {
  name = aName;
  student = new Student(aNumberForStudent, this); }
```

The second constructor *Mentor(String aName, int aNumberForStudent)* takes all the required parameters for both objects; allowing both objects to be created *effectively* instantaneously. Please note that in this case, there is no *setStudent* interface and the logic to verify that the student is valid is provided directly in the constructor. We do not include a *setStudent* or *setMentor* interface because the one-to-one constraint means you cannot re-assign a mentor or a student, as the *replaced* object would then be an orphan; violating the one-to-one constraint.

In the *One* to *Mandatory Many* case a Mentor must have more than one Student (m..n, n, or m..*) and a Student must have exactly one Mentor.

Table 10 Constructor Signature Patterns When Both Association Ends are Mandatory

Multiplicity	Constructor Implementation Multiplicity
1 -- m..*	1 -- *
1 -- m..n	1 -- 0..n
1 -- n	1 -- 0..n

The Mentor's constructor is implemented as though the lower bound was zero; allowing the objects to exist in an invalid state immediately following its construction. To verify the status of an object in such a case, we add an additional method `isNumberOfStudentsValid`, which checks if the number of students is valid.

The *mandatory-many* to *mandatory-many* constructor is similar, is that *both* constructors are initiated without the constraint being satisfied, and an additional method isNumberOfMentorsValid() is provided. The implementation as proposed for these latter two cases allows for handling of cases that are in fact rare. Our implementation provides the developer with the necessary tools to query the association to verify the satisfaction of the constraint.

6.3 Get Method Code Generation Patterns

Table 11 outlines an the interface to access an association end available to a Mentor based on the multiplicity end of the Student. The method pattern is based on a generic association end name (*name*), and the association end's type (*type*).

Table 11 Method Signature Patterns for Get Methods

Mult. Constraint	Pattern	Example
UB = 1	getName() : Type	getStudent() : Student
UB > 1	getName(int index) : Type getNames() : Iterator<Type> indexOfName(Type aName) : int numberOfNames() : int hasNames() : boolean	getStudent(int index) : Student getStudents() : Iterator<Student> indexOfStudent(Student aStudent) : int numberOfStudents() : int hasStudents() : boolean

The *getStudent* implementation is shown below.

```
public Student getStudent() { return student; }
```

The difference between *mandatory one* (1) and *optional one* (0..1) is that the student member might be null in the optional case; whereas the 1 multiplicity end will never be null.

When the upper bound multiplicity is greater than 1, there are four common accessor methods as shown below.

```
public Student getStudent(int index)
 { return students.get(index);}
public Iterator<Student> getStudents()
 { return students.iterator(); }
public int numberOfStudents()
 { return students.size(); }
public boolean hasStudents()
 { return students.size() > 0; }
public int indexOfStudent(Student aStudent)
 { return students.indexOf(aStudent); }
```

Although one has access to all associated students, one is not able to alter the association by manipulating a list retrieved using the *get* methods shown above. To change the number of elements one must use the available *add* methods as shown below. The reason for this is to prevent the caller of API methods from being able to violate the multiplicity constraints or corrupt the referential integrity. Other implementers of 'many' associations simply pass the collection of objects around, however, we explicitly ensure this never happens.

6.4 Set Method Code Generation Patterns

Next, we consider an interface to add, remove and set links of an association end. Again, we will be adding Student instances to a Mentor object based on various multiplicity constraints. Table 12 describes the generated interface.

Table 12 Method Signature Patterns for Set Methods

Multiplicity Constraint	Pattern	Example
UB = 1	setName(Type aName)	setStudent(Student aStudent)
UB > 1	addName(Type aName)	addStudent(Student aStudent)
	removeName(type aName)	removeStudent(Student aStudent)

The implementation of set methods is considerably more complex than get methods. First, set methods must undo any existing links between objects and establish the new links. Second, the methods must ensure referential integrity: when creating one end of a binary association they must create the other end as well.

Let us begin with the case where the upper bound is one. When the relationship is optional, the following scenarios must be considered.

If adding a new link, there must be code to set the inverse link as well. Conversely, if the inverse link has already been set, then it must not be set again. For example, if adding a Student to a Mentor, the code must be sure to add a Mentor to the Student (but only once).

If replacing or removing an existing link, both directions of the link must be removed. For example, if a Mentor can only have one Student, then when assigning a Mentor to a new Student, the implementation must unassign that Mentor from the existing Student.

When creating a new link, the multiplicity constraints on both ends must be satisfied. If a Mentor can only have four Students, then a Student is not allowed to add a Mentor such that the Mentor would now be linked to five Students.

Finally, whenever removing an existing link, the multiplicity constraints on the existing objects must be satisfied. If a Mentor must have at least two Students, then the implementation must not allow a Student to set itself to a new Mentor if the existing mentor is at its two-Student minimum.

Two examples are outlined below; one where UB = 1, and the other where UB > 1. First the implementation of *setMentor* from in Student class as part of the 0..1 Mentor -- 0..1 Student association.

```
public void setMentor(Mentor newMentor) {
  if (newMentor == null) {
    Mentor existingMentor = mentor;
    mentor = null;
    if (existingMentor != null &&
        existingMentor.getStudent() != null) {
      existingMentor.setStudent(null); }
    return; }
  Mentor currentMentor = getMentor();
  if (currentMentor != null &&!currentMentor.equals(newMentor)) {
    currentMentor.setStudent(null); }
  mentor = newMentor;
  Student existingStudent = newMentor.getStudent();
  if (!equals(existingStudent)) { newMentor.setStudent(this); } }
```

Next are the implementations of addStudent, and removeStudent for the Mentor class as part of the 0..1 Mentor -- * Student association.

```
public boolean addStudent(Student aStudent) {
  if (students.contains(aStudent)) { return false; }
  Mentor existingMentor = aStudent.getMentor();
  if (existingMentor == null) {
    students.add(aStudent);
    aStudent.setMentor(this);
  } else if (!existingMentor.equals(this)) {
    existingMentor.removeStudent(aStudent);
    addStudent(aStudent);
  } else { students.add(aStudent); }
  return true; }

public boolean removeStudent(Student aStudent) {
  if (!students.contains(aStudent)) { return false; }
  else {
    students.remove(aStudent);
    aStudent.setMentor(null);
    return true; } }
```

There are 42 different association combinations (28 different classes, 7 for directed and an additional 7 for symmetric). Each implementation follows the general guidelines shown above, and each combination can explored online at [9].

6.5 Patterns for Generated Support Methods

In addition to establishing relationships between objects, we include methods to query the minimum and maximum bounds of a relationship. Due to space constraints, we omit the full details of the support methods, but Table 13 highlights the interface.

Table 13 Interface for support methods

Multiplicity	Interface
m..n, m..*	minimumNumberOfStudents() : int
m..n, 0..n	maximumNumberOfStudents() : int
n	requiredNumberOfStudents() : int

7 Related Work

Several studies [12-16] propose approaches to formalizing the semantics of associations. They generally agree on the interpretation of the associations, but do not address uniqueness and ordering of associations.

Other studies refer to two types of associations; static and dynamic [17, 18]. Static associations, a view we adopt, represent structural relationships between objects, where the association is enforced throughout the lifetime of both objects. *Dynamic* (or *contextual* associations) are enforced only during the interactions of the two objects. Miliev [19] proposes yet another view of associations: *intentional* associations that encapsulate the intention of association of each participating object. Milicev highlights deficiencies in the traditional semantics of associations and multiplicities that can be overcome by the introduction of an intentional perspective on associations.

Acknowledging deficiencies in automated code generation of UML associations and multiplicities, Wang and Shen [20] propose a run-time verification approach for UML association constraints. Østerbye [21] proposes supporting association referential integrity with a reusable class library that ensures the consistency of the relationship is maintained.

Executable UML (eUML) [6] aims is to provide a (yet to be approved) specification of an unambiguous subset of executable UML (using model compilers). The Umple language also behaves as a model compiler and provides a concrete implementation of a subset of UML. However unlike Executable UML, Umple integrates with standard object oriented languages, and supports a wider range of multiplicity, as well as a variety of other features not present in executable UML.

Umple has been under continuous development since 2007. Experimentation with users has revealed that the comprehensibility levels of model oriented code is superior to the equivalent object oriented code [22-24]. Umple has also been used and evaluated in open source projects [25].

8 Conclusion

This paper discussed problems with generating code for UML associations, and proposed Umple as a solution. We identified the 42 combinations of multiplicity for association ends and analyzed their impact on code generation. We reviewed the code generated by five modeling tools, and found that none dealt with multiplicity constraints or referential integrity. This may be one reason why code generation is not as widely used in practice as might be expected.. As a result developers must modify generated code by hand, which is awkward and error-prone. We provided an overview of the Umple language for associations and its model compiler that addresses the above issues.

References

1. Costal, D., Gómez, C.: On the use of association redefinition in UML class diagrams. In: Embley, D.W., Olivé, A., Ram, S. (eds.) ER 2006. LNCS, vol. 4215, pp. 513–527. Springer, Heidelberg (2006)
2. Object-Oriented Software Engineering: Practical Software Development using UML and Java. McGraw-Hill (2005)
3. Badreddin, O., Forward, A., Lethbridge, T.C.: Exploring a Model-Oriented and Executable Syntax for UML Attributes. Accepted in SERA 2013 (2013)
4. Object-Oriented Software Engineering: Practical Software Development using UML and Java. McGraw Hill (2001)
5. UmpleOnline, `http://www.try.umple.org` (accessed 2013)
6. Executable UML: A Foundation for Model-Driven Architectures. Addison-Wesley, Boston (2002)
7. Umple Language, `http://cruise.site.uottawa.ca/umple/` (accessed 2013)
8. Forward, A., Lethbridge, T.C., Brestovansky, D.: Improving program comprehension by enhancing program constructs: An analysis of the umple language, pp. 311–312 (2009)
9. Umple language online, `http://cruise.site.uottawa.ca/umpleonline/` (accessed 2013)
10. Norton, D.: Open-Source Modeling Tools Maturing, but Need Time to Reach Full Potential, Gartner, Inc., Tech. Rep. G00146580 (April 20, 2007)
11. Wikipedia Listing of UML modeling tools, `http://en.wikipedia.org/wiki/List_of_UML_tools` (accessed 2013)
12. Bourdeau, R.H., Cheng, B.H.C.: A formal semantics for object model diagrams. IEEE Trans. Software Eng. 21, 799–821 (1995)
13. Diskin, Z., Dingel, J.: Mappings, maps and tables: Towards formal semantics for associations in UML2. In: Wang, J., Whittle, J., Harel, D., Reggio, G. (eds.) MoDELS 2006. LNCS, vol. 4199, pp. 230–244. Springer, Heidelberg (2006)
14. France, R.: A problem-oriented analysis of basic UML static requirements modeling concepts. ACM SIGPLAN Notices 34, 57–69 (1999)
15. Overgaard, G.: A formal approach to relationships in the unified modeling language. In: Proceedings PSMT (1998)
16. Overgaard, G.: Formal specification of object-oriented ModellingConcepts. PhD Thesis, Dept. of Teleinformatics, Royal Inst. of Technology, Stockholm, Sweden (November 2000)
17. Stevens, P.: On the interpretation of binary associations in the Unified Modelling Language. Software and Systems Modeling 1, 68–79 (2002)
18. Genova, G., Llorens, J., Fuentes, J.M.: UML associations: A structural and contextual view. Journal of Object Technology 3, 83–100 (2004)

19. Miliev, D.: On the semantics of associations and association ends in UML. IEEE Trans. Software Eng., 231–258 (2007)
20. Wang, K., Shen, W.: Runtime checking of UML association-related constraints. In: Proceedings of the 5th International Workshop on Dynamic Analysis (2007)
21. Osterbye, K.: Design of a class library for association relationships. In: Proceedings of the 2007 Symposium on Library-Centric Software Design, pp. 67–75 (2007)
22. Badreddin, O.: Empirical Evaluation of Research Prototypes at Variable Stages of Maturity. In: ICSE Workshop on User Evaluation for Software Engineering Researchers, USER (to appear, 2013)
23. Badreddin, O., Lethbridge, T.C.: Combining experiments and grounded theory to evaluate a research prototype: Lessons from the umple model-oriented programming technology. In: User Evaluation for Software Engineering Researchers (USER). IEEE (2012)
24. Badreddin, O., Forward, A., Lethbridge, T.C.: Model oriented programming: an empirical study of comprehension. In: Proceedings of the 2012 Conference of the Center for Advanced Studies on Collaborative Research. IBM Corp. (2012)
25. Badreddin, O., Lethbridge, T.C., Elassar, M.: Modeling Practices in Open Source Software. In: OSS 2013, 9th International Conference on Open Source Systems (to appear, 2013)

How Process Enactment Data Affects Product Defectiveness Prediction – A Case Study

Damla Aslan, Ayça Tarhan, and ve Onur Demirörs

Abstract. The quality of a software product is highly influenced by the software process used to develop it. However, abstract and dynamic nature of the software process makes its measurement difficult, and this difficulty has supported the assessment insight of indirectly measuring the performance of software process by using the characteristics of the developed product. In fact, enactment of the software process might have a significant effect on product characteristics and data, and therefore, on the use of measurement and analysis results. In this article, we report a case study that aimed to investigate the effect of process enactment data on product defectiveness in a small software organization. We carried out the study by defining and following a methodology that included the application of Goal-Question-Metric (GQM) approach to direct analysis, the utilization of a questionnaire to assess usability of metrics, and the application of machine learning methods to predict product defectiveness. The results of the case study showed that the accuracy of predictions varied according to the machine learning method used, but in the overall, about 3% accuracy improvement was achieved by including process enactment data in the analysis.

Keywords: software defect prediction; machine learning; process enactment, software measurement, defectiveness.

Damla Aslan
Simsoft Computer Technologies Co., Ltd, Technopolis of METU, Ankara, Turkey
e-mail: damla.sivrioglu@simsoft.com.tr

Ayça Tarhan
Department of Software Engineering, Hacettepe University, Ankara, Turkey
e-mail: atarhan@cs.hacettepe.edu.tr

ve Onur Demirörs
Informatics Institute, METU, Ankara, Turkey
e-mail: demirors@ii.metu.edu.tr

R. Lee (Ed.): *SERA*, SCI 496, pp. 151–166.
DOI: 10.1007/978-3-319-00948-3_10

1 Introduction

Product defectiveness is a suitable measure for the assessment of software quality. Defect data usage for quality evaluation is common among studies in literature since accessibility to product defect data by issue tracking and configuration management tools are easy. These tools are important information sources with their huge databases that store descriptions of the defects detected in software, including detection dates, resolution status, and etc.

The process reference models such as CMMI [22] suggest after second maturity level the mapping between the product and process data and also suggest taking into account this mapping for process improvement. However, the accessibility and collection of process data are difficult to use in software measurement and analyses [1, 2] because of abstract and dynamic nature of the software process. Therefore, we propose that using process enactment data might be a good idea to obtain process traces and combine them with defect data. By this way process characteristics can be realized and the patterns in the data can be recognized. Besides, we can take process dynamics into consideration by directly using its enactment data instead of indirectly using derived metrics such as defect density.

In this study, we aimed to investigate how the usage of process enactment data affects software defectiveness prediction results. We used defect open duration as dependent variable and applied machine learning algorithms for the purpose of classifying defect data and defect tracking process enactment data to predict defect open duration. The data used in the study belonged to a software development project of a small software organization called Simsoft. WEKA tool [14] was used to conduct the analysis. To assess the benefit of including process enactment data in the analysis, we compared the reliability of the datasets with and without process enactment data.

The remainder of this paper is organized as follows. Section two provides an overview of studies that consider software process data in defect analysis and prediction. Section three explains the methodology and its application in the case study. Section four provides the results of the analysis and a discussion on them. Finally, section five provides overall conclusions.

2 Related Studies

Catal and Diri reviewed software defect prediction studies in a systematical way [11]. The review states that the studies with using class-level, process-level and component-level measures are not sufficient. Besides, machine learning methods are suggested since they give better results than statistical analysis methods and expert view. Tarhan and Demirors [18, 19] emphasized the importance of process differences in software projects. They developed and applied some assets such as Metric Usability Questionnaire, Process Execution Record, and Process Similarity Matrix for data collection and evaluation. Sivrioglu and Tarhan [3] analyzed defect data with both statistical and machine learning methods. The results indicated that machine learning techniques were more accurate than the ones of statistical techniques. At the end of the study they suggested to use process enactment data for more accurate results.

It is slightly possible to find prediction studies by using process data in literature. Jalote et al. [15] explained a defect prediction approach by performing quantitative quality management and statistical process charts. Wahyudin et al. [16] presented a defect prediction model by using statistical hypothesis with a combination of product and process measures. Lee et al. [2] developed a prediction model with micro interaction metrics which are supposed as process-related metrics. In their study, they made comparisons among the accuracy results of the model of code metrics, the model of history metrics, and the combination of them. Fenton et al. [17] suggested Bayesian Belief Networks machine learning technique as prediction model. Process data is given for their model. Dhiauddin [4] generated a prediction model for testing phase in his master thesis. With this model he discovered the strong factors that contributed to the number of testing defects. Zeng and Rine [5] have estimated defect fix effort by using dissimilarity matrix and Self Organizing Maps (Kohonen Networks) which is a type of Neural Networks method. With this data mining technique the data have been clustered for prediction. Model performance has been evaluated by magnitude of relative error (MRE) values of 6 grouped data sets. The input attributes of the model are defect fix time in hour unit, defect severity, the activity during which the defect is detected, system mode, defect category and SLOC (source lines of code) changed. Defect severity, detection activity, system mode and defect category attributes can be considered as contextual metrics. Weiss et al [6] have used the defects life-time phases gone through issue tracking tool as the attributes for defect fix effort prediction. They compared two types of Nearest Neighbor approaches called as with (α-kNN) and without thresholds (kNN). They used text mining for grouping the data before kNN analysis. Hassouna and Tahvildari [7] have improved Weiss' study by adding 1. data enrichment to infuse additional issue information into the similarity-scoring procedure, 2. majority voting to exploit many of the similar historical issues repeating effort values, 3. adaptive threshold to automatically adjust the similarity threshold to ensure that they obtain only the most similar matches and 4. binary clustering to form clusters when the similarity scores are very low phases. Hewett and Kijsanayothin [8] have penned down a comprehensive study regarding defect repair time prediction. Firstly, they have applied five different empirical machine learning approaches to two individual data sets with and without attribute selection.

It is seen that researchers claim the benefits of using process measures, machine learning methods, some data collection and grouping methods for generating defect prediction models one by one. However, none of them uses several of these methods together in empirical studies. Combining product data with process data to generate defect prediction models by using qualitative and quantitative techniques, we believe, is a promising research topic.

3 Methodology

Figure-1 shows the methodology that was developed and applied in the case study.

To carry out the analysis in the direction of our purpose, we applied Goal-Question-Metric (GQM) approach [13] as the first step. The GQM table constructed to direct the analysis is given in Table 1.

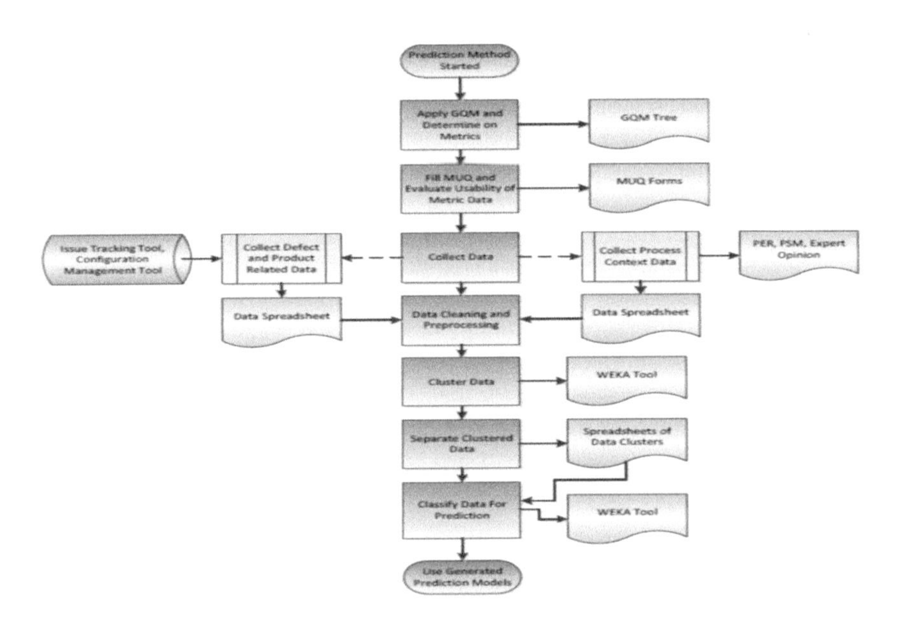

Fig. 1 Methodology

Table 1 GQM Table of the Case Study

GOAL	QUESTION	ANALYSIS METHOD	DERIVED METRIC	BASE METRICS AND DATA
To understand if there is an effect of process enactment on software product defectiveness prediction.	How much impact has process enactment on defect open duration prediction?	Bayesnet, Logistic, C4.5 Tree, Multilayer Perceptron Machine Learning Techniques	Defect Data: open duration (closed date-created date)	**Defect and Product Data:** detected module name, closed date, created date, detected test type, product version, product SLOC, product complexity, reproducibility, detected project phase
			Defect Data: open duration (closed date-created date)	**Defect and Product Data:** detected module name, closed date, created date, detected test type, product version, product SLOC, product complexity, reproducibility, detected project phase **Process Enactment Data:** defect management process attributes

In the case study, we used metric data of the defects detected in software qualification tests of a project completed in a small software development organization called Simsoft. The company has 30 employees that contain Software Engineers, Modeling and Graphics Designers, and Quality Assurance Support Personnel. It has developed software projects for a large amount of institutes, especially for defense industry by now. In the project of which defect data was used in this study, 6 personnel worked for 7-months project duration to develop a software product of a Computer Service Configuration Item (CSCI) with 5 modules. At the end of the development, the project ended with 23 KLOC (Kilo Lines of Code) of C++ code implemented for 955 requirements, and with 296 defects detected during qualification tests run by test specialists.

Table 2 Metric Usability Questionnaire for "created date" Metric of Defect Records

Indicators			Answers	Rating	Expected
Measure Identity				**N**	
	Q1	Which entity does the measure measure?	Process		
	Q2	Which attribute of the entity does the measure measure?	Defect record's time		
	Q3	What is the scale of the measurement data?	Nominal		Ratio, Absolute
	Q4	What is the unit of the measurement data?	Time		
	Q5	What is the type of the measurement data? (integer, real, etc.)	Date		
	Q6	What is the range of the measurement data?	00.00.0000 00:00		
Data Existence				**F**	
	Q7	Is measurement data existent?	Yes		
	Q8	What is the amount of overall observations?	296	⍰	Available > 20
	Q9	What is the amount of missing data points?	0		
	Q10	Are data points missing in periods?	0		
	Q11	Is measurement data time sequenced?	Yes		
Data Verifiability				**F**	
	Q12	When is measurement data recorded in the process?	At start		
	Q13	Is all measurement data recorded at the same place in the process?	Yes	⍰	Yes
	Q14	Who is responsible for recording measurement data?	Test Specialist		
	Q15	Is all measurement data recorded by the responsible body?	Yes	⍰	Yes
	Q16	How is measurement data recorded? (on a form, report, tool, etc.)	Tool		
	Q17	Is all measurement data recorded the same way?	Yes	⍰	Yes
	Q18	Where is measurement data stored?	The tool's database		
	Q19	Is all measurement data stored in the same place?	Yes	⍰	Yes
Data Dependability				**P**	
	Q20	What is the frequency of generating measurement data?	Asynchronous		
	Q21	What is the frequency of recording measurement data?	Synchronous		
	Q22	What is the frequency of storing measurement data?	Synchronous		
	Q23	Are the frequencies for data generation, recording, and storing different?	No	⍰	No
	Q24	Is measurement data recorded precisely?	Yes	⍰	Yes
	Q25	Is measurement data collected for a specific purpose?	No	⍰	Yes
	Q26	Is the purpose of measurement data collection known by process performers?	No	⍰	Yes
	Q27	Is measurement data analyzed and reported?	No		Yes
	Q28	Is measurement data analysis results communicated to process performers?	No		Yes
	Q29	Is measurement data analysis results communicated to management?	No		Yes
	Q30	Is measurement data analysis results used as a basis for decision making?	No		Yes

The defect data from Issue Tracking Tool (ITT) and product size data from Configuration Management Tool (CMT) were collected manually. ITT contains detailed information such as issue status, issue created date, issue updated date, issue reproducibility, which are basically related to Issue Management Process. Since the aim of the study was to analyze defect prediction process, we extracted the data which belonged to "Defects" detected rather than "Changes" implemented in software. Other process metrics such as project phase when the issue was detected, and the test type in which the issue was detected were collected manually. The product metrics such as product version size and complexity, however, were obtained indirectly from the tool. We say "indirectly" because these metrics were calculated with LocMetrics tool [12], and related data was elicited from the information recorded in the tool by using the product version where the issue was detected.

After identifying metrics and gathering metric data, metric usability analysis for each base metric was performed to determine if the metric was suitable and its data was sufficient for our study. This analysis was carried out by using an asset called Metric Usability Questionnaire (MUQ) as shown in Table 2. It is an asset that was defined by one of the authors and has been utilized in more than 10 case studies to evaluate the usability of metrics and data for quantitative software management [14]. The MUQ was filled for each base metric, and it was seen that selected metrics were only "partially usable" for the analysis.

Table 3 Metric Descriptions

Metrics	Metric Description	Measurement Scale
Defect Open Duration	The time starting with the creation of the defect and finishing with the closure of the defect. Calculated by the difference of defect closed date and defect created date. Unit is number of days.	Absolute
Source Component	The component name in which the defect detected. Component name can be component-A, component-B, component-C, component-D or component-E. Recorded by the Issue Tracking Tool automatically when the tester enters the defect info.	Nominal
Created Date	The date when the defect is detected. Filled by the Issue Tracking Tool automatically when the tester records the defect.	Interval
Closed Date	The date when the defect is closed. Filled by the Issue Tracking Tool automatically when the project manager changes the status of the defect as "Closed".	Interval
Test Type	The name of test type during which the defect is detected. Entered by tester into the Issue Tracking Tool.	Nominal
Product Version	The version of the software product in which the defect is detected. Entered by tester into the Issue Tracking Tool.	Ordinal
SLOC (Source Lines of Code)	The size of the product version where the defect is detected. Collected from Configuration Management Tool by using LocMetrics tool.	Absolute
Complexity	The McCabe complexity of the product version where the defect is detected. Collected from Configuration Management Tool by using LocMetrics tool.	Absolute
Reproducibility	The repeatability of the defect detected. Entered by tester into the Issue Tracking Tool.	Nominal
Project Phase	The project phase where the defect is detected. Collected manually by domain expert.	Nominal

The descriptions of defect and product metrics specified in the GQM table (in Table 2) are given in Table 3. The data for these metrics extracted from ITT and CMT were gathered in an Excel sheet. "Defect open duration" metric was calculated by subtracting "created date" from "closed date" during data cleaning and preparation phase.

To collect process enactment data, Process Execution Record (PER) shown in Figure 2 was utilized. This is an asset to capture internal process attributes such as inputs, activities, outputs, roles, and tools and techniques taking place during process enactment [15]. A PER was filled retrospectively for each defect recorded in the ITT with defect management process professionals. At the end of process enactment data collection phase, we had process enactment data for each one of the 296 defects detected during software qualification tests performed in the scope of the software project.

Process Execution Record
(Internal Attributes)

Process Name:	Issue Management	Recorded On:	26.03.2012
Process Execution No:	N/A	Recorded By:	Damla Sivrioğlu

1. **Inputs:** Please list the inputs to the process execution.

No	Name	Description
1	Defects	
2	Change requests	

2. **Outputs:** Please list the outputs from the process execution.

No	Name	Description
1	Updated product version	

3. **Activities:** Please list in sequence the activities that were performed while executing the process.

No	Name	Description
1	Assign defect	
2	Defect resolution (Fix issues)	
3	Not verified for second time	
4	Defect verification	
5	Close defect	

4. **Roles:** Please list the roles that were allocated responsibilities in process execution.

No	Name	Description
1	Project Manager	Track issues, Fix issues
2	Configuration Manager	Track issues
3	Developer	Fix issues
4	Modelling and Graphics Designer	Fix issues
5	Tester	Open issues

5. **Tools and Techniques:** Please list the tools and techniques that are used to support process execution.

No	Name	Description
1	Redmine	Issue tracking tool
2	Excel	Version Description List is in Excel format.
3	SVN	Configuration management tool

Fig. 2 Process Execution Record

Process Similarity Matrix (PSM) given in Table 4 is another asset which was organized as a spreadsheet in Excel file and used to gather process attribute values for all process enactments [15]. This matrix enables one to see the similarities and differences in process enactments at a glance, and this information is valuable to cluster process executions prior to data analysis or to derive causal relationships while interpreting data analysis results. Process attributes specified in PERs were placed vertically and process enactments (regarding the management of every defect recorded in ITT) were placed horizontally in this matrix. Each cell in the PSM was filled by entering either "1" or "0" depending on weather the process attribute was applicable for regarding process execution or not. After the PSM is completed, the differences in columns were examined and the clustering of process enactments was performed manually. Prior to this, the data collected by the PSM was reviewed and similar colons were removed to prevent the usage of redundant data in the analysis. We observed that dmA1, dmA5, dmR1, dmR3, and dmR5 displayed the same patterns. We kept only dmR3 from these six attributes as the representative. We also observed that dmI1 and dmI2 did not differ in values among the executions, and we did not include dmI2 in our analyses.

Table 4 Process Similarity Matrix

		Process Attributes															
		1.1 <Input 1>	1.2 <Input 2>	2.1 <Output 1>	3.1 <Activity 1>	3.2 <Activity 2>	3.3 <Activity 3>	3.4 <Activity 4>	3.5 <Activity 5>	4.1 <Role 1>	4.2 <Role 2>	4.3 <Role 3>	4.4 <Role 4>	4.5 <Role 5>	5.1 <Tools and Techniques 1>	5.2 <Tools and Techniques 2>	5.3 <Tools and Techniques 3>
Process Executions	**Defect No**	dmI1	dmI2	dmO1	dmA1	dmA2	dmA3	dmA4	dmA5	dmR1	dmR2	dmR3	dmR4	dmR5	dmT1	dmT2	dmT3
PE1	**1**	1	0	1	1	1	0	1	1	1	1	1	0	1	1	1	1
PE2	**2**	1	0	0	1	1	0	0	1	1	0	1	0	1	1	1	1
PE3	**3**	1	0	1	1	1	0	1	1	1	1	1	0	1	1	1	1
PE4	**4**	1	0	1	1	1	0	1	1	1	1	1	0	1	1	1	1
PE5	**5**	1	0	1	1	1	1	1	1	1	1	1	0	1	1	1	1
PE6	**6**	1	0	1	1	1	0	1	1	1	1	1	0	1	1	1	1
PE7	**7**	1	0	1	1	1	0	1	1	1	1	1	0	1	1	1	1
PE8	**8**	1	0	1	1	1	0	1	1	1	1	1	0	1	1	1	1
PE9	**9**	1	0	1	1	1	1	1	1	1	1	1	0	1	1	1	1
PE10	**10**	1	0	1	1	1	0	1	1	1	1	1	0	1	1	1	1
PE11	**11**	1	0	1	1	1	0	1	1	1	1	1	0	1	1	1	1
PE12	**12**	1	0	0	1	0	0	0	1	1	0	1	0	1	1	1	1
PE13	**13**	1	0	1	1	1	0	1	1	1	1	1	0	1	1	1	1
PE14	**14**	1	0	1	1	1	0	1	1	1	1	1	0	1	1	1	1
...	**...**																

Table 5 Process Clusters and Process Attribute Patterns

Process Attributes Pattern (PAP)	2.1 <Output 1> dmO1	3.2 <Activity 2> dmA2	3.3 <Activity 3> dmA3	3.4 <Activity 4> dmA4	4.2 <Role 2> dmR2	4.3 <Role 3> dmR3	4.4 <Role 4> dmR4
Cluster Name	**c0**						
PAP1	1	1	0	1	1	1	0
PAP2	1	1	1	1	1	1	0
Cluster Name	**c1**						
PAP1	1	1	0	1	1	0	1
PAP2	1	1	1	1	1	0	1
PAP3	1	1	1	1	0	0	1
PAP4	0	0	1	1	1	0	1
Cluster Name	**c2**						
PAP1	1	1	0	1	0	1	0

Table 5 (*continued*)

Cluster Name	**c3**						
PAP1	1	1	0	1	1	0	0
Cluster Name	**c4**						
PAP1	1	1	1	1	1	0	0
Cluster Name	**c5**						
PAP1	0	1	0	0	0	1	0
Cluster Name	**c6**						
PAP1	0	0	0	0	0	1	0
PAP2	0	0	1	0	0	1	0

data_with_process_context_c0
data_with_process_context_c1
data_with_process_context_c2
data_with_process_context_c3
data_with_process_context_c4
data_with_process_context_c5
data_with_process_context_c6

Fig. 3 Clustered Data Files

As the next step we combined defect, product, and process enactment data in an Excel file, and clustered the data by using K-Means and Euclidean Distance clustering techniques. We obtained seven clusters including one or more Process Attribute Patterns (PAPs) as shown in Table 5. The characteristics of the clusters shown in the table are described below in terms of process attributes.

Cluster 0 included process executions through which an updated product version was obtained as output, defect resolution and defect verification activities were implemented, and configuration manager and developer performed their roles. However, modeling and graphics designer did not perform his role.

Cluster 1 included process executions through which defect verification activity was implemented, and modeling and graphics designer performed his role. However, developer did not perform his role.

Cluster 2 included process executions through which an updated product version was obtained as output, defect resolution and defect verification activities were implemented, and developer performed his role. However, configuration manager, and modeling and graphics designer did not perform their roles.

Cluster 3 included process executions through which an updated product version was obtained as output, defect resolution and defect verification activities were implemented, and configuration manager performed his role. However developer, and modeling and graphics designer did not perform their roles.

Cluster 4 included process executions through which an updated product version was obtained as output; defect resolution, not verified for second time and defect verification activities were implemented; and configuration manager performed his role. However developer, and modeling and graphics designer did not perform their roles.

Cluster 5 included process executions through which defect resolution activity was implemented, and developer performed his role. However configuration manager, and modeling and graphics designer did not perform their roles.

Cluster 6 included process executions through which no activities documented in PERs were implemented, and only developer performed his role. In only one of 296 executions, the "not verified for second time activity" was implemented. It means that in one defect management process execution, the defect in resolved status could not be verified during second test repetition by the test specialist.

We separated the data in the Excel file into separate files in accordance to the clusters identified, and prepared a separate .csv file for each cluster accordingly. At the end of the clustering, we obtained the files shown in Figure 3. Each of these files included defect, product, and process enactment date of the related defects.

WEKA tool [14] was used to conduct data analyses. The main purpose of using Weka is to discover patterns between process enactment and defect open duration metric. We applied Multilayer Perceptron, Bayesian Belief Networks, Logistic Regression, and C4.5 Decision Tree (J48) machine learning techniques [21] for each cluster separately by keeping "defect open duration" metric as class attribute (dependent variable). To define this quantitative variable as class attribute, it was transformed to nominal scale by using Weka discretization method.

4 Analysis Results

To analyze our GQM goal, we compared the results of two different data sets. The first data set included only defect and product data without process enactment data. The second data set included both defect and product data with process enactment data.

According to the analysis of first data set, the following results were obtained. The set included 296 data points which were sufficient to obtain confident prediction results. Multilayer perceptron gave the best performance values compared with other machine learning approaches. Other performance values of the models are provided in Table 6.

- Multilayer perceptron machine learning technique validated with 10-folds gave 95% correctly classified instances.
- Bayesian networks machine learning technique validated with 10-folds gave 85% correctly classified instances.
- Logistic machine learning technique validated with 10-folds gave 82% correctly classified instances.
- J48 decision tree machine learning technique validated with 10-folds gave 92% correctly classified instances.

According to the analysis of the second data set, the following results were obtained. Correctly classification performance values of the generated models for cluster 0 are given below. The other performance values of the models and the clusters are provided in Table-6. Bayesian networks gave the best performance values compared with other machine learning approaches.

- Multilayer perceptron machine learning technique validated with 10-folds gave 96% correctly classified instances.
- Bayesian networks machine learning technique validated with 10-folds gave 97% correctly classified instances.
- Logistic machine learning technique validated with 10-folds gave 95% correctly classified instances.
- J48 decision tree machine learning technique validated with 10-folds gave 96% correctly classified instances.

Table 6 shows the analysis results of the comparative data sets. Since clusters 3, 4 and 5 included low number of data, we could not apply machine learning techniques to them. In the overall, we observed that the analysis results of clustered data sets (clusters 0 and 2) with process enactment were more accurate than the data set without process enactment. When looked into the details of Table 6, we had the results as described in the following paragraphs.

The average of correctly classified instances values of the methods applied to cluster 0 data was 95.98%. On the other hand the average of correctly classified instances values of the methods applied to data without process enactment was 88.51%. The correctly classified rate was 7.47% higher in cluster 0 than the result of the data set that did not include process enactment. The average of root mean squared error values of the methods applied to cluster 0 data was 11.73%. On the other hand the average of root mean squared error values of the methods applied to data without process enactment was 19.29%. The root mean squared error was 7.55% lower in cluster 0 than the result of the data set that did not include process enactment.

The average of correctly classified instances values of the methods applied to cluster 1 data was 83.10%. The correctly classified rate was 5.41% lower in cluster 1 than the result of the data set that did not include process enactment. The average of root mean squared error values of the methods applied to cluster 1 data was 25.09%. The root mean squared error was 5.81% higher in cluster 1 than the result of the data set that did not include process enactment.

The average of correctly classified instances values of the methods applied to cluster 2 data was 90.00%. The correctly classified rate was 1.49% higher in cluster 2 than the result of the data set that did not include process enactment. The average of root mean squared error values of the methods applied to cluster 2 data was 23.43%. The root mean squared error was 4.14% higher in cluster 2 than the result of the data set that did not include process enactment.

The average of correctly classified instances values of the methods applied to cluster 6 data was 100.00%. The correctly classified rate was 11.49% higher in cluster 6 than the result of the data set that do not include process enactment. The average of root mean squared error values of the methods applied to cluster 6 data was 1.80%. The root mean squared error was 17.49% lower in cluster 6 than the result of the data set that does not include process enactment.

The analysis results received from cluster 1 are not promising. Because the average of correctly classified instances value is lower than the result value of the data set that does not include process enactment data. However if we divide it into further clusters (PAPs 1-4 in c1) shown in Table 5, the prediction performance might be improved. In other words, we consider that the reason of this situation is the noise in various Process Attribute Patterns.

Table 6 Results from Comparative Data Sets

Number of instances (data points)	Data set	Method	Correctly Classified Instances	Incorrectly Classified Instances	Kappa statistic	Mean absolute error	Root mean squared error	Relative absolute error
112	Cluster 0 Data (With Process Enactment)	Multilayer Perceptron	96.43%	3.57%	94.86%	1.70%	10.37%	6.06%
		Bayesnet	97.32%	2.68%	96.16%	1.40%	10.45%	4.98%
		Logistic	94.64%	5.36%	92.28%	2.14%	14.64%	7.63%
		J48	95.54%	4.46%	93.55%	2.15%	11.47%	7.64%
71	Cluster 1 Data (With Process Enactment)	Multilayer Perceptron	84.51%	15.49%	79.06%	7.35%	24.39%	61.41%
		Bayesnet	80.28%	19.72%	73.61%	8.31%	27.57%	69.87%
		Logistic	81.69%	18.31%	75.58%	7.19%	26.46%	23.84%
		J48	85.92%	14.08%	80.95%	7.41%	21.94%	24.57%
70	Cluster 2 Data (With Process Enactment)	Multilayer Perceptron	95.71%	4.29%	92.13%	3.61%	14.75%	9.76%
		Bayesnet	91.43%	8.57%	83.48%	5.53%	21.94%	14.96%
		Logistic	90.00%	10.00%	81.04%	6.54%	25.37%	17.70%
		J48	82.86%	17.14%	64.87%	17.21%	31.64%	46.55%
296	Data Without Process Enactment	Multilayer Perceptron	94.93%	5.07%	93.38%	2.40%	13.14%	7.80%
		Bayesnet	85.14%	14.86%	80.54%	5.79%	20.81%	18.86%
		Logistic	82.43%	17.57%	76.90%	7.00%	26.16%	22.7%
		J48	91.55%	8.45%	88.87%	5.63%	17.03%	18.35%

5 Lessons Learned

GQM provided a systematic way to direct the analysis and to determine the data that would be collected. MUQ supported data cleansing phase, and enabled us to assess and characterize the data. PER and PSM helped us to collect process enactment data since it was not straightforward to gather it from tools.

The metrics utilized in our study were selected by applying the GQM approach and characterized by using MUQs. It was observed that the metrics are “partially usable” rather than being “fully usable” for the analysis. This was due to the weakness of “data dependability” attribute of the MUQ, more specifically to the lack of feedback related to metric data collection and usage within the project and the company. Actually, this is one of the essential problems that the organizations face with while practicing measurement and analysis. The weakness of this attribute might have caused the noise in the data and have increased the error rates in predictions; however, these proposals need to be validated by additional studies. The weakness of data dependability attribute was reported to company management, and initiating a small scale measurement program was recommended as the first step.

The module name in which the defect is detected needs to be stored in Issue Tracking Tool (ITT). Since it had not been stored there, this data was collected manually by the developer. SLOC and complexity metrics should be collected on product version basis. Since these metrics had been collected on monthly basis, the product version information mapped to these product metrics was not available. This data was collected manually by using Version Description List Document.

It was observed that the history data stored by ITT was beneficial to collect process enactment data. We collected process enactment data by filling PERs retrospectively with process professionals to identify process attributes. These process attributes can be identified easier by reviewing history data in tool database since all process activity alternatives are stored with their dates and the personnel who perform the activity. For example, when any personnel updates the defect status as "verified", the tool constitutes a record that "Defect status was updated by <personnel name> on <date>." in database. This process history data was elicited manually and used to fill PSM for each defect record, in other words, for each process execution. Elicitation of this data from the tool cost 40 man-hours.

Since machine learning analysis techniques are pattern oriented measurement methods, the raw defect data alone cannot be sufficient either for defect prediction or bringing recommendations for product and process improvement. Therefore, if process enactment data is gathered and added to defect data, the data can be clustered according to similar process attributes and machine learning techniques can give more accurate results when the researchers have sufficient data points.

The factors that have an impact on software product defectiveness can be considered in two categories [15]: (outer) environmental factors and (inner) process execution. Process execution means the applied process in the company while developing the product (e.g., development life cycle). Environmental factors mean the factors which affect product defectiveness but cannot be controlled by human beings at the time of process execution (e.g., developer skills). As a constraint in our study, we included only the (inner) process attributes.

We could not obtain successful results for each cluster when we compared the data sets with process enactment data to those without it. We explain the insights below.

- For cluster 1 we could not obtain promising results due to the the noise in cluster patterns that was shown in Table 5. To avoid this noise and achieve more accurate prediction for cluster 1, one more clustering can be performed within cluster-1 data. This clustering operation might also show one or more process attribute patterns.
- For cluster 2, although average correctly classified instances was high, we obtained a high average error value. The reason of this might be the low error rate in J48 (C4.5) decision tree method. This machine learning method needs more data points for a more accurate prediction than the other methods.

- The questions that we wanted to answer in this study were "Does process enactment help software defectiveness prediction?" and "How much impact has process enactment on defect open duration prediction?". As we observed from the results of cluster 0 and cluster 2 given in Section IV, process enactment improves software defectiveness prediction performance up to 12% (with at least 2%) if you have sufficient data points and homogeneous clusters (which are not noisy with PAPs).

As the last point, the process clusters obtained by using K-Means and Euclidean Distance clustering techniques indicated the variations in process enactments and therefore might have arisen important issues for process improvement (e.g., the reasons of variations). In this study, we did not focus on identifying the opportunities for process improvement specifically (unless they arose naturally), and left them out of the discussion since our aim was to investigate the effects of process enactment data on defectiveness prediction. However, clustering of process attribute values by machine learning methods might also serve this specific purpose.

6 Conclusion and Future Work

This study has been carried out to introduce a methodology to use process enactment data for defect prediction and analyze the effect of process enactment on defect prediction. We have practiced a prediction model by combined application of GQM, MUQ, PER, PSM, process attribute clustering, and machine learning techniques. We investigated the validity of our model by comparing error values of two different datasets. The first data set contained only defect and product data. The second data set contained defect, product, and process enactment data together. We had a number of supporting as well as threatening results. Despite the threats, since predicting defects in the development life-cycle is crucial to delivering quality products, using qualitative and quantitative techniques together for defect prediction deserves attention as a research topic.

The methodology can be used in other domains as well, since the only requirements for machine learning analysis are process enactment data, result metrics and one dependable variable. In any business area you can find this information and you apply the method offered in this study.

This study was performed with the data of only one project of a single company. Therefore, in some data clusters the number of data points was not sufficient for machine learning classification techniques. Similar studies with several similar projects' data can be performed in the future to have a more solid base for the validation. As another future work, the outer process factors can be investigated and different collection methods might be discovered for them. Automatic data collection will reduce data collection efforts and increase the motivation to use the prediction models.

Acknowledgement. We thank to Simsoft team, especially Aydın Okutanoğlu, Veysi İşler, and Şafak Burak Çevikbaş for allowing to use company data in our research study and sharing of their expert views.

References

1. Koru, A.G., Liu, H.: Building Effective Defect-Prediction Models in Practice. IEEE Software 22(6) (November/December 2005)
2. Lee, T., Nam, J., Han, D., Kim, S., In, H.P.: Micro Interaction Metrics for Defect Prediction. In: ESEC/FSE 2011 Proceedings of the 19th ACM SIGSOFT Symposium and the 13th European Conference on Foundations of Software Engineering (2011)
3. Sivrioğlu, D., Tarhan, A.: Defectiveness Analysis According To Software Module Features: A Case Study (Yazılım Modül Özelliklerine Göre Hatalılık Analizi: Bir Durum Çalışması) Original is Turkish (February 2012)
4. Dhiauddin, M.: Defect Prediction Model For Testing Phase. Master Thesis, Universiti Teknologi Malaysia, Faculty of Computer Science and Information System (May 2009)
5. Zeng, H., Rine, D.: Estimation of Software Defects Fix Effort Using Neural Networks. In: COMPSAC 2004 Proceedings of the 28th Annual International Computer Software and Applications Conference - Workshops and Fast Abstracts, USA, vol. 02, pp. 20–21 (2004)
6. Weiss, C., Premraj, R., Zimmermann, T., Zeller, A.: How Long will it Take to Fix This Bug? In: MSR 2007 Proceedings of the Fourth International Workshop on Mining Software Repositories, USA, p. 1 (2007)
7. Hassouna, A., Tahvildari, T.: An Effort Prediction Framework for Software Defect Correction. Information and Software Technology 52, 197–209 (2010)
8. Hewett, R., Kijsanayothin, P.: On Modeling Software Defect Repair Time. Empir. Software Eng. 14, 165–186 (2008, 2009)
9. Runeson, P., Höst, M.: Guidelines for conducting and reporting case study research in software engineering. Empirical Software Eng. 14, 131–164 (2009)
10. Florac, A.W., Park, R.E., Carleton, A.D.: Practical Software Measurement: Measuring for Process Management and Improvement. Guidebook: CMU/SEI-97-HB-003 (1997)
11. Çatal, Ç., Diri, B.: A Systematic Review of Software Fault Prediction Studies. Expert Systems with Applications 36, 7346–7354 (2009)
12. `http://www.locmetrics.com/` (last access date: April 11, 2012)
13. Basili, V.R., Caldiera, G., Rombach, H.D.: Goal Question Metric Paradigm. In: Encyclopedia of Software Engineering – 2 Volume Set (1994) ISBN#1-54004-8
14. `http://www.cs.waikato.ac.nz/~ml/weka/` (last access date: April 11, 2012)
15. Jalote, P., Dinesh, K., Raghavan, S., Bhashyam, R., Ramakrishnan, M.: Quantitative Quality Management through Defect Prediction and Statistical Process Control
16. Wahyudin, D., Schatten, A., Winkler, D., Tjoa, A.M., Biffl, S.: Defect Prediction using Combined Product and Project Metrics a Case Study from the Open Source "Apache" MyFaces Project Family. In: 34th Euromicro Conference on Software Engineering and Advanced Applications, SEAA 2008, September 3-5, pp. 207–215 (2008)
17. Fenton, N., Krause, M., Neil, P.: A Probabilistic Model for Software Defect Prediction. For submission to IEEE Transactions in Software Engineering

18. Tarhan, A., Demirörs, O.: Apply Quantitative Management Now. IEEE Software 29(3), 77–85 (2012), doi:10.1109/MS.2011.91
19. Tarhan, A., Demirörs, O.: Investigating the Effect of Variations in Test Development Process: A Case from a Safety-Critical System. Software Quality Journal, doi:10.1007/s11219-011-9129-8
20. Boetticher, G.D.: Nearest Neighbor Sampling for Better Defect Prediction
21. Witten, I.H., Frank, E.: Data Mining Practical Machine Learning Tools and Techniques, 2nd edn. Elsevier (2005)
22. CMMI Product Team, CMMI for Development, Version 1.3, Technical Report, SEI (2010)
23. Sivrioğlu, D.: A Method for Product Defectiveness Prediction with Process Enactment Data in a Small Software Organization. Master Thesis, Middle East Technical University, Informatics Institute (June 2012)

Modeling Business and Requirements Relationships for Architectural Pattern Selection

Javier Berrocal, José García-Alonso, and Juan Manuel Murillo

Abstract. In analysis of the business and the system requirements, the identified elements are modeled using notations that fully describe their characteristics. Nevertheless, implicit relationships often exist between different types of elements that subsequently have to be identified and explicitly represented during the design of the system. This requires an in-depth analysis of the generated models on behalf of the architect in order to interpret their content. Misunderstandings that take place during this stage can lead to an incorrect design and difficult compliance with the business goals. Here we present a series of profiles that explicitly represent these relationships during the initial development phases, and which are derived to the system design. They are reusable by the architect, thereby decreasing the risk of their misinterpretation.

1 Introduction

The initial phases of software product development require a careful analysis of the business to which the application is addressed. In particular, the analysis will identify information such as the goals and processes of the business. This information will then form the basis on which to identify and specify the requirements, both functional and non-functional, of the system to develop [15], [8], [25].

The information generated from such analyses is detailed in specific artefacts. For example, business goals are documented in a *Goals Model*, and the business processes in a *Business Processes Model* [26]. The same applies to the system requirements [9]. The functional requirements are documented in the *Use Cases Model* and the non-functional in the *Supplementary Specification Document* [20]. In this way the characteristics of each type of information are detailed in a specific artefact without interfering with other elements.

Javier Berrocal · José García-Alonso · Juan Manuel Murillo
University Of Extremadura, Avda. de la Universidad s/n, 10003, Cáceres, Spain
e-mail: {jberolm,jgaralo,juanmamu}@unex.es

R. Lee (Ed.): *SERA*, SCI 496, pp. 167–181.
DOI: 10.1007/978-3-319-00948-3_11

Later, at the system design phase, all this information must be analysed and consolidated. The architect must know how non-functional requirements influence the functional ones in order to select the architectural tactics and patterns that best facilitate the fulfilment of both requirements.

The consolidation of this information is complicated and costly. Firstly, the architect has to know the content of every artefact and, secondly, he or she has to identify the relationships between the elements. Because each document is focused on specific types of elements, the relationships between them are usually detailed implicitly, leaving the architect the task of discovering and inferring the network of relationships. Thus for example, in the *Business Processes Model* the Goals that impose restrictions or limitations that affect some of the modeled tasks are not detailed. Instead, it is the architect who, depending on his or her experience and ability, has to detect this type of information.

There have been various studies aimed at making these relationships explicit. In [1], the authors link business goals with business processes in order to identify areas of the business processes that should be improved to meet these needs. In [21], the authors define how to model business operational restrictions in business processes in order to define controls to handle these restrictions. In the present work, we examine how to document these relationships so that they can be reused in subsequent phases of development.

In particular, we define how to make explicit the relationships both between elements of the business and between the requirements, the objective being that the architect can subsequently use them for the design of the system. To that end, we have defined an extension for BPMN 2.0 [7], to support the modeling of relationships between business elements, and a set of extension for the UML 2 Use Case, Activity and Sequence Diagrams [27], to support the modeling of relationships between functional and non-functional requirements. Modeling these relations, the chances of their misinterpretation is reduced. These annotations are also used to automatically preselect the architectural patterns that can be applied to meet the requirements. This is possible thanks to the definition of a set of rules which, when applied to the models and annotations, facilitate this selection. The result is to narrow down the solution space that the architect has to explore.

The paper is organized as follows. Section 2 presents the motivation of the work, section 3 describes the process for documenting the relationships, section 4 reviews some related work, and section 5 presents the conclusions and future works.

2 Motivation

Developing a system to provide maximum value to a firm requires a thorough analysis of its business. Some methodological approaches, such as Ullah [26] or BMM [6], detail how to perform and document this analysis. They define, for example, how business goals are modeled using Goals, and how business processes are documented with BPMN notation [7].

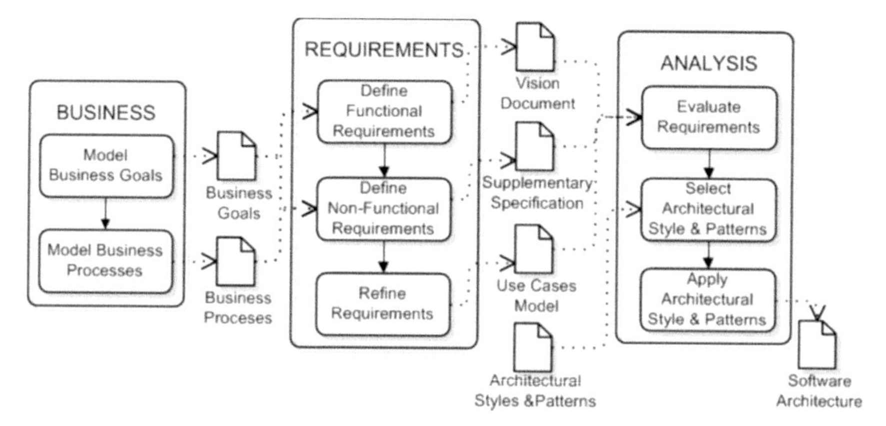

Fig. 1 Main activities of the design process

Performing this analysis in the case of an online shop, for instance, one could identify, among other things, the business goals "facilitate the implementation of any changes in the workflow or in the rules of the business", and "transmit quickness by checking whether an order is correct in less than 0.7 seconds". All these objectives are documented in the *Business Goals* artefact, as shown in figure 1. Similarly, with the analysis of the business, the business processes are identified and modeled, documenting them in the *Business Processes* artefact. Figure 2 shows part of the business process to handle orders.

As each element is documented in artefacts and with specific annotations, all of its features are perfectly detailed and encapsulated. This does not, however, facilitate the modeling of the relationships between elements of different types, making it difficult to document the influence of one element on another. Thus, for the online shop of figure 2, before a customer validates his order, he has to be given an indication of possible mistakes together with a series of suggestions of products that might interest him. The performance goal is to convey quickness during the validation of the order. With an initial analysis of this goal and the business process, one could identify that only the task "Check Order" has to meet that goal. The identification of this relationship is easy, nonetheless, for the user to really perceive this rapidity, the task "Search for Additional Products" must also satisfy that objective. The identification of this second relationship requires a deeper analysis of the business. During the business analysis phase, the identification of these relationships may be relatively simple. However, if they are kept undocumented, their identification at later stages of development, such as during the design of the architecture, can be very expensive and dependent on the experience of the architect.

In [21], the authors define an extension of BPMN to annotate the tasks of business processes with *operational restrictions*. With this extension, in processing orders, the tasks "Check Order" and "Search for Additional Products" would be annotated with *Operating Conditions* indicating that the response time should be less than 0.7 seconds. To get a complete picture of the restrictions, it would be desirable

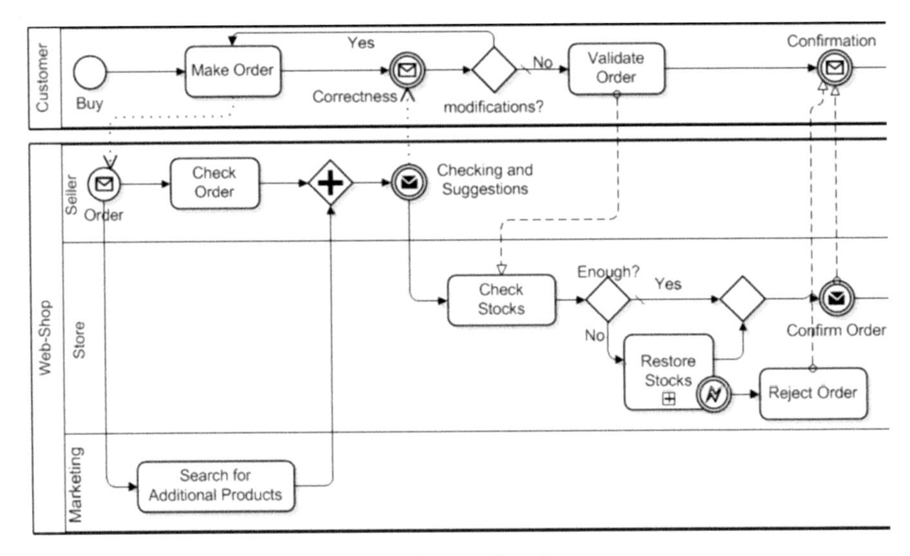

Fig. 2 Business process to handle orders in an online shop

to also model the conditions of the non-operational objectives. Thus, in the order business process, one could also model which process tasks are constrained by the objective "facilitate the implementation of new business rules".

As shown in figure 1, one defines from the business information both the non-functional and functional requirements. Thus, from the online shop's business goals, two non-functional requirements are defined: "the system has to be easy to change" and "the response time in checking orders has to be shorter than 0.7 seconds". These requirements are detailed in the *Supplementary Specification Document*. Similarly, from the business process of figure 2, one identifies such functionalities as "making orders" or "checking the validity of each order". These functionalities are first detailed with a Use Cases Diagram in the *Vision Document*, as figure 3 shows. Subsequently, the use cases are specified in the artefact *Use Cases Model* with sequence and activity diagrams.

Again, both types of requirement are detailed with different diagrams and in specific artefacts that allow their features to be better detailed, but making hard to document the relationships between them. Thus, in the online shop's Use Cases Diagram, figure 3, one can not specify that the "Check Order" use case is related to a non-functional requirement that constrains the time in which the checking has to be done. These are relationships that have to be taken into account for such activities as the system design, so that they will have to be identified by the architect. Correct identification of these relationships is highly dependent on the architect's experience in correctly interpreting both requirements.

In [13], the authors add notes to the use case diagrams, detailing in natural language the non-functional requirements that each use case must satisfy. For example, in the online shop's use case diagram, a note would be linked to the "Check Order"

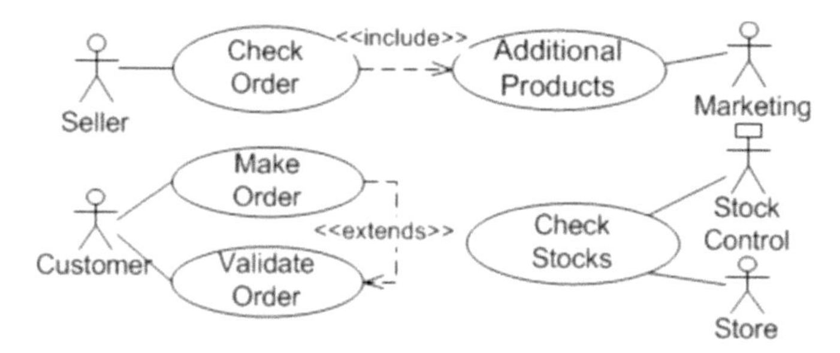

Fig. 3 Use cases extracted from the order process

use case, indicating that some of its activities must be executed in less than 0.7 seconds. This information will thus provide invaluable documentation for the architect. Nonetheless, for it to have even greater value, it would be desirable if it could be processed by tools that guide the architect in the design of the system.

The detailed requirements are then analysed during the system design process. To decide which pattern to apply, the architect must know perfectly the system's requirements and their relationships. If the latter are not fully specified, they will have to be identified by means of an in-depth analysis of the artefacts. For the online shop, the architect should analyse the performance non-functional requirement and all the use cases in order to identify those relationships. Whereas an initial analysis will show that the "Check Order" use case must satisfy that requirement, to identify the "Additional Products" use case as also having to satisfy it will need a more detailed analysis of both the requirements and the business – a laborious task that requires great experience.

Any misidentification of relationships may cause an incorrect pattern to be chosen, which would make it difficult to fulfil the business goals. For the online shop, once it has been identified that the entire system must be maintainable and that only two use cases need to take performance into account, the architect may decide to apply the *Layer* pattern together with some tactic to achieve the desired performance [16]. If these relationships are identified incorrectly, it may be decided to apply a pattern that is oriented more to performance but that does not pay special attention to maintainability, thus hindering the fulfilment of this objective.

Works such as [3] and [19] facilitate the choice of architectural patterns in documenting the different patterns and how they affect each requirement and the relationships between requirements. In this way, the architect is guided in the selection of the most suitable patterns. For these techniques to be fully effective, the architect needs to have a clear picture of the relationships between system requirements. The present work focuses on making these relationships explicit so that the architect can identify them clearly and unambiguously.

To document the relationships between the elements of the business, we define an extension for the Business Process Modeling Notation (BPMN 2.0). In addition, to model the relationships between requirements, we define profiles for the UML 2 Use Case, Activity, and Sequence Diagrams. Finally, we show how the architect can use these requirement relationships together with a set of rules to narrow down the possible patterns which may be used for the system design. As a result, there is less chance of misinterpreting the relationships, and the architect's task in selecting patterns is made easier.

3 Documenting the Relationships

In order to guide the specification of relationships during the early stages of development, we have added two new activities to the design process of figure 1. Furthermore, another activity has been added to guide the architect in the analysis of those relationships. These activities, together with their supporting tools, are:

- *Model the context information.* In this activity, the business processes are annotated with information on the relationships between process tasks and different elements of the business. An extension of BPMN 2.0 notation is used to be able to model this information.
- *Define the requirement relationships.* In this activity, the relationships between the system requirements are specified. To this end, we have defined a set of profiles that allow the relationships to be modeled in UML diagrams.
- *Analyse the requirement relationships.* During this activity, the requirements and their relationships are analysed with the objective of delimiting the possible patterns that can be applied. We have defined a set of rules to facilitate this task. These rules, together with the annotated models, are automatically applied by a tool.

Figure 4 shows highlighted the activities added to the design process illustrated in figure 1, together with the artefacts that are modified as a result. The following subsections describe each of these activities and the corresponding profiles.

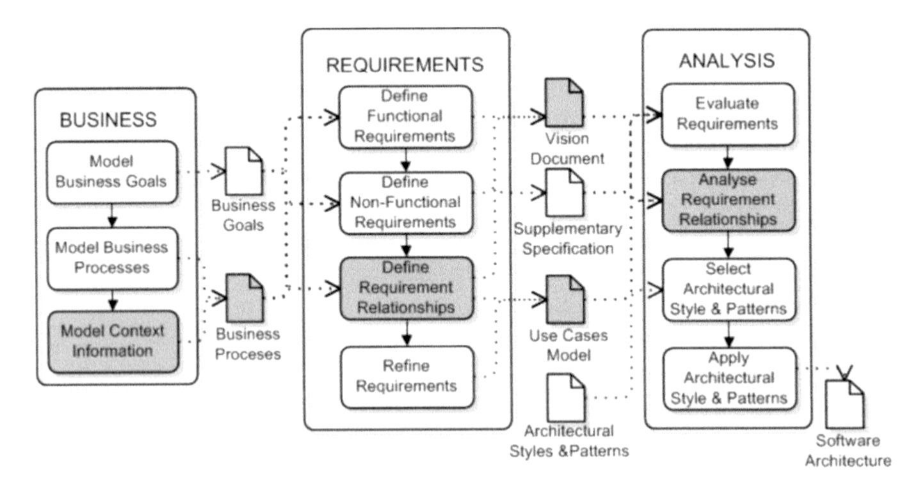

Fig. 4 Design process indicating the new tasks and modified artefacts

3.1 *Model the Context Information*

In analysing the organization's business, one identifies and models information about its objectives, needs, and processes. The objective of this activity is to detail the relationships between these elements. The relationships are annotated on the business processes modeled with BPMN. These models are then used as a basis on which to model the relationships since BPMN provides a number of communication facilities [7] that are reused to facilitate discussion about the relationships.

To model these relationships, the BPMN notation was extended with a profile. This extension defines the *Quality Attribute*, *Legacy System*, and *Business Use Case* stereotypes.

A new element is needed to model the relationships between business goals and business processes so that they can be documented. This element is the *Quality Attribute* stereotype, which allows one to group tasks of the processes that must meet certain of the business's quality objectives or requirements. In figure 5, the virtual shop process has been annotated with three business goal relationships. The first specifies that the entire business process is related to the goal of ease of change. The second and third indicate that the "Check Order" and "Search for Additional Products" tasks are constrained by the objective which limits the time in which they must be completed. These annotations make it easy for the architect to see which of the process's specific tasks must fulfil each goal.

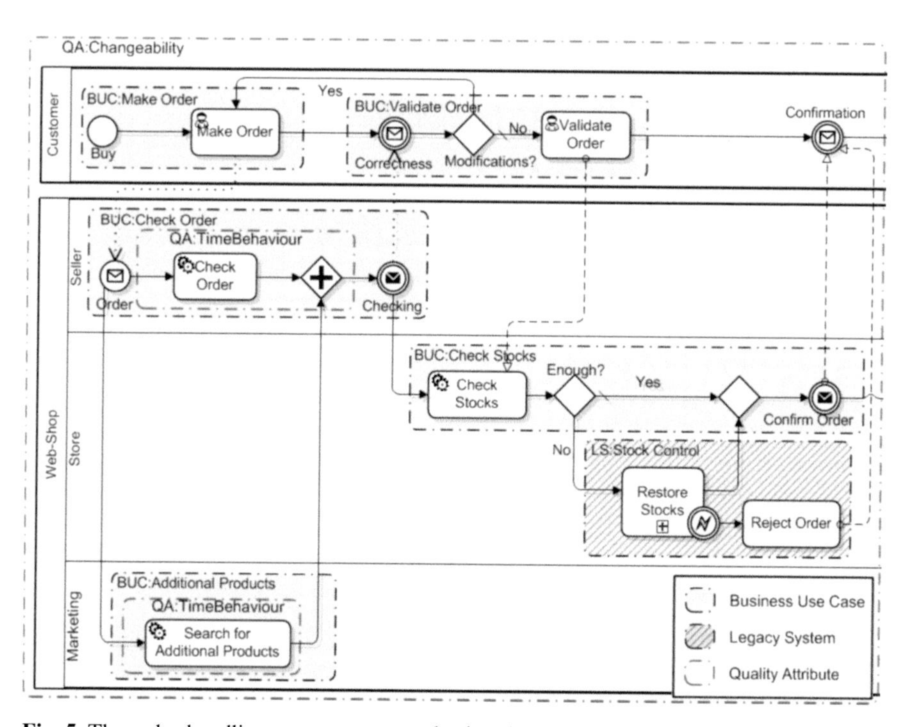

Fig. 5 The order-handling process annotated using the defined profile

It is also necessary to know whether certain tasks, even though are modeled in a lane because a certain role is responsible of them, are already being supported by a legacy system. The *Legacy System (LS)* stereotype allows such tasks to be grouped. The resulting groupings are then used to derive the relationships between the new system and the legacy systems. In figure 5 for instance, this stereotype is used to indicate that the "Restore Stocks" sub-process is already being supported by the "Stock Control" legacy system. In this way, one can identify when the new system should invoke the legacy system to restore stock.

Finally, when business processes are used to identify the functionalities of a system, natively there exists no element to show an approximation of the functionalities which will support the process and each of its task. The "Business Use Case" (BUC)[1] stereotype, which is based on the Cockburn's *Coffee break* rule [10] and on the *Step* concept defined in [12], allows process tasks to be grouped to represent this information. These groupings, in addition to facilitate the derivation of use cases, are reused in the "Define Requirement Relationships" activity to derive the relationships between requirements. Five Business Use Cases are modeled in the online shop's business process. The first BUC, for example, groups the tasks the user performs to place the order. In this way, the business expert and the requirements engineer can easily see and discuss the business tasks covered by each functionality.

By means of the above stereotypes, the engineer can reflect the relationships between different elements of the business, documenting them and facilitating their consideration in future development phases. In addition, these annotations may also be automatically dealt by tools that aid certain development activities.

3.2 *Define the Requirement Relationships*

The functional and non-functional requirements are defined on the basis of the information modeled in the business. The functional requirements are modeled and specified as use cases in the *Vision Document* and in the *Use Cases Model*, and the non-functional requirements in the *Supplementary Specification Document*.

The relationships between the functional and non-functional requirements are documented and modeled in this activity. These relationships are documented at two levels of granularity. At the first level, the non-functional requirements that must be satisfied by each use case are detailed. At the second level, which of each use case's tasks or actions must satisfy each restriction are detailed.

To model the relationship between non-functional requirements and use cases, we have defined a profile for the UML 2 Use Case Diagrams. This profile defines stereotypes that extend the "Extension Points" to enable these relationships to be modeled. It was decided to extend this metaclass because it maintains the readability of the diagram when a large number of relationships are modeled.

---

[1] The term "Business Use Case" is used here, even though it is similar to the System Use Case term [10], because it groups business tasks that should be refined in order to detail the system tasks.

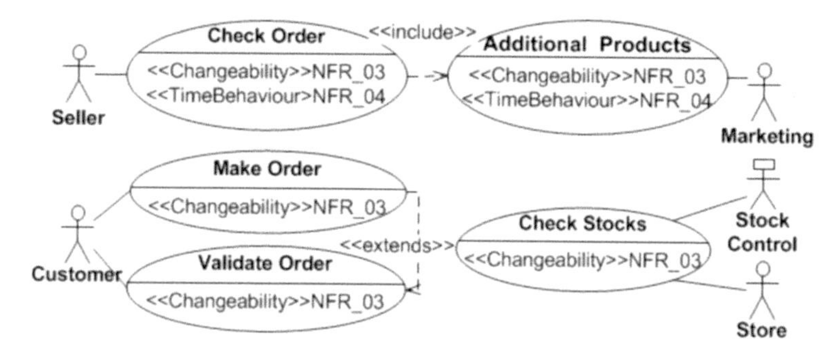

Fig. 6 Annotated use case diagram extracted from the online shop process

Figure 6 shows the use case diagram extracted from the information annotated in the business process illustrated in figure 5. The diagram shows five use cases, all annotated with the "Changeability" non-functional requirement, and the "Check Order" and "Additional Products" use cases must also fulfil the "Time Behaviour" requirement. Thus, the relationships between each use case and non-functional requirement are easily documented and visualized. In addition, these annotations can be reused by tools that assist development.

Besides being able to deduce these relationships manually, we have also defined a series of patterns to guide the engineer in their identification. These patterns are based on the information annotated in the business processes to derive the relationships between requirements. For example, since the tasks covered by "BUC:Check Order" of the online shop are annotated with the "Time Behaviour" requirement, one of these patterns indicates that there also exists a relationship between the use case derived from that BUC and the indicated quality requirement. In [4], one can find more detailed information on these patterns.

In order to be able to detail at a finer granularity the relationships between use cases and non-functional requirements, we have also defined extensions to the Activity and Sequence Diagrams. These extensions define stereotypes that group the actions that have to fulfil each non-functional requirement.

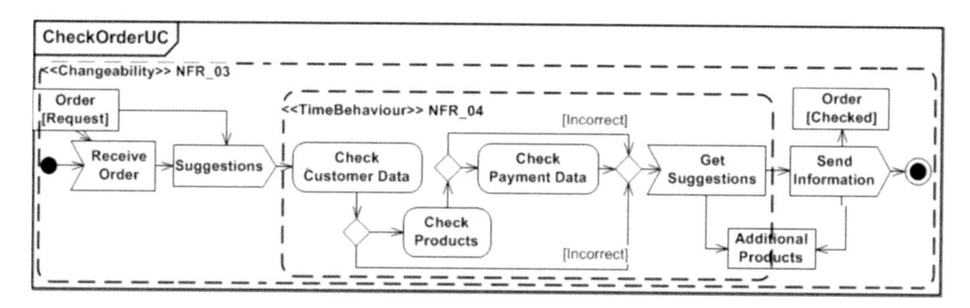

Fig. 7 Activity diagram of the Check Order use case

Figure 7 shows the activity diagram for the online shop's "Check Order" use case. As in the use case diagram, this diagram has two relationships annotated with non-functional requirements. The first is that all of the use case's actions must satisfy the "Changeability" requirement. The second indicates that just the "Check Customer Data, Check Products, Check Payment, and Get Suggestions" actions must satisfy the "Time Behaviour" requirement.

The above profiles allow one to reflect the relationships between use cases and non-functional requirements to the point of detailing exactly which actions each non-functional requirement must satisfy. In this way, with little effort the architect can identify and analyse that information in the system design phase, thereby reducing the risk of failure due to the misidentification of some relationship.

The stereotypes of the presented profiles were defined taking into account the ISO/IEC 9126 quality model [18], and permitting the possibility of detailing properties for each quality requirement. One can thus apply rules to filter, search, or reason the use cases constrained by specific non-functional requirements.

3.3 Analyse the Requirement Relationships

For designing the system, the architect should have formed a complete picture of the requirements and how they relate to each other. This information is needed to select the most appropriate patterns and tactics. Each pattern may affect different requirements positively or negatively. For example, as shown in table 1 derived from [16] and [5], the *Layer* pattern affects positively the adaptability and ease of change requirements, but negatively the performance requirement.

Analysing all the patterns and how they affect the non-functional requirements, those that best facilitate the fulfilment the system requirements could be selected. Nevertheless, whether all the pattern's benefits, or its liabilities, are obtained largely depends on the functionalities on which it is applied. Thus, for example, in order for the *Pipe and Filter* pattern to provide the performance and maintenance benefits, the system functionalities should be implemented to exploit the parallel processing. Similarly, for the *Layer* pattern to provide the stated benefits, the system must have a certain size. Otherwise, the complexity introduced for the separation between layers harm instead of benefit to the system maintainability.

Table 1 Benefits and liabilities of the architectural patterns

Pattern	Benefits	Liabilities
Layer	Security Maintainability Adaptability Developed by multiple teams	Efficiency Development Complexity
Pipes And Filter	Maintainability Efficiency	Usability Security Reliability

In this activity, the architect analyses the relationships between functional and non-functional requirements through the diagrams annotated with the profiles detailed above. The annotated use case diagrams allow the architect to obtain a high level vision of the use cases together with the non-functional requirements that each of them has to fulfil. The activity diagrams give a finer grain view of which actions have to support each non-functional requirement. This way, one can easily assess whether the application of a pattern to a given set of use cases facilitates or hinders the fulfilment of the annotated non-functional requirements. In addition, the architect can also evaluate whether applying the desired pattern on that set of use cases the desired benefits are really obtained.

For the online shop for example, it is readily seen that the entire system must fulfil the ease of change requirement, while only the "Check Order" and "Additional Products" use cases need to take performance into account when executing their actions. Taking these non-functional requirements and the patterns indicated in table 1 into account, the architect could decide to apply the *Pipes and Filter* pattern, since it benefits both non-functional requirements. However, since for processing an order there has to be interactions between the customer and the online shop, and that most of the functionalities should be executed sequentially, this pattern does not provide all the desired benefits. Therefore, because all the use cases must satisfy the changeability requirement and that the system has an appropriate number of functionalities, the architect may decide to apply the *Layer* pattern to fulfil it. Observing the annotated diagrams, he or she is aware of which actions would be positively affected regarding the maintainability advantage provided by the pattern, and which would be negatively affected by the by the performance liability. Thus, to achieve the desired performance, the application of a variant of the *Layer* pattern may be evaluated, such as, for example, bypassing some of the layers as indicated in [17]. More information about this process can be found in [5].

In addition, the annotated diagrams are reused by a tool that assists the architect in selecting patterns. This tool, first, documents the different architectural patterns that a firm uses, together with their variants and how they affect each non-functional

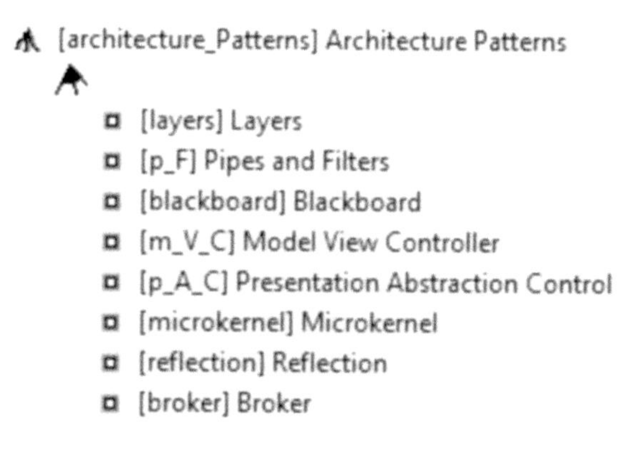

Fig. 8 Fragment of the feature model documenting architecture patterns

requirement. Feature models [11] are used for this purpose. Figure 8 shows a fragment of the feature model documenting architecture patterns. This diagram models the common top-level architecture patterns. By examining each pattern, one can see how it affect the functional and non-functional requirements, and its architectural variability. Second, the tool contains a series of rules which, by evaluating the annotated use cases, preselects a set of architectural patterns that might be applied, thus assisting the architect in finally deciding on the patterns to apply. More information about this process can be found in [14].

4 Related Work

Many studies describe techniques for relating business processes and business goals for various purposes. In [1] for example, the authors define a process to identify, decompose, and model quality requirements using a goal tree. These goals are then attached to the business process models. The resulting relationships are used to identify areas of the processes that are candidates for improvement. Such annotations could also be used, together with the methodological approach presented in this work, to derive the relationships between the system requirements.

In [21], the authors extend BPMN with two new elements ("Operating Condition" and "Control Case"). The former is used to annotate the processes' operating restrictions, and the latter to define control mechanisms that handle those restrictions. This extension is used to detail the relationships between business processes and operational goals. Again, this extension could be used together with the present proposals to detail the relationships with not only operational but also non-operational goals.

Other works have focused on facilitating the derivation of system requirements from business processes. In [12], [24], and [23], the authors define mappings that use business processes modeled with Activity Diagrams or BPMN as the basis to identify actors, use cases, and some of the relationships between use cases. Likewise, in [28], functional requirements, represented as Task Descriptions, are derived from business processes modeled with BPMN. These proposals allow a set of functional requirements to be defined that are aligned with the business. For this alignment to be perfect, the present work complements the aforementioned proposal by providing a mechanism to extract not just the functional requirement relationships but also the non-functional.

In [22], the authors extend the BPMN notation to model security concepts in business processes. In addition, they define a set of transformations which, based on these annotated processes, derive the system's use cases and the functionalities needed to provide the security being modeled. In particular, they document the relationships between business processes and security goals to then derive the corresponding relationships with the system requirements. These extensions may be used together with the present proposals to also address the relationships with other goals and non-functional requirements.

There has been work aimed at improving requirement specification in order to provide the architect with more information [2]. Thus, IESE NFR [13] sets out a guide for the identification, analysis, and documentation of non-functional requirements. Once identified, they are then linked to the functionalities that have to satisfy them. To this end, notes are added to the use cases and their scenarios. In addition, each non-functional requirement is refined until metrics can be associated with it. These metrics are used to evaluate whether or not a particular architectural design can fulfil the requirements. The IESE NFR methodological approach can be used together with the work presented here to, firstly, apply the BPMN extension to obtain a set of requirements and relationships aligned with the business, and, second, to use the rules, for the selection of patterns, and the metrics, they define, as a guide in the design of the architecture.

Other work has been specifically oriented to facilitating the design of the architecture. For example, Quality-Driven Architecture Development [19] models architectural patterns, and their variants, with quality models based on feature models. The architect uses this model to identify patterns or variants that will satisfy the system's quality attributes. The selection of these architectural patterns is made entirely by the architect on the basis of his or her knowledge of the system requirements and the relationships between them. The present work also uses feature models to document architectural patterns and tactics, but with the selection of patterns being narrowed down by the relationships annotated on the business process and requirements models.

Finally, the studies of Harrison and Avgeriou [16], [17] first examines how architectural patterns influence positively and negatively the quality requirements, and second how architectural patterns and tactics influence each other. The feature model and the pattern selection rules of the present work take these studies as their basis.

5 Conclusions and Future Works

Designing the architecture of a software system is a very complex activity that requires great experience. Software architects need to know perfectly all the functional and non-functional requirements and the relationships between them to design the architecture that best allows them to be satisfied. The identification of this information requires architects with extensive experience and skill, since any misinterpretation of the requirements or their relationships may lead to the design of an architecture that not only does not facilitate compliance with some of them, but may even actively make such compliance difficult.

In order to reduce the effort of interpreting and identifying this information, we have here presented extensions to model the relationships both between the various elements of the business and between the system requirements. The relationships between business elements are modeled through an extension for BPMN 2.0. The relationships between requirements are modeled by mean of extensions for the UML 2 Use Case, Activity and Sequence Diagrams. We are currently working on applying

MDD techniques to define transformations between models that semi-automate the derivation of a system's requirements and their relationships from the information annotated in the business.

Furthermore, these relationships, together with a model documenting the architectural patterns, are used by a set of rules that pre-select the patterns to apply. This model and the rules can be used by a firm to generate repositories of knowledge about its architectural style. These repositories can then be used during the development of new applications, thus taking advantage of the knowledge that has been acquired in previous projects.

Acknowledgements. This research was supported by the Spanish Ministry of Science and Innovation under Project TIN2012-34945, by the Department of Employment, Enterprise, and Innovation of the Government of Extremadura under Project GR10129, and by the European Regional Development Fund (ERDF).

References

1. Aburub, F., Odeh, M., Beeson, I.: Modelling non-functional requirements of business processes. Inf. Softw. Technol. 49, 1162–1171 (2007), doi:10.1016/j.infsof.2006.12.002
2. Avgeriou, P., Grundy, J., Hall, J.G., Lago, P., Mistrík, I. (eds.): Relating Software Requirements and Architectures. Springer (2011)
3. Bachmann, F., Bass, L., Klein, M., Shelton, C.: Designing software architectures to achieve quality attribute requirements. IEE Proceedings Software 152(4), 153–165 (2005), doi:10.1049/ip-sen:20045037
4. Berrocal, J., García-Alonso, J., Murillo, J.M.: Patrones para la extracción de casos de uso a partir de procesos de negocio. In: II Taller de Procesos de Negocio e Ingenieŕ-a de Servicios, pp. 1–11 (2009)
5. Berrocal, J., García-Alonso, J., Murillo, J.M.: Facilitating the selection of architectural patterns by means of a marked requirements model. In: Babar, M.A., Gorton, I. (eds.) ECSA 2010. LNCS, vol. 6285, pp. 384–391. Springer, Heidelberg (2010)
6. BMM: Business motivation model version 1.1, `http://www.omg.org/spec/BMM/`
7. BPMN: Business process modeling notation version 2.0, `http://www.bpmn.org/`
8. Cardoso, E., Almeida, J., Guizzardi, G.: Requirements engineering based on business process models: A case study. In: 13th Enterprise Distributed Object Computing Conference Workshops, EDOCW 2009, pp. 320–327 (2009), doi:10.1109/EDOCW.2009.5331974
9. Chung, L., do Prado Leite, J.C.S.: On non-functional requirements in software engineering. In: Borgida, A.T., Chaudhri, V.K., Giorgini, P., Yu, E.S. (eds.) Conceptual Modeling: Foundations and Applications. LNCS, vol. 5600, pp. 363–379. Springer, Heidelberg (2009)
10. Cockburn, A.: Writing Effective Use Cases, 1st edn. Addison-Wesley Longman Publishing Co., Inc., Boston (2000)
11. Czarnecki, K., Helsen, S., Eisenecker, U.W.: Formalizing cardinality-based feature models and their specialization. Software Process: Improvement and Practice 10(1), 7–29 (2005)

12. Dijkman, R.M., Joosten, S.M.M.: Deriving use case diagrams from business process models. Tech. Rep. TR-CTIT-02-08, University of Twente (2002)
13. Dörr, J.: Elicitation of a complete set of non-functional requirements. Ph.D. thesis, University of Kaiserslautern (2011)
14. García-Alonso, J., Berrocal, J., Murillo, J.M.: Modelado de la variabilidad en arquitecturas multicapa. Jornadas de Ingeniera del Software y Bases de Datos (JISBD), 895–900 (2011)
15. Grau, G., Franch, X., Maiden, N.A.M.: Prim: An i*-based process reengineering method for information systems specification. Inf. Softw. Technol. 50(1-2), 76–100 (2008)
16. Harrison, N.B., Avgeriou, P.: Leveraging architecture patterns to satisfy quality attributes. In: Oquendo, F. (ed.) ECSA 2007. LNCS, vol. 4758, pp. 263–270. Springer, Heidelberg (2007)
17. Harrison, N.B., Avgeriou, P.: How do architecture patterns and tactics interact? a model and annotation. Journal of Systems and Software 83(10), 1735–1758 (2010)
18. International Standard Organization (ISO/IEC): Informational technology – product quality: Quality model. International Standard ISO/IEC 9126 (2001)
19. Kim, S., Kim, D.K., Lu, L., Park, S.: Quality-driven architecture development using architectural tactics. J. Syst. Softw. 82, 1211–1231 (2009), doi:10.1016/j.jss.2009.03.102
20. OpenUP: Open unified process (2013), `http://epf.eclipse.org/wikis/openup/`
21. Pavlovski, C.J., Zou, J.: Non-functional requirements in business process modeling. In: Proceedings of the Fifth Asia-Pacific Conference on Conceptual Modelling, APCCM 2008, vol. 79, pp. 103–112. Australian Computer Society, Inc., Darlinghurst (2008)
22. Rodríguez, A., de Guzmán, I.G.R., Fernández-Medina, E., Piattini, M.: Semi-formal transformation of secure business processes into analysis class and use case models: An mda approach. Information & Software Technology 52(9), 945–971 (2010)
23. Siqueira, F., Silva, P.: Transforming an enterprise model into a use case model using existing heuristics. In: Model-Driven Requirements Engineering Workshop (MoDRE), pp. 21–30 (2011)
24. Stolfa, S., Vondrak, I.: Mapping from business processes to requirements specification. Tech. rep., CUniversitat Trier (2006)
25. Traetteberg, H., Krogstie, J.: Enhancing the usability of bpm-solutions by combining process and user-interface modelling. In: Stirna, J., Persson, A. (eds.) PoEM 2008. LNBIP, vol. 15, pp. 86–97. Springer, Heidelberg (2009)
26. Ullah, A., Lai, R.: Modeling business goal for business/it alignment using requirements engineering. Journal of Computer Information Systems 51(3), 21–28 (2011)
27. UML: Unified modeling language, `http://www.uml.org/`
28. de la Vara, J.L., Sánchez, J.: BPMN-based specification of task descriptions: Approach and lessons learnt. In: Glinz, M., Heymans, P. (eds.) REFSQ 2009. LNCS, vol. 5512, pp. 124–138. Springer, Heidelberg (2009)

Introducing Critical Thinking to Software Engineering Education

Oumout Chouseinoglou and Semih Bilgen

Abstract. Software and its development processes are changing continuously pervading our daily life, new and diverse techniques and approaches are being proposed and the software industry is eager to adopt the ones that will provide competitive advantage. The diversity of these new techniques and approaches and the diversity of clients and contexts in the software industry, requires software developers to have the ability to judge correctly and to discriminate successfully among these. These skills need to be taught to software developers in the course of their formal undergraduate education. However, traditional approaches in software engineering education (SEEd) are mostly inadequate in equipping students with these unusual and diverse skills. This study, as part of a larger study aiming to develop a model for assessing organizational learning capabilities of software development organizations and teams, proposes and implements a novel educational approach to SEEd combining different methodologies, namely lecturing, project development and critical thinking. The theoretical background and studies on each approach employed in this study are provided, together with the rationales of applying them in SEEd. Student opinions and instructor observations demonstrate that the proposed course structure is a positive step towards the aforementioned goals.

Keywords: Software engineering education, critical thinking, practicum, SQ4R.

1 Introduction

Software systems evade our daily life in an increasing pace with diverse applications and the need for quality software is eminent. Different methodologies and

Oumout Chouseinoglou
Statistics and Computer Science Department, Başkent University, 06810, Ankara, Turkey
e-mail: umuth@baskent.edu.tr

Semih Bilgen
Electrical and Electronics Engineering Department, Middle East Technical University, 06531, Ankara, Turkey
e-mail: semih-bilgen@metu.edu.tr

R. Lee (Ed.): *SERA*, SCI 496, pp. 183–195.
DOI: 10.1007/978-3-319-00948-3_12

approaches in software engineering are being employed to provide the necessary levels of software quality, however, the quality of developed software is directly related to the supply of capable and up-to-date software developers [1], who have to cope with technical as well as non-technical issues [2], and who have to discriminate among the criteria for success by identifying the good solutions for the problem at hand [3]. In other words todays software developers need to "think out of the box" [3]. On the other hand, software engineers, in accordance with the characteristics of the engineering domain, are expected to reconcile conflicting constraints and to make deliberate selections among alternative designs with their judgments based on deep knowledge of the discipline [1], need to have social skills, and must be capable of evaluating competing values [4]. Thus, our belief is that software developers and engineers need to be equipped with the aforementioned skills and capabilities through their education with the use of specifically constructed software courses.

However, traditional and generic approaches in software engineering education (SEEd) are insufficient in helping students to keep their knowledge current, cannot prepare them for the intricacies of the domain and fail to produce the supply and quality of developers required by the industry [1][5]. Traditional learning approaches have been proven to be ineffective as students are not actively involved in the learning process but are merely passive listeners [6]. SEEd should first identify the critical ingredients that result in competence in the field and then the instructional models that will transform students to effective practitioners should be developed [7]. Moreover, new approaches should not only require students to study theory using text books but also should educate them to "learn how to learn" through state-of-the-art analyses [8], allow them to stay up-to-date regardless of rapid change, prepare them for different and new roles [1], and provide them with the experience of non-technical issues and practical know-how [2]. Similarly, in a detailed review of current trends in SEEd [6] the importance of self-directed learning (the ability to learn on their own) and higher order cognitive skills of application, analysis, evaluation and synthesis is emphasized. Surveying the current trends in SEEd within the pedagogical context, the authors of [9] point to the increasing importance of practice-based education and alternative ways of teaching such as empirical methods where students will acquire knowledge for evaluating and proposing technology and processes. SEEd is moving from lecture-format courses to team projects where students are expected to exercise the ideas they are learning [1][6], the so called practicum, a positive change bridging the academia-industry gap [6]. The efforts of transforming the existing knowledge to a curriculum have produced two important milestones for the SEEd, namely the Guide to the Software Engineering Body of Knowledge, which manifests the general perceptions on what a software engineer with bachelor's degree and four years of experience should know, and the Software Engineering 2004, which suggests curriculum guidelines for undergraduate software engineering degree programs [4]. However, experiences in project based SEEd have shown that students with little industrial involvement (i.e., undergraduate students) aren't mature enough to appreciate the importance of many software engineering topics [4], they lack the intuition to understand problems, ambiguity and hidden constraints of real projects and therefore find it challenging to apply their acquired knowledge to a project [3], and focus more on programming issues and less on the development process and the associated software engineering issues [6]. Therefore, students often fail to appreciate the importance of tasks that software practitioners continuously conduct (e.g., requirements engineering, project management, cost estimation) [4], as they consider them theoretical and of very little use in future [6].

In [7], SEEd courses are investigated with respect to three distinctive knowledge categories, namely declarative (knowledge from textbooks), procedural (knowledge by doing) and metacognitive knowledge (planning, monitoring process and progress, changing when appropriate and reflecting). Software domain studies make it clear that each knowledge category must be addressed explicitly in instruction. This paper describes the practice of combining the approach of critical thinking, a metacognitive approach, to a project based SEEd course (traditionally encompassing the declarative and procedural approaches) with the aim of increasing the understanding capabilities of students with respect to software engineering practices, allowing them to appreciate these practices and equipping them with skills to evaluate and judge alternative approaches that they will face in their professional careers. The rest of the paper is organized as follows: First, we briefly review the related practices and the theoretical background on the subject. Section 3 gives the details of the developed course as a comparative study, and Section 4 outlines the lessons learned. The last section concludes the paper and overviews the planned future work.

2 Related Work

2.1 *Software Engineering Practicum*

Acknowledging that the skills to be effective software engineers are not limited to declarative knowledge and claiming that expertise is domain-specific and can only be acquired in the context in which it will be practiced, practicum is a procedural approach which uses a realistic environment, usually in the form of a project for an actual client, where students learn the skills they will use in the future by applying their knowledge in a real-world setting [7][10][11]. Especially it is important that these practicum approaches be team based, according to the CMM statement that in higher levels of maturity individual activities transform to team activities [12]. Practicum is closely related to constructivism, a learning theory that is learner centered, which states that students learn better if they construct knowledge for themselves and regards learning as a process of active construction [13]. The set of constructivist instructional principles and the skills needed by software engineers to solve real-world problems within the constructivist approach are given in detail in [13].

Numerous successful cases of SEEd courses incorporating theoretical knowledge and practical experience with the use of semester-long projects have been discussed in the literature. Surveys and descriptions of courses reflecting the realities and focusing on specific areas of software engineering such as requirements engineering, supply chain development and global software development are presented in [14] and [15]. In [16] it is argued that the proposed approach taught inexperienced graduate students many software engineering principles, such as software verification and validation. An initiative towards restructuring an undergraduate software engineering class from lecture-based to lab-oriented by focusing on learning and personality types and emphasizing practical tools is given in [17]. A review of textbooks addressing the difficulties of learning by doing in the SEEd domain and the challenges faced by the universities is available in [2], and a university-wide example of learner centered approach by solving real problems is provided in [18]. In [13], the author surveys a vast number of approaches employed to render software education more realistic, pinpointing the most important ones that have provided valuable contributions. However, it is argued that the majority of these approaches do not explicitly integrate pedagogical innovations, they do not sufficiently take into consideration the human aspect of the learning process such as the emotions, behavior and thoughts of students and finally they do not expose students to all phases of software development [13]. Despite these successful cases in the literature, the practicum approach has several drawbacks, pointed out in detail in [7]: instructors use little effort on identifying the skills that a project should bring to the students, practicum does not help students to develop the skills necessary for deliberate practice and students mostly are able to apply what they have learned only to very similar situations [7]. In [11], observed issues with practicum are poor testing, ineffective teams, no documentation, no use of metrics, no measures of quality and failure of students to transfer knowledge from the formal curriculum to the practicum project.

Another successful practicum in SEEd is the CSCI577ab Software Engineering course [19][20] which is further described in detail in Section 3 as it has been pivotal in our study.

2.2 *Personality and Learning Types*

In [17] it is discussed that students can be classified with respect to their personality types using the Myers-Briggs type indicator and to their learning styles using Felder-Silverman model and a software engineering course should be constructed in a way that should appeal to most students. In Myers-Briggs personality types the students can be characterized with respect to four dimensions, namely as introvert vs. extrovert, sensing vs. intuitive, thinking vs. feeling and judging vs. perceiving. On the other hand, with respect to the Felder-Silverman learning styles, the students can be classified as active vs. reflective, sensing vs. intuitive, visual vs. verbal and sequential vs. global. The details of how these dimensions should be assessed are given in detail in [17]. However, it is evident that SEEd courses need to be designed in a fashion to contain elements that would address the majority of students. We believe that critical thinking is an approach that can address a wide number of students with different personality and learning types.

2.3 *Critical Thinking in Education*

The importance of judgment and decision making in engineering is studied in detail in [3], where it is clearly shown that methods, tools, processes, skills, heuristics, and other tools and techniques can be utilized in the search of good and cost-effective candidate solutions but cannot replace judgment. According to the engineering principle of striking a balance between conflicting goals, SEEd students should be taught judgment and the commitment to use it [3]. This, we believe, is closely related to critical thinking, a metacognitive approach. A survey of the conceptions of critical thinking is given in [21] where it is defined as "reasonable, reflective thinking focusing on task, people or belief"; involving abilities such as "identifying a problem and its associated assumptions, clarifying and focusing the problem, and analyzing, understanding and making use of inferences, inductive and deductive logic, as well as judging the validity and reliability of the assumptions, sources of data or information available". In [7] it is stated that when students are asked to think in a metacognitive fashion at every stage of a problem-solving process, not only they accomplish this task but they also develop a deeper understanding about it and achieve better performance on following problems. Moreover, studies in programming domain have demonstrated that students perform better both on declarative and procedural tasks when they reflect on what they are learning. This suggests that the metacognitive activity is the main reason for producing the improved performance [7].

In [3], a critical thinking approach for SEEd is presented where supplemental materials (mostly books) are incorporated to the course by having the groups read and report on the ideas from these materials via critical analysis and interpretation, and having the students to identify their association with the course. According to the authors [3] this activity highlights the value of engineering judgment by emphasizing critical evaluation and helps students learn to recognize external but relevant material, evaluate techniques on their own, and recognize how to use different techniques together. The course moves from the traditional memorize-and-recite-back to the critical application of content. The related theoretical framework and the application details of this approach are provided in detail in [3]. In [7] extreme programming practices are assessed within the metacognitive approach by focusing on how each practice may facilitate the acquisition of the metacognitive skills that are required for the development of enhanced competence in students.

In this study the Survey, Question, Read, Recite, Review, and wRite (SQ4R) technique is proposed as a critical thinking method. Details of SQ4R and how it was implemented are discussed in Section 3.

3 Comparative Study

3.1 *General Structure of the Course*

The İST478 Current Topics in Information Technologies course, which is offered in the Department of Statistics and Computer Science, Başkent University, Turkey, is a course with flexible content and focuses on addressing current topics in information systems, software engineering and programming. In the 2011-2012 Spring term in which this study was conducted, İST478 was given as a Software Engineering Team Project Practicum course, as explained below, and was developed with the initial aims of: (a) to teach students the practical techniques and tools that are used in professional software development through regular lecture sessions and a practicum conducted in teams, and (b) to provide a test bed and a pilot study to validate whether a model for assessing the organizational learning capabilities of software development teams, namely AiOLoS (Assessing Organizational Learning of Software Development Organizations), developed by the authors [23], is actually applicable in real life teams. AiOLoS has been developed with the main aims of (a) providing a framework for comparison between software organizations with respect to their organizational learning capabilities, (b) allowing software organizations to identify their deficiencies and shortcomings, (c) offering the means for the measurement of the realized improvement in organizational learning and (d) providing a starting point for software process improvement. AiOLoS consists of three major process areas that map to the three major objectives of an Learning Software Organization [22], namely obtaining, using and passing knowledge. AiOLoS proposes that the organizational learning activity can be assessed with respect to 12 core processes that elaborate the 3 major process areas. The details of AiOLoS are available in [23] whereas an exploratory case study of AiOLoS conducted on the İST478 course is given in [24]. As explained in [9], when pilot studies such this one are carried out with students, they are required to have pedagogical value and both the researchers and the students have to perceive that value. Therefore, in order to enhance the pedagogical value of the course a critical thinking approach within the perspective of metacognitive knowledge was employed. However, as this critical thinking approach is also novel and was developed based on different practices in the literature not previously applied in the SEEd domain in conjunction, a comparative study was performed in order to assess its applicability and usefulness.

The subjects consisted of 15 undergraduate and 4 graduate level students who were enrolled in the İST478 course. All graduate level students and 6 of the 15 undergraduate level students had taken an introductory software engineering course. Four software development groups were formed of varying sizes, with each graduate student being assigned as a team leader (project manager) to each group. Moreover, each group had at least two students who had previously taken an introductory software engineering course. In order to achieve fairness in the workload, each group was assigned the development of systems similar in size and context, but with significant requirement and development differences. Specifically, each group was assigned the development of a score tracking software respectively for chess, tennis, basketball and football.

The course followed a customization of the outline provided by CSCI577ab Software Engineering [19], a graduate software engineering course at University of Southern California, being offered since 1996. CSCI577ab focuses on software plans, processes, requirements, architectures, risk analysis, feasibility analysis, software product creation, integration, test, and maintenance with an emphasis on quality software production [20]. Moreover, CSCI577ab has been used as an experimental test-bed to deploy various research tools and approaches for validation of new methods and tools, leading to twelve PhD dissertations until 2008. As stated in [16], partially employing an already defined course outline and building the novel approaches of our study on top of that outline, ensures that our study is in accordance with published, well-grounded work and may encourage other instructors to use and employ the methodology that is proposed in this research with less effort. İST478 followed the Incremental Commitment Spiral Model (ICSM) [25][26][27], a new generation process model developed specifically for CSCI577ab and the architected agile approach for software development. İST478 covered the full system development life cycle of ICSM, which consisted of the Exploration phase, Valuation phase, Foundations phase, Development phase, and Operation phase. The deliverable deadlines and the items to be delivered for each of these phases were predefined. The tasks and artifacts to be developed by the students in İST478 were based on specific templates and they were described in detail in the Incremental Commitment Spiral process model - Electronic Process Guide (ICSM-EPG) [28]. Table 1 provides the list of conducted phases and the artifacts delivered by groups in each phase.

3.2 Implementing the SQ4R Approach

SQ4R [29] is a metacognitive approach to facilitate students' comprehension and memory specifically when reading science texts. Moreover, SQ4R has been implemented successfully in courses from a variety of areas, such as poetry analysis [30] or arts teaching [31]. SQ4R teaches learners to "attack" content in five sequential steps [32]: (i) Survey and (ii) Question where self-questioning and predicting occurs, (iii) Reading where learners check if their predictions are accurate, (iv) Recording where learners take notes regarding the content, Reciting where learners fill the gaps in their understanding based on their notes and finally (v) wRiting where learners write a brief summary to reflect what they have understood from the subject [33].

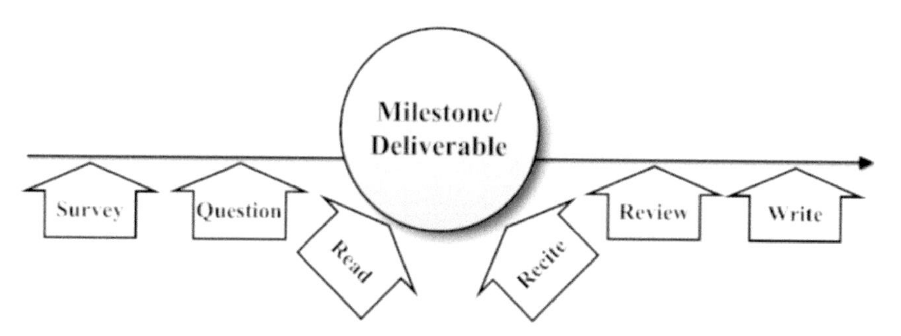

Fig. 1 The implemented SQ4R approach

Table 1 The ICSM phases followed in this study

Phase	Deliverable	Artifact
Exploration	Customer Interaction Package	Customer Interaction Report
Valuation	Valuation Commitment Package	Customer Interaction Package + Life Cycle Plan Operational Concept Description Feasibility Evidence Description
Foundation	Foundation Commitment Package	Valuation Commitment Package + System and Software Architecture Description System and Software Requirements Description Prototype Report Supporting Information Document
Development	Development Commitment Package	Foundation Commitment Package + Quality Management Plan Acceptance Test Plan and Cases Iteration Plan
Transition	Transition Readiness Package	Development Commitment Package + Iteration Assessment Report Training Plan User Manual Transition Plan Test Procedures and Results Functioning Product

Two randomly selected groups (groups 2 and 3) were assigned a differentiated development method of the ICSM which incorporated the SQ4R, to enhance their learning experience. The two groups implementing SQ4R were provided with prior knowledge of the phase they were conducting, the artifacts they were expected to develop and the deliverables to submit. During SQ4R, before working on and developing the deliverable, the students were given the deliverable name and were asked to conduct a small "survey" on the subject. After the survey, the team members were asked to write a brief reflection paper where they "questioned" and discussed why they thought the phase and the related deliverables are of importance for the software development process. Then all teams were given the guidelines and templates of the deliverables to be developed. The teams, while developing the deliverables, "read" the documents provided by the instructor and team members would "recite" to each other what they have understood on the material provided by the instructor. After the submission of the deliverable, the members of the teams undertaking SQ4R would conduct a "review" session with the instructor where they discussed their understanding of the process they have concluded/undertaken and the deliverable they have submitted. Finally they would write a closure paper, where they discussed what they have done, if they have understood it, what their initial thoughts and final thoughts were on the process, if they would change some or all parts of the deliverable or process, and their final comments/proposals. Figure 1 displays the SQ4R approach which was undertaken by the two randomly assigned groups in all five phases (depicted as "Milestone" in Fig. 1) of the software development lifecycle of İST478 course.

Similar to the experience in [34], during the course period one of the groups submitted no acceptable documents and deliverables (Group 4) and subsequently the members failed the course; thus no metrics or data were collected from this group. The authors in [34] argue that this experience is one of the most important lessons learned while conducting experiments in SEEd courses: surprises happen, and evaluations rarely turn out exactly as planned.

Among these three groups, only Group 1 did not undertake the SQ4R approach. The results of the AiOLoS research regarding the organizational learning capabilities of the assessed three teams are given in [24].

4 Lessons Learned / Experience and Evaluation

As explained in detail in [34], evaluation in the domain of education and especially SEEd is a challenging undertaking, as it is almost impossible to adequately isolate the effects of a new or proposed educational technique, there are difficulties in getting a statistically significant number of subjects, assessment of software engineering skills is less straightforward with respect to other disciplines, and comparative evaluations are difficult to be conducted due to the immaturity of the domain of software engineering. Due to these reasons in SEEd a new technique is usually intended to be a supplement to a curriculum.

The developed İST478 course was aimed to be as close as possible to reality and all five phases of the ICSM model were completed in a period of 16 weeks. All three groups that attended the course completed and submitted a working software artifact. However, as the developed software products were not meant for real-life usage, no payments or similar incentives existed; the incentives of students for developing the projects according to the expectations of the ICSM-EPG and SQ4R were credit points. Students also received credit points for attending the lectures.

After the conclusion of the course, the teams undertaking the SQ4R approach, a total of 11 students (both undergraduate and graduate), were asked to evaluate and assess the SQ4R approach and provide their opinions regarding the model and its results. The team members were asked five questions regarding the developed SQ4R model and they submitted their results using a Likert Scale. The questions and the Likert scores of the answers are given in Table 2. The major threat to the validity of that evaluation was the instructor-student relationship that existed between the assessor and the assessed team members. This relationship could force the students to alter their answers in the questionnaires to more favorable ones, believing that such answers would contribute to their grades. In order to resolve this, the students were informed that they would not be graded based on the answers they provide. Moreover, the survey answers were collected after the submission of the grades, so that students would not feel compelled to provide answers that do not depict their true opinions about the SQ4R model.

The frequency of the results regarding the answers given in the opinion questionnaires were:

- 8 out of 11 believed that the SQ4R approach mostly helped them to learn the course topics better (mode value being Mostly, median value being 4 out of 5),
- 6 out of 11 believed that the SQ4R approach mostly helped them to apply the course topics better to their project (mode value being Mostly, median value being 4 out of 5),
- 9 out of 11 believed that the SQ4R approach mostly provided them with an advantage in the development of the project (mode value being Mostly, median value being 4 out of 5),
- 6 out of 11 believed that the time spent for conducting the SQ4R approach mostly was worth it (mode value being Mostly, median value being 4 out of 5),
- 6 out of 11 believed that the SQ4R approach mostly would contribute to passing the acquired knowledge to their later professional life (mode value being Mostly, median value being 4 out of 5).

Moreover, the students felt that the SQ4R experience enhanced their academic curiosity by the questions they were asking and the level of their participation in classroom activities. Even though it was not measured with questionnaires, another important observation was that the attitude of team members towards the task at hand would usually shift positively after the conclusion of the SQ4R for that task. On the other hand, the majority of the students would complain regarding the extra work the SQ4R required. However, we believe that these complaints were mostly related to the fact that one development team was not undertaking the SQ4R approach.

Table 2 Student opinions regarding SQ4R approach

Question	Fully	Mostly	Somewhat	Very Little	Not at all
Q1) Do you think the SQ4R helped you to learn the course topics better?	2	8		1	
Q2) Do you think the SQ4R approach helped you to apply the topics better to your project?	3	6	2		
Q3) Do you think the SQ4R approach provided you an advantage in the development of your project?	1	9		1	
Q4) Do you think the extra time spent for conducting the SQ4R was worth it?		6	2	3	
Q5) Do you think the SQ4R approach will contribute to you passing the acquired knowledge to your later professional life?	4	6	1		

5 Conclusion

This study has been a part of a larger study aiming to develop the AiOLoS model for assessing organizational learning capabilities of software development organizations and teams. Within that perspective, a pilot study was constructed to test whether AiOLoS can be actually implemented in software development teams. In conjunction, in order to propose a pedagogical contribution, a novel SEEd course structure has been introduced, implementing several different methodologies, namely lecturing, practicum with ICSM and critical thinking.

It is obvious that the number of subjects who participated in the comparative study described in this paper and the overall structure of the study are not conducive to statistically significant and definitive results, especially as one of the development teams did not continue the course. As confirmed in [34], the anecdotal usage alone of an education technique does not provide much information, instead a multi-angled evaluation approach which is partially rooted in educational theory, can be a useful solution.

Nevertheless, we believe that the combination of lecture and practicum, the use of a well-grounded software process model (ICSM), and the implementation of a critical thinking based learning approach to the process has contributed to the development of a successful learning environment. It is our opinion that, with the aforementioned amalgamation of approaches and methods, the course has (a) constituted a step forward towards the realization of an overall SEEd course that will arm students with the critical thinking and judgment capabilities the engineering approach requires, and (b) appealed to a wide variety of students with different personality types and learning styles. The student opinions given in Table 2 and the observations of the instructor are supplements of these claims. We have interpreted these results as an indication that the proposed approach of utilizing creative thinking in SEEd provides students with both the knowledge and the judgment capabilities that the software engineering industry requires.

We are currently analyzing the obtained results and the lessons learned from that first experience with this course in order to develop a more structured and well-defined course, that will be used to better evaluate and compare the educational attainments of the proposed model. It is our belief that by harvesting further data from students in the newly constructed course we will have a better understanding of the learning needs of SEEd students and we will be able to find solutions in filling the gap between the requirements of software industry and SEEd. Furthermore, we are also investigating the case of proposing the critical thinking approach to professional software development organizations as a technique of increasing the organizational learning of teams and individuals and enhancing AiOLoS in order to embody the critical thinking capabilities of software practitioners.

References

1. Shaw, M.: Software engineering education: a roadmap. In: Proceedings of the Conference on the Future of Software Engineering. ACM (2000)
2. Gnatz, M., Kof, L., Prilmeier, F., Seifert, T.: A practical approach of teaching software engineering. In: Proceedings of the 16th Conference on Software Engineering Education and Training, CSEE&T 2003, pp. 120–128. IEEE (2003)

3. Shaw, M., Herbsleb, J., Ozkaya, I., Root, D.: Deciding what to design: Closing a gap in software engineering education. In: Inverardi, P., Jazayeri, M. (eds.) ICSE 2005. LNCS, vol. 4309, pp. 28–58. Springer, Heidelberg (2006)
4. Van Vliet, H.: Reflections on software engineering education. IEEE Software 23(3), 55–61 (2006)
5. Blake, B.M.: A student-enacted simulation approach to software engineering education. IEEE Transactions on Education 46(1), 124–132 (2003)
6. Garg, K., Varma, V.: A study of the effectiveness of case study approach in software engineering education. In: Proceedings of the 20th Conference on Software Engineering Education & Training, CSEET 2007. IEEE (2007)
7. Williams, L., Upchurch, R.: Extreme programming for software engineering education? In: The Proceedings of the 31st Annual Frontiers in Education Conference. IEEE (2001)
8. Boehm, B.: A view of 20th and 21st century software engineering. In: Proceedings of the 28th International Conference on Software Engineering. ACM (2006)
9. Carver, J., Jaccheri, L., Morasca, S., Shull, F.: Issues in using students in empirical studies in software engineering education. In: Proceedings of the Ninth International Software Metrics Symposium. IEEE (2003)
10. Katz, E.P.: Software engineering practicum course experience. In: Proceedings of the 23rd IEEE Conference on Software Engineering Education and Training (CSEE&T). IEEE (2010)
11. Bareiss, R., Katz, E.P.: An exploration of knowledge and skills transfer from a formal software engineering curriculum to a capstone practicum project. In: Proceedings of the 24th IEEE-CS Conference on Software Engineering Education and Training (CSEE&T). IEEE (2011)
12. Favela, J., Feniosky, P.M.: An experience in collaborative software engineering education. IEEE Software 18(2), 47–53 (2001)
13. Hadjerrouit, S.: Learner-centered web-based instruction in software engineering. IEEE Transactions on Education 48(1), 99–104 (2005)
14. Gotel, O., Kulkarni, V., Neak, L.C., Scharff, C., Seng, S.: Introducing global supply chains into software engineering education. In: Meyer, B., Joseph, M. (eds.) SEAFOOD 2007. LNCS, vol. 4716, pp. 44–58. Springer, Heidelberg (2007)
15. Gotel, O., Kulkarni, V., Say, M., Scharff, C., Sunetnanta, T.: A global and competition-based model for fostering technical and soft skills in software engineering education. In: Proceedings of the 22nd Conference on Software Engineering Education and Training (CSEE&T 2009). IEEE (2009)
16. Hayes, J.H.: Energizing software engineering education through real-world projects as experimental studies. In: Proceedings of the 15th Conference on Software Engineering Education and Training (CSEE&T 2002). IEEE (2002)
17. Layman, L., Cornwell, T., Williams, L.: Personality types, learning styles, and an agile approach to software engineering education. ACM SIGCSE Bulletin 38(1), 428–432 (2006)
18. Nikolov, R., Ilieva, S.: Building a research university ecosystem: the case of software engineering education at Sofia University. In: Proceedings of the 6th Joint Meeting of the European Software Engineering Conference and the ACM SIGSOFT Symposium on the Foundations of Software Engineering. ACM, Cavtat near Dubrovnik (2007)
19. Boehm, B., Koolmanojwong, S.: Software Engineering I - Fall 2011. USC Viterbi School of Engineering (August 12, 2011), `http://greenbay.usc.edu/csci577/fall2011/index.php` (cited June 30, 2012)

20. Koolmanojwong, S., Boehm, B.: Using Software Project Courses to Integrate Education and Research: An Experience Report. In: Proceedings of the 22nd Conference on Software Engineering Education and Training (CSEE&T 2009), Hyderabad, India (2009)
21. Pithers, R., Soden, R.: Critical thinking in education: A review. Educational Research 42(3), 237–249 (2000)
22. Ruhe, G.: Learning Software Organisations. In: Chang, S.K. (ed.) Handbook of Software Engineering and Knowledge Engineering, vol. 1, pp. 663–678. World Scientific Publishing (2001)
23. Chouseinoglou, O., Bilgen, S.: Assessing Organizational Learning in Software Development Organizations. Technical Report. METU/II-TR-2012-02, Department of Information Systems, Middle East Technical University, Ankara, Turkey (2012), `http://www.baskent.edu.tr/~umuth/METU-II-TR-2012-02.pdf`
24. Chouseinoglou, O., Bilgen, S.: A Model for Assessing Organizational Learning in Software Development Organizations. In: Winckler, M., Forbrig, P., Bernhaupt, R. (eds.) HCSE 2012. LNCS, vol. 7623, pp. 251–258. Springer, Heidelberg (2012)
25. Boehm, B., Lane, J.A.: Using the Incremental Commitment Model to Integrate System Acquisition, Systems Engineering, and Software Engineering. CrossTalk the Journal of Defense Software Engineering, 4–9 (October 2007)
26. Pew, R.W., Mavor, A.S.: Human-System Integration in the System Development Process: A New Look. National Academy Press (2007)
27. Boehm, B.: Some future software engineering opportunities and challenges. In: The Future of Software Engineering, pp. 1–32. Springer, Heidelberg (2011)
28. USC-CSSE: Instructional Commitment Spiral Model - Software Electronic Process Guide. USC Viterbi School of Engineering (2008), `http://greenbay.usc.edu/IICMSw/index.htm` (cited June 30, 2012)
29. Thomas, E.L., Robinson, A.H.: Improving Reading in Every Class: A Sourcebook for Teachers. Allyn & Bacon, Boston (1982)
30. Casebeer, E.F.: SQ4R in the Analysis of Poetry. College Composition and Communication 19(3), 231–235 (1968)
31. Applegate, M.D., Quinn, K.B., Applegate, A.J.: Using metacognitive strategies to enhance achievement for at-risk liberal arts college students. Journal of Reading 38(1), 32–40 (1994)
32. Glynn, S.M., Muth, D.K.: Reading and writing to learn science: achieving scientific literacy. Journal of Research in Science Teaching 31(9), 1057–1073 (1994)
33. Yakupoglu, F.: The effects of cognitive and metacognitive strategy training on the reading performance of Turkish students. Practice and Theory in Systems of Education 7(3), 353–358 (2012)
34. Navarro, E.O., Van Der Hoek, A.: Comprehensive evaluation of an educational software engineering simulation environment. In: Proceedings of the 20th Conference on Software Engineering Education and Training, CSEET 2007. IEEE (2007)

Activity Diagrams Patterns for Modeling Business Processes*

Étienne André, Christine Choppy, and Gianna Reggio

Abstract. Designing and analyzing business processes is the starting point of the development of enterprise applications, especially when following the SOA (Service Oriented Architecture) paradigm. UML activity diagrams are often used to model business processes. Unfortunately, their rich syntax favors mistakes by designers; furthermore, their informal semantics prevents the use of automated verification techniques. In this paper, (i) we propose activity diagram patterns for modeling business processes, (ii) we devise a modular mechanism to compose diagram fragments into a UML activity diagram, and (iii) we propose a semantics for the produced activity diagrams, formalized by colored Petri nets. Our approach guides the modeler task (helping to avoid common mistakes), and allows for automated verification.

1 Introduction

Business processes are collections of related and structured activities or tasks, producing a specific service or product. Being able to model and to analyze business processes is of paramount importance, not only for the design of such processes, but also in the field of the software development whenever the SOA (Service Oriented Architecture) paradigm [9] is followed. The most common modeling notations for

Étienne André · Christine Choppy
Université Paris 13, Sorbonne Paris Cité, LIPN, F-93430, Villetaneuse, France
e-mail: {Etienne.Andre,Christine.Choppy}@lipn.univ-paris13.fr

Gianna Reggio
DIBRIS, Genova, Italy
e-mail: gianna.reggio@unige.it

* This work is partially supported by project #12 "Méthode de modélisation des systèmes dynamiques" (CREI, Université Paris 13, Sorbonne Paris Cité).

R. Lee (Ed.): *SERA*, SCI 496, pp. 197–213.
DOI: 10.1007/978-3-319-00948-3_13

business processes are the BPMN[1] and the UML [1] activity diagrams. We consider in this paper the UML since it offers also many other diagrams (classes, state machine, etc.), providing an integrated way to model all the aspects of a business as the used data and the participant entities; also it may be used in all the other phases of the software development. Furthermore, there is no relevant difference between the readability of the UML and of the BPMN (see, e.g., [17]).

Although UML diagrams are widely used, they suffer from some drawbacks. Indeed, since UML specification is documented in natural language, inconsistencies and ambiguities may arise. First, their rich syntax is quite permissive, and hence favors common mistakes by designers. Second, their informal semantics in natural language prevents the use of automated verification techniques, that could help detecting errors as early as the modeling phase.

Our contribution is twofold. First, we define precise activity diagrams for modeling business processes. These precise activity diagrams are based on patterns, that can be inductively composed so as to build complex activity diagrams. Our approach also takes classes into account. We have selected a minimal subset of the useful UML activity diagram constructs (viz., sequence, fork, join, choice, merge, loops). This paper does not consider accept and timed event, which is the subject of ongoing work. Second, we give a semantics to these patterns, by translating them into Colored Petri Nets (CPNs) [12] in a modular way. Petri net is a natural formalism as result of the translation: the UML specification explicitly mentions them, and the informal semantics of activity diagrams is given in terms of token flows.

Related Works. The first issue we address is that of an adequate notation and approach for business process modeling. [19, 6] compare different styles of activity diagrams (precise, "ultra-light") in experiments. The workflow pattern initiative [20] issued a collection of workflow patterns for business modeling. These patterns address the modeling of control, data, etc., and are expressed in Petri nets.

Another issue is to propose a formal associated semantics to UML diagrams using a formal notation, which is important to allow for automated verification [10]. This has been addressed in quite a variety of works using automata, different kinds of Petri nets, etc., so we mention only a few. Instantiable Petri nets are the target of transformation of activity diagrams in [13], and this is supported by tool BCC (Behavioral Consistency Checker); however they do not consider data, whereas we do. In [8, 4], the issue is performance evaluation, from activity diagrams and others (use case, state diagrams, etc.) to stochastic Petri nets. In [21] and [2], various syntactic features of UML state machines are translated into CSP# and colored Petri nets, respectively. Also note that [11] proposes an operational semantics of the activity diagrams (for UML 2.2). Börger [5] and Cook *et al.* [7] present other formalizations of the workflow patters of [20] using formalisms different from Petri nets, viz., Abstract State Machines and Orc, respectively. In [16], patterns for specifying the system correctness are defined using UML statecharts, and then translated into timed automata. The main differences with our approach are that the authors mainly focus on real-time properties, and the patterns of [16] do not seem to be

[1] `http://www.bpmn.org/`

hierarchical: the "composition" of patterns in [16] refers to the simultaneous verification of different properties in parallel. In [15], a reactive semantics is defined for a subset of UML activities, which makes it a precise design language for reactive systems. The same authors also define in [14] an automated compositional mechanism for UML activities together with an interface (a so-called External State Machine), seen as building blocks.

Outline. Section 2 presents our UML-based modeling for business processes (static view, activity diagram, etc.), details the activity diagram features we consider, and describes how to compose them in a modular way. Then, we provide a translation of the considered activity diagrams into colored Petri nets in Section 3 (activity diagram) and Section 4 (static view). We use as a running example an electronic commerce system EC. Section 5 concludes, gives some hints on our implementation, and sketches future directions of research.

2 Business Process Modeling

2.1 Precise Business Process Models

Business processes are collections of related and structured activities or tasks, producing a specific service or product. In this section, we consider *precise* models of business processes. The word "precise" means here that we define such models in a sharper way than usual; the word is used in several related works on models (see, e.g., [19]). A precise model of a business process consists of (1) the *static view*, i.e., a class diagram defining the types of all the entities in the process; (2) the list of the *process participants* and of the used data typed using the classes and the datatypes in the static view; and (3) an *activity diagram* representing the process behavior.

The process participants are entities taking part in a process, and can be classified as: (i) business worker, if they correspond to human beings acting in the process, (ii) system, if they are software or hardware systems with a role in the process, and (iii) business object, when they are passive entities used in the activities of the workers and of the systems. The classes in the static view may be stereotyped by <<worker>>, <<system>> and <<object>> to explicit which kind of entities they model. A class with these stereotypes is called an *entity class*.

The operations of the classes stereotyped by either <<worker>> or <<system>> represent the atomic activities that they are able to perform in the business process. These classes may have also some auxiliary operations stereotyped by <<aux>> not modeling any activity.

The operations of the classes stereotyped by <<object>> represent the atomic activities that may be performed over them. The constructor operations for any class have the stereotype <<create>>.

We consider an e-commerce EC as a running example of a precise business process. Fig. 1 presents its static view, while Fig. 2 presents its activity diagram and the list of the participants with the used data.

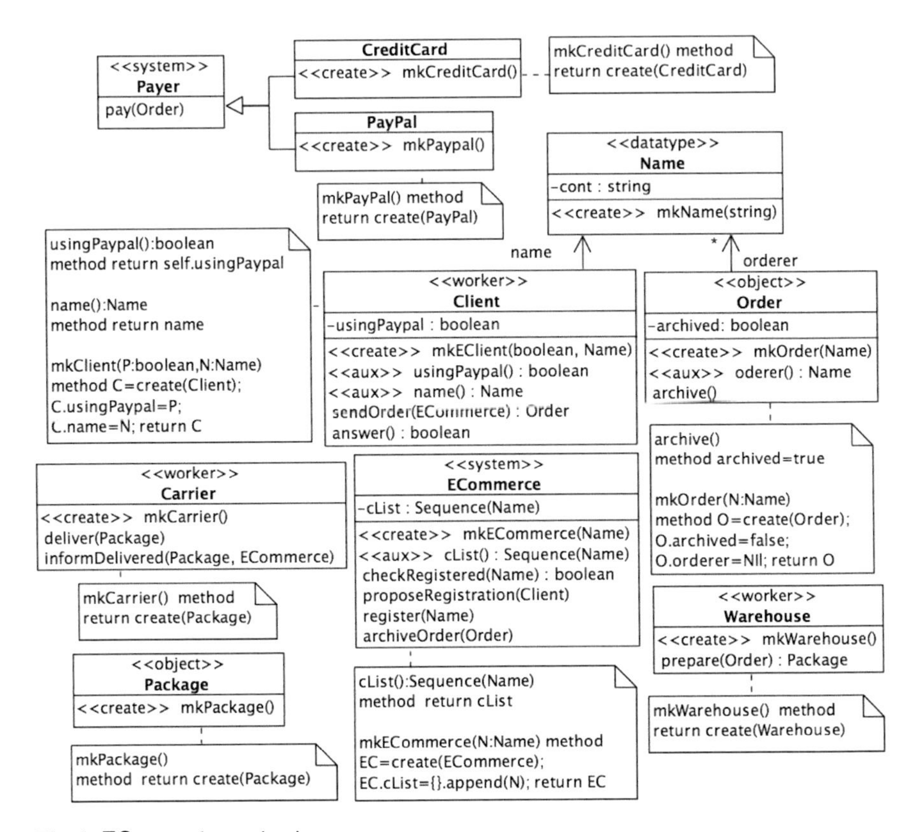

Fig. 1 EC example: static view

The EC business process has seven participants, and two of them, ORDER and PACK, are created during the process execution. Two boolean values, ANS and RES, are set during the process execution. It is important to note that the listed participants and data are not specific individuals, but roles that can be instantiated in many different ways. If a participant/data is marked by <<out>>, then it means that it is created/defined during the process execution.

The static view should be complemented with methods defining the meaning of the operations of the datatypes, of the classes stereotyped by <<object>>, and of any operation stereotyped by <<aux>> or <<create>>. In Fig. 1, the various methods are reported in notes attached to the corresponding classes. The behavior of the classes stereotyped by <<worker>> or <<system>> will be defined by state machines, where all events are calls of their operations not stereotyped by <<aux>>. In the case of the EC process, these state machines are not shown here. They have a simple "daisy form", with a unique state and with a transition leaving and entering this state for any operation. This corresponds to say that the instances of these classes may perform anytime any atomic activity represented by an operation.

The following subsection describes how the business process behavior is modeled by a precise activity diagram.

2.2 *Precise Activity Diagrams*

2.2.1 UML Activity Diagrams

We first briefly recall UML activity diagrams [1]. They feature in particular an *initial node* (e.g., the top node in Fig. 2), and two kinds of final nodes: *activity final*, that terminate the activity globally ("final1" and "final2" in Fig. 2), and *flow final*, that terminate the local flow [1, p.340]. They also feature *choice* (e.g., "dec1"), i.e., the ability to follow one path among different possibilities, depending on guards, and *merge* (e.g., "Merge1"), i.e., the converse operation. They also feature *fork*, i.e., the ability to split the flow into different subactivities executed in parallel (e.g., the large line below "Merge1"), and *join*, i.e., the converse operation (the large line below "Merge3").

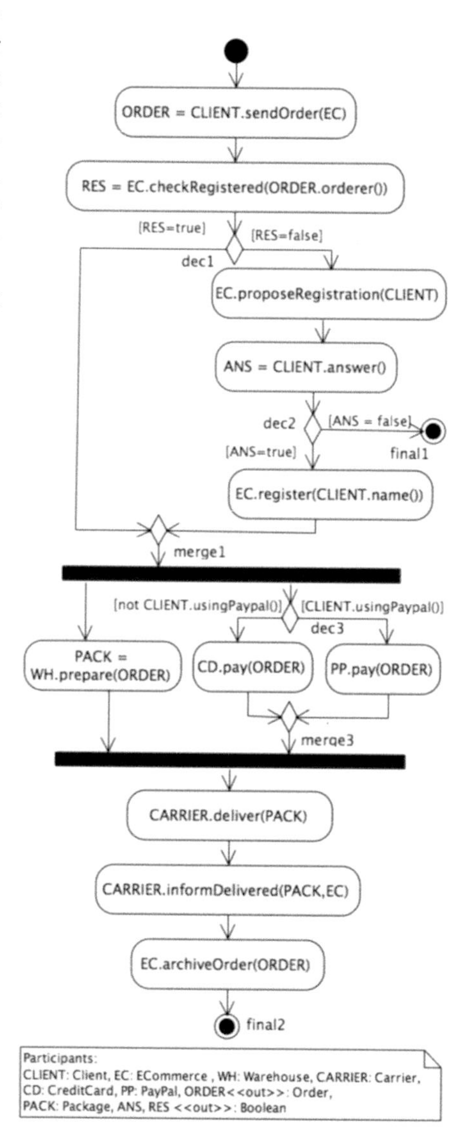

Fig. 2 EC example: activity diagram

2.2.2 Activity Diagrams Patterns

General Scheme for Patterns. We now introduce precise activity diagrams. Whereas UML activity diagrams provide a lot of freedom in the syntax, we give here precise rules for building activity diagrams in an iterative and modular way. First, from years of experience in the area of modeling, we believe that some of the syntactic features of UML activity diagrams are not often used in practice, or are ambiguous, and are then discarded here. Second, some constructions can reflect ill-formed diagrams. For example, we make here compulsory that a fork must always be eventually followed by a join, except in very particular cases.

Hence, following these patterns can help the designer to avoid common mistakes (see, e.g., [18]). Providing these precise activity diagrams with a semantics will be the subject of Section 3. Note that, different from software engineering design patterns, that can be inserted into freely written code, precise activity diagrams are exclusively made of activity diagram patterns composed with each other.

Inductive Rules. We assume the static view and the list of the participants of the business process are already defined. Now, the set ***PACT*** of the precise activity diagrams is inductively defined below using a set of rules. Each rule defines an activity diagram pattern. For each activity diagram pattern in ***PACT***, we define a *begin node* and an *end node*. Either the begin or the end node may be undefined, but not both. When composing the activity diagram fragments, we denote by $\perp$ the fact that a fragment has no end node.

In the following ***EXP*** denotes the set of the OCL (Object Constraint Language) expressions built on the participant names, the operations of the datatypes defined in the static view, and the operations of the entity classes appearing in the static view stereotyped by <<aux>>. Such expressions are without side-effects on the process since the stereotype <<aux>> requires an operation to be a query.

Rules 1–4 define simple patterns, whereas rules 5–8 define complex patterns by composing fragments built using the patterns. We also compare our patterns with those of [20], when applicable.

Rule 1: Initial. The initial node ● belongs to ***PACT***, and its begin node is undefined, while its end node is itself.

Rule 2: Activity final. ◉ belongs to ***PACT***, and its begin node is itself, while its end node is undefined.

Rule 3: Flow final. ⊗ belongs to ***PACT***, and its begin node is itself, while its end node is undefined.

Rule 4: Action. If X is a participant of the process, Exp, $\text{Exp}_1, \ldots, \text{Exp}_n$ belong to ***EXP***, and op is an operation of a class stereotyped either by <<worker>>, <<system>>, <<object>> in turn not stereotyped by <<aux>> or <<create>>, then $\text{X} := \text{Exp}$ (4a) , $\text{X} := \text{Exp.op}(\text{Exp}_1, \ldots, \text{Exp}_n)$ (4b), and $\text{Exp.op}(\text{Exp}_1, \ldots, \text{Exp}_n)$ (4c) belong to ***PACT***, and their begin and end nodes coincide with themselves.

Rule 5: Sequence. This pattern corresponds to pattern 5 ("sequence") in [20]. If A_1 (with a defined end node) and A_2 (with a defined begin node) belong to ***PACT***, then A_1 → A_2 belongs to ***PACT***, and has the begin node of A_1 and the end node A_2, if they exist. Note that A_1 and A_2 represent here activity diagrams fragments inductively defined using our set of rules. The begin node of A_1 (resp. end node of A_2) is not depicted: this means it can either be defined or not. These conventions will be used throughout this section.

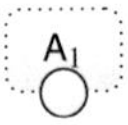

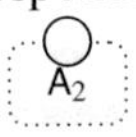

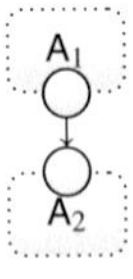

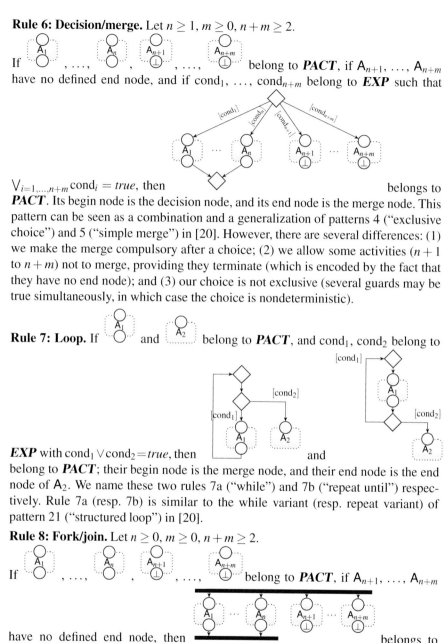

Rule 6: Decision/merge. Let $n \geq 1$, $m \geq 0$, $n+m \geq 2$.

If 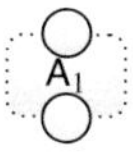

, …, 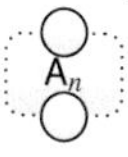

, 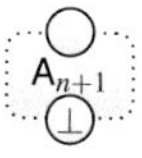

, …, 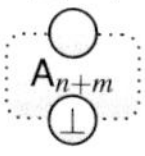

belong to ***PACT***, if $\mathsf{A}_{n+1}, \ldots, \mathsf{A}_{n+m}$ have no defined end node, and if $\text{cond}_1, \ldots, \text{cond}_{n+m}$ belong to ***EXP*** such that $\bigvee_{i=1,\ldots,n+m} \text{cond}_i = \textit{true}$, then

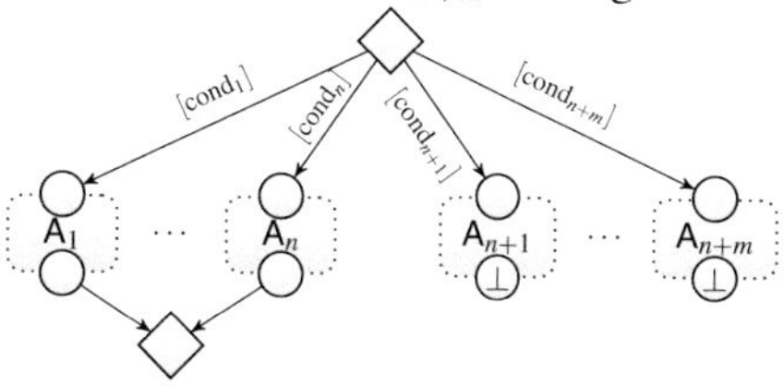

belongs to ***PACT***. Its begin node is the decision node, and its end node is the merge node. This pattern can be seen as a combination and a generalization of patterns 4 ("exclusive choice") and 5 ("simple merge") in [20]. However, there are several differences: (1) we make the merge compulsory after a choice; (2) we allow some activities ($n+1$ to $n+m$) not to merge, providing they terminate (which is encoded by the fact that they have no end node); and (3) our choice is not exclusive (several guards may be true simultaneously, in which case the choice is nondeterministic).

Rule 7: Loop. If 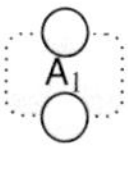

and 

belong to ***PACT***, and cond_1, cond_2 belong to ***EXP*** with $\text{cond}_1 \vee \text{cond}_2 = \textit{true}$, then

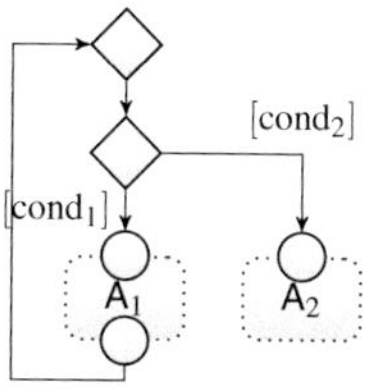

and

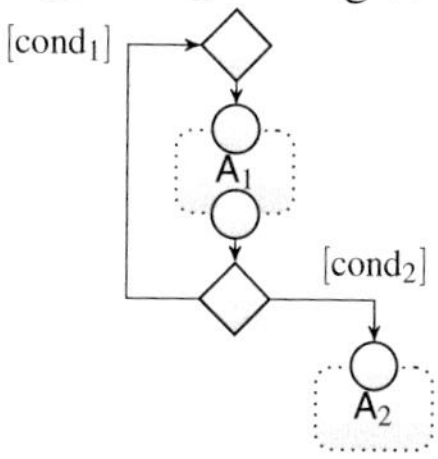

belong to ***PACT***; their begin node is the merge node, and their end node is the end node of A_2. We name these two rules 7a ("while") and 7b ("repeat until") respectively. Rule 7a (resp. 7b) is similar to the while variant (resp. repeat variant) of pattern 21 ("structured loop") in [20].

Rule 8: Fork/join. Let $n \geq 0$, $m \geq 0$, $n+m \geq 2$.

If 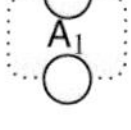

, …, 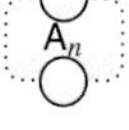

, 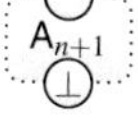

, …, 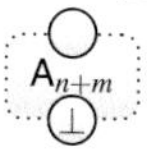

belong to ***PACT***, if $\mathsf{A}_{n+1}, \ldots, \mathsf{A}_{n+m}$ have no defined end node, then

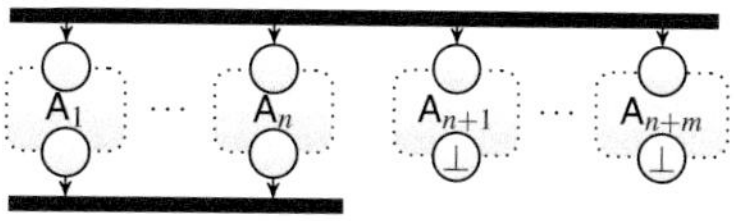

belongs to ***PACT***. Its begin node is the fork node, and its end node is the join node if $n > 0$, otherwise it is undefined. This pattern can be seen as a combination and a generalization of patterns 2 ("parallel split") and 3 ("synchronization") in [20]. However, we make the join compulsory after a fork; and we allow some activities ($n+1$ to $n+m$) not to join, providing they terminate.

3 Translation of the Activity Diagram

In the remaining of the paper, we consider the translation into a CPN of the business process models introduced in Section 2. On the one hand, the translation of the static view and of the lists of the participants of a business process will result in a set of declarations of types and of functions over them defining a special type State, whose values represent the current situation of the process participants and of the process data during the process execution. On the other hand, the translation of the activity diagram will result in a CPN. This CPN will use the type declarations and functions in its inscriptions.

We first recall the formalism of CPNs (Section 3.1), and then introduce the translation of the activity diagram (Section 3.2). The translation of the static view will be the subject of Section 4.

3.1 Colored Petri Nets with Global Variables

We briefly recall here colored Petri net (CPNs) [12] (for a precise definition, see [3]). CPNs are an extension of Petri nets with color sets, or types. In CPNs, places, tokens and arcs have a *type*. In Fig. 3(a), place p_1 has type $\mathbb{N}$, whereas p_2 has type $\mathbb{N} \times \mathbb{B}$. Arcs can be labeled with *arc expressions* modifying the (colored) token (e.g., $(i, true)$ in Fig. 3(a)). Transitions can have a *guard*, hence enabling the transition only if the guard is true (e.g., $[i \neq 2]$). We use for arc inscriptions and guards the syntax of CPN ML, an extension of the functional programming language Standard ML, and used by CPN Tools [12].

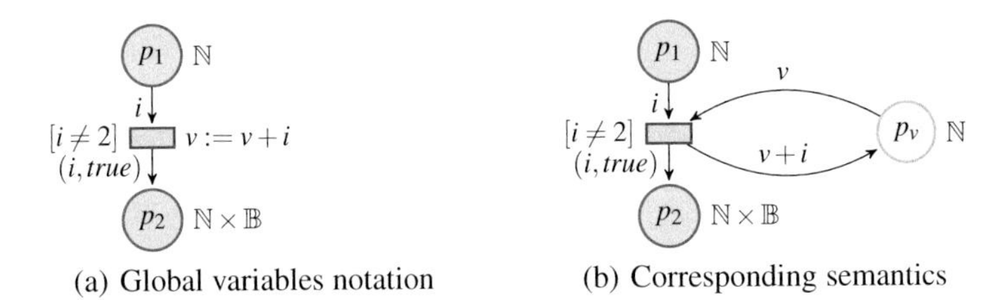

(a) Global variables notation (b) Corresponding semantics

Fig. 3 Example of a use of global variables

We use here the concept of *global variables*, a notation that does not add expressive power to CPNs, but renders them more compact. Global variables can be read in guards and updated in transitions. Some tools (such as CPN Tools) support these global variables. Otherwise, one can simulate a global variable using a "global" place, in which a single token (the type of which is the variable type) encodes the current value of the variable. An example of use is given in Fig. 3(a). The variable v (of type $\mathbb{N}$) is a global variable updated to the expression $v + i$. This CPN construction is equivalent to the one in Fig. 3(b). The case where a global variable is read in a guard is similar, with the difference that v is not modified.

3.2 *Translation*

The translation of the precise activity diagrams belonging to ***PACT*** (defined in Section 2.2) will be given compositionally following the rules defined there.

3.2.1 Assumptions

We make the following choice: each translated activity diagram fragment must start and finish with a place, so that the composition of the translations of the subparts is straightforward: it suffices to connect the places the same way as for the nodes we defined for the activity diagram patterns.

We define two global variables go: BOOL and s (see Section 4). In particular, variable go records whether the CPN should still execute, or should be completely stopped. This go variable is used to encode the activity final pattern (rule 2); if such a state is entered, then the whole process must immediately stop. Here, we assume that, for each transition of the CPN, the guard includes a check [go=true] (for sake of conciseness, this variable will not be depicted in our graphics). This go variable is initialized with true, and will be set to false when entering the CPN transition encoding the activity final state (see Fig. 4(c)).

Note that all edges and places have type "UNIT", i.e., the same type as in place/transition nets (we omit that type in Fig. 4 for sake of conciseness). Nevertheless, our CPN is still colored because of the use of global variables, guards in transitions, and functions updating the variables in transitions.

3.2.2 Translation of the Rules

We now give in Fig. 4 the translation of the rules from Section 2.2.2. The translation of each activity diagram pattern will result in a CPN fragment having the shape of Fig. 4(a). Modular composition is performed using the begin and end nodes, using the same way as for activity diagram patterns in Section 2.2.2.

Rule 1: Initial. The initial state is encoded into an initial place, containing the only initial token of the resulting CPN, followed by a transition assigning InitState to the global variable *s* (InitState will be detailed in Section 4). Finally, an outgoing place allows connection with the next component.

Rule 2: Activity final. An activity final pattern is translated into a transition updating the global variable go to false. Hence, since each transition has an implicit guard checking that go=true, the execution of the CPN is immediately stopped.

Rule 3: Flow final. A flow final pattern is translated into a simple place; hence local execution is terminated, without any consequence on the rest of the system.

Rule 4: Action. Recall from Section 2.2 that this rule translates the actions using three different schemes (i.e., Rules 4a, 4b, and 4c).

Rule 5: Sequence. We translate A_1 and A_2 inductively, and we directly merge the end node of A_1 with the begin node of A_2.

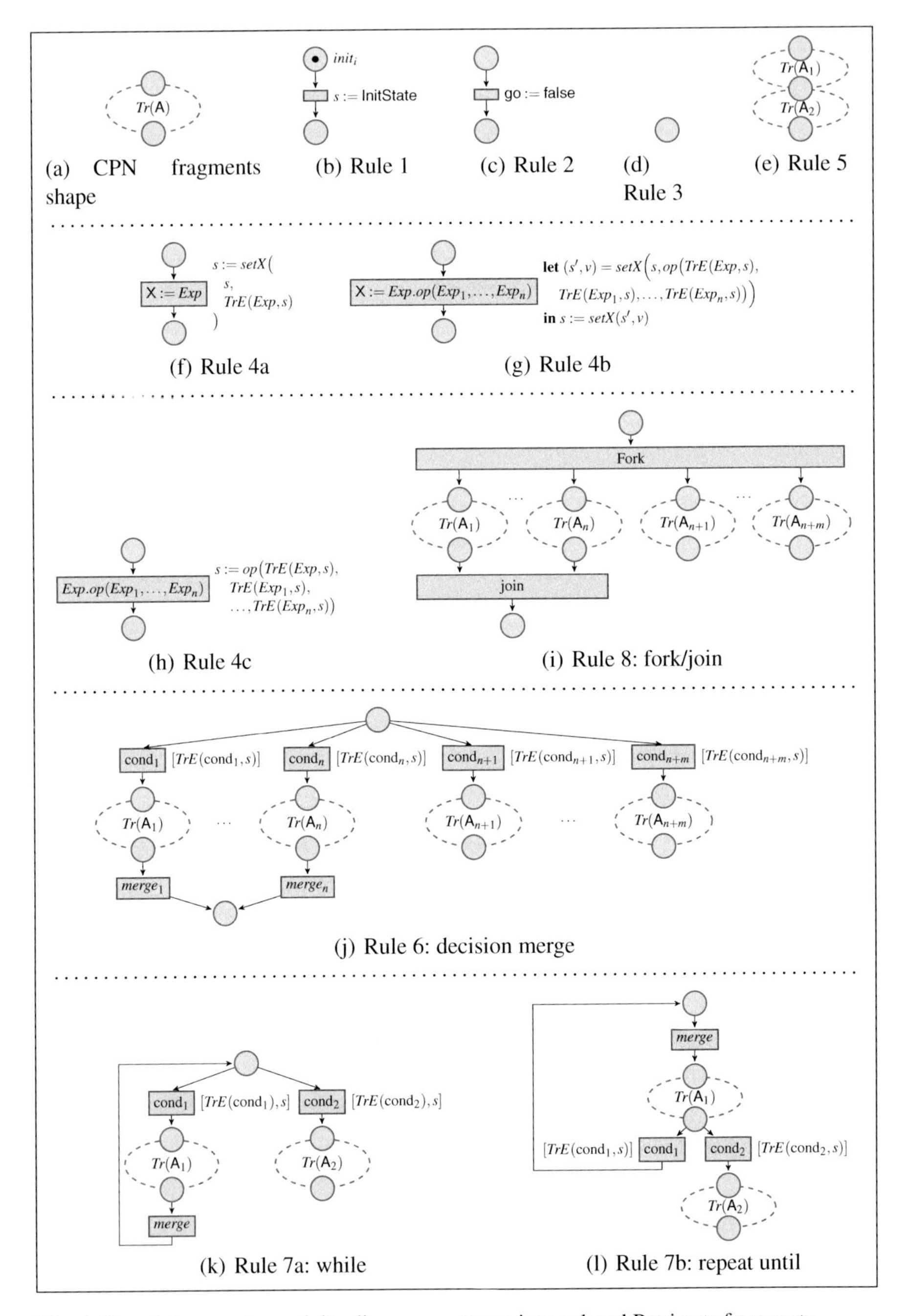

Fig. 4 Translating precise activity diagrams patterns into colored Petri nets fragments

Rule 6: Decision/merge. Here, (only) one of the transitions will fire (depending on the guards[2]). If the corresponding activity has an end node (activities 1 to n), then the process continues afterwards from the outgoing place below; otherwise (activities $n+1$ to $n+m$), it is stopped when the activity stops.

Rule 7: Loop. The translation of the while loop (resp. repeat until loop) is given in Fig. 4(k) (resp. Fig. 4(l)).

Rule 8: Fork/join. The translation is quite straightforward. The $n+m$ activities are subject to a fork; then, only the n first activities are merged later.

A full translation of the activity diagram in Fig. 2 is available in [3].

4 Translation of the Static View and of the Participant List

In this section, we translate the static view and the participant list into a set of CPN ML declarations. In particular, we translate the type (color set) State together with a set of declarations of auxiliary types and of functions needed to handle them, used by the CPN defined in Section 3. Recall that the values of State represent the current situation of the process participants and of the process data during the execution of the process itself.

We first present the part of the translation generating the definition of State (Section 4.1). Then we give the translation of the expressions (Section 4.2). We terminate with the part concerning the definition of the initial state (Section 4.3), the particular value of State representing the situation at the beginning of the process execution. We use the EC example to illustrate our approach throughout the section. The complete model can be found in [3].

In the following E_1: T_1, ..., E_n: T_n are the participants of the business process, $\mathsf{Class}_1, \ldots, \mathsf{Class}_m$ are all the entity classes introduced by the static view (i.e., those stereotyped by <<object>>, <<worker>> or <<system>>), and $\mathsf{Datatype}_1$, ..., $\mathsf{Datatype}_h$ are all datatypes included in the static view.

4.1 State Definition

As mentioned earlier, the values of type State represent all possible states of the process participants during the process execution. State is defined by the list of type and function declarations shown in Fig. 5(a). The first n components of State are used to record the associations between the names of the participants (E_1, ..., E_n) and the CPN ML value identifying them; whereas, given Class a class, then classes: *CompType*(Class) is the component of State recording all existing instances (objects) of the class Class with their current states. Function *CompType* returns the proper types for the various components of State. *Comp* generates all the functions and type declarations needed to handle the State component corresponding either to

[2] If several guards are true simultaneously, the choice is nondeterministic, according to the CPN semantics.

Decls(Datatype$_1$) ... *Decls*(Datatype$_h$); *Decls*(Class$_1$) ... *Decls*(Class$_m$)
Comp(E$_1$: T$_1$) ... *Comp*(E$_n$: T$_n$); *Comp*(Class$_1$) ... *Comp*(Class$_m$)
colset State = **record**
E$_1$: *CompType*(T$_1$) * ... * E$_n$: *CompType*(T$_n$)
class$_1$s: *CompType*(Class$_1$) * ... * class$_m$s: *CompType*(Class$_m$); ;

(a) State translation

let att$_1$: T$_1$, ..., att$_k$: T$_k$ be the attributes of Class
colset ClassID = int;
if Class is stereotyped by <<object>> then
colset ClassState = **record** att$_1$: *TrType*(T$_1$) * ... * att$_k$: *TrType*(T$_k$);
otherwise
colset ClassControl = **with** s$_1$ | ... | s$_h$;
colset ClassState = **record** att$_1$: *TrType*(T$_1$) * ... * att$_k$: *TrType*(T$_k$) * control: ClassControl;
where s$_1$, ..., s$_h$ are the states of the state machine associated with Class

(b) Definition of *Decls*(Class)

let att$_1$: T$_1$, ..., att$_k$: T$_k$ be the attributes of Datatype
colset DatatypeVal = **record** att$_1$: *TrType*(T$_1$) * ... * att$_k$: *TrType*(T$_k$);
for any op(T$_1$, ..., T$_n$): T operation of Datatype
op: *TrType*(T$_1$) * ... * *TrType*(T$_n$) → *TrType*(T)
these operations must be defined by looking at the associated methods in the static view

(c) Definition of *Decls*(Datatype)

fun setE: State × *TrType*(T) → State

(d) Definition of *Comp*(E: T)

colset Classes = **list product** ClassID * ClassState;
upClass: Classes * ClassID * ClassState → Classes
getClass: Classes * ClassID → ClassState
for any op(T$_1$, ..., T$_n$) operation of Class
op: State * ClassID * *TrType*(T$_1$) * ... * *TrType*(T$_n$) → State
for any op(T$_1$, ..., T$_n$): T operation of Class not marked by <<aux>>
op: State * ClassID * *TrType*(T$_1$) * ... * *TrType*(T$_n$) → (State * *TrType*(T))
for any op(T$_1$, ..., T$_n$): T operation of Class marked by <<aux>>
op: State * ClassID * *TrType*(T$_1$) * ... * *TrType*(T$_n$) → *TrType*(T)

(e) Definition of *Comp*(Class)

Fig. 5 Translation of the static view

a process participant or to all the instances of a class, whereas *Decls* generates the data structures and the relative functions needed to represent a class/dataype.

We give below the definition of State in the case of the EC example.

```
colset State = record
       CLIENT : ClientID      *    EC : ECommerceID *        WH : WarehouseID *
      CARRIER : CarrierID     *    CC : CreditCardID *        PP : PaypalID    *
        ORDER : OrderID       *  PACK : PackageID    *       ANS : BOOL        *
          RES : BOOL          * clients : Clients    * eCommerces : ECommerces *
   warehouses : Warehouses    * carriers : Carriers  * creditCards : CreditCards *
      paypals : Paypals       * orders : Orders      *    packages : Packages;
```

Function *Decls* (defined in Fig. 5(b) and 5(c)) transforms a class/datatype present in the static view into the set of CPN ML type and function declarations needed to represent its values and to handle them. The values of a datatype Datatype are represented by the type DatatypeVal, i.e., a record having a component for each attribute of Datatype. A class Class determines a set of objects having an identity, typed by ClassID, and a local state typed by ClassState. The local state is a record having a component for each attribute of Class and, in the case of active objects and extra component corresponding to the control state, typed by ClassControl, and defined by the state machine associated with Class.

In the EC example, the WarehouseState is defined as follows:
colset WarehouseState = **record** control: WarehouseControl;

As all identifiers, the WarehouseID is an integer: **colset** WarehouseID = int;

And the WarehouseControl is an enumerated type with (in this case) only one value: WarehouseControl = **with** Warehouse0;

Function *Comp* (defined in Fig. 5(d) and 5(e)) transforms a process participant declaration (resp. a class) in the static view into a set of the type and function declarations needed to define and handle component State recording the participant state (resp. the states) of all class instances. The set of the states of the instances/objects of a class is realized by a list of pairs, made of an object identity and an object state.

For example, type Warehouses is defined as a list of pairs of WarehouseID and WarehouseState: **colset** Warehouses = **list product** WarehouseID * WarehouseState.

The function corresponding to an operation op of a class in the static view is defined by looking either at the method associated with op in the static view, in case of business object classes and of <<aux>> operation of workers and system classes, whereas for the other operations of the workers and system classes they are defined using the state machines associated with that class. By looking at the state machine transitions, it will be possible to know how these operation calls modify the attribute values and the control state. In particular, our mechanism defines functions set to set a value inside a record (e.g., "State.set_CLIENT *s id*" sets field CLIENT to *id* in state *s*), as well as functions to get a value from the record, and to update it. The definitions of these set, get and upd functions are omitted here; their definition for the EC example can be found in [3].

Finally, the *TrType* function translates a UML type into its corresponding CPN ML type. Native types (string, boolean, integer) are translated to CPN ML types (viz., STRING, BOOL, int respectively). Then, we have $TrType(\mathsf{Class}) = \mathsf{ClassID}$, where Class is the name of a UML class of the static view. And $TrType(\mathsf{Datatype}) = \mathsf{DatatypeVal}$, where Datatype is the name of a UML datatype.

4.2 Expressions

We give here the translation of the expressions of ***EXP*** into CPN ML expressions, since they will appear in the activity diagrams as conditions on the arcs leaving the merge nodes, as well as in the action nodes. We define below by cases the translation function $TrE(\mathsf{Exp}, s)$, that associates a CPN ML with an OCL expression Exp, given the current state s.

- $TrE(\mathsf{X}, s) = \#\mathsf{X}(s)$, if X is a participant of the process, (#X is the CPN ML operation selecting a record type component), e.g. CLIENT is translated to $\#\mathsf{CLIENT}(s)$;
- $TrE(C, s) = C$, if C is a primitive data type constant;
- $TrE(\mathsf{op}(\mathsf{Exp}_1, \dots, \mathsf{Exp}_n), s) = \mathsf{op}'(TrE(\mathsf{Exp}_1, s), \dots, TrE(\mathsf{Exp}_n, s))$, if op is an operation of a primitive type, op' will be either op itself or it will be defined case by case in case of name mismatch between the operations on the UML primitive types and the corresponding ones of CPN ML;
- $TrE(\mathsf{op}(\mathsf{Exp}_1, \dots, \mathsf{Exp}_n), s) = \mathsf{op}(TrE(\mathsf{Exp}_1, s), \dots, TrE(\mathsf{Exp}_n, s))$, if op is an operation of a datatype defined in the static view;
- $TrE(\mathsf{Exp.op}(\mathsf{Exp}_1, \dots, \mathsf{Exp}_n), s) = \mathsf{op}(s, TrE(\mathsf{Exp}, s), TrE(\mathsf{Exp}_1, s), \dots, TrE(\mathsf{Exp}_n, s))$, if op is an operation of a class defined in the static view of kind query.

For example, the translation of the guard [RES = true] in Fig. 2 using function TrE results in the CPN ML expression $[\#\mathsf{RES}(s) = true]$. And the OCL expression CARRIER.deliver(PACK) is translated to $\mathsf{deliver}(s, \#\mathsf{CARRIER}(s), \#\mathsf{PACK}(s))$.

4.3 Initial Process Execution State

In order to translate a business process into CPNs, and specifically define the initial execution state of the process itself, we also need a specific list of individual participants. Recall that the names in the participant list part of the process model are roles, not specific individuals.

If n is the number of participants and data not marked by <<out>>, we call a *business process instantiation* a list of n ground OCL expressions defined using the data type defined in the static view, and the constructors of the classes in the static view itself (operations stereotyped by <<create>>).

Given the business process instantiation (i.e., a list of ground expressions G_1, ..., G_n), the function *Initialize* returns the CPN ML expression defining the initial state, where the participants not marked by <<out>> are initialized with the values determined by the process instantiation. The other ones are initialized with some standard default values depending on their type (e.g., 0 for int, *false* for booleans, nil for list types, etc.), and the components corresponding to the objects of the various classes just contain the states of the objects appearing in the process instantiation. Hence, we have: **val** InitState = *Initialize*($C_1, \ldots, C_n$); *Initialize* is defined using *TrE* (details can be found in [3]).

5 Conclusion and Future Work

In this work, we define precise business models, where the activity diagrams are inductively defined using a set of patterns combined in a modular way. Hence, we characterize a set of commonly used behaviors in activity diagrams. Moreover, our patterns provide the designer with guidelines, thus avoiding common modeling errors. Our second contribution is to provide the activity diagrams built using these patterns with a formal semantics using colored Petri nets, hence allowing the use of automated verification techniques.

Implementation. Following our algorithm, we implemented (manually) the EC example into the CPN Tools model checker [12]. This results in a CPN containing 24 places, 25 transitions and about 500 lines of CPN ML code; the detailed CPN description is available in [3], and the CPN Tools model is available online[3]. Such an implementation allows for automated verification techniques; among the properties are for example the fact that the various final nodes may be reached in any case, and hence that the process is well-formed. Automatizing the translation process from a precise activity diagram to a CPN using model-driven methods and technologies does not raise any particular theoretical problem, and is the subject of ongoing work.

Future Works. Among directions for future research is the comparison of our semantics given in terms of CPNs where the process execution state is modeled by colored tokens, with existing (partial) semantics, such as [15] and [11] (a source of inspiration for our work). Furthermore, integrating accept and timed events to our approach is an interesting direction of research. Finally, we aim at finding the properties relevant for the business process, and providing guidelines to prove them.

Also note that the resulting CPN (including the functions) may be simplified in some cases. First, some places and transitions added by the translation may be unnecessary. This is the case, e.g., of a decision/merge pattern with only one activity on the left side, and one on the right side ($n = m = 1$). In that case, the only activity synchronizing in the merge is the left one; hence, the transition "$merge_1$" in Fig. 4(j), as well as the place below, are unnecessary. Second, some functions could be simplified for similar reasons. These simplifications, that are beyond the scope of this paper, could help to speed up the automated verification of the resulting CPN.

Acknowledgment. We wish to thank Michael Westergaard for his kind help when using CPN Tools, and anonymous reviewers for their helpful comments.

[3] `http://lipn.univ-paris13.fr/~andre/activity-diagrams-patterns/`

References

1. OMG unified language superstructure specification (formal). version 2.4.1 (August 06, 2011), http://www.omg.org/spec/UML/2.4.1/Superstructure/PDF/
2. André, É., Choppy, C., Klai, K.: Formalizing non-concurrent UML state machines using colored Petri nets. ACM SIGSOFT Software Engineering Notes 37(4), 1–8 (2012)
3. André, É., Choppy, C., Reggio, G.: Activity diagrams patterns for modeling business processes (report version) (2013), http://lipn.fr/~andre/adp/
4. Bernardi, S., Merseguer, J.: Performance evaluation of UML design with stochastic well-formed nets. Journal of Systems and Software 80(11), 1843–1865 (2007)
5. Börger, E.: Modeling workflow patterns from first principles. In: Parent, C., Schewe, K.-D., Storey, V.C., Thalheim, B. (eds.) ER 2007. LNCS, vol. 4801, pp. 1–20. Springer, Heidelberg (2007)
6. Di Cerbo, F., Dodero, G., Reggio, G., Ricca, F., Scanniello, G.: Precise vs. ultra-light activity diagrams – An experimental assessment in the context of business process modelling. In: Caivano, D., Oivo, M., Baldassarre, M.T., Visaggio, G. (eds.) PROFES 2011. LNCS, vol. 6759, pp. 291–305. Springer, Heidelberg (2011)
7. Cook, W.R., Patwardhan, S., Misra, J.: Workflow patterns in Orc. In: Ciancarini, P., Wiklicky, H. (eds.) COORDINATION 2006. LNCS, vol. 4038, pp. 82–96. Springer, Heidelberg (2006)
8. Distefano, S., Scarpa, M., Puliafito, A.: From UML to Petri nets: The PCM-based methodology. IEEE Transactions on Software Engineering 37(1), 65–79 (2011)
9. Erl, T.: SOA Principles of Service Design. The Prentice Hall Service-Oriented Computing Series from Thomas Erl (2007)
10. France, R.B., Evans, A., Lano, K., Rumpe, B.: Developing the UML as a formal modelling notation. In: Computer Standards and Interfaces: Special Issues on Formal Development Techniques, pp. 297–307. Springer (1998)
11. Grönniger, H., Reiß, D., Rumpe, B.: Towards a semantics of activity diagrams with semantic variation points. In: Petriu, D.C., Rouquette, N., Haugen, Ø. (eds.) MODELS 2010, Part I. LNCS, vol. 6394, pp. 331–345. Springer, Heidelberg (2010)
12. Jensen, K., Kristensen, L.M.: Coloured Petri Nets – Modelling and Validation of Concurrent Systems. Springer (2009)
13. Kordon, F., Thierry-Mieg, Y.: Experiences in model driven verification of behavior with UML. In: Choppy, C., Sokolsky, O. (eds.) Monterey Workshop 2008. LNCS, vol. 6028, pp. 181–200. Springer, Heidelberg (2010)
14. Kraemer, F.A., Herrmann, P.: Automated Encapsulation of UML Activities for Incremental Development and Verification. In: Schürr, A., Selic, B. (eds.) MODELS 2009. LNCS, vol. 5795, pp. 571–585. Springer, Heidelberg (2009)
15. Kraemer, F.A., Herrmann, P.: Reactive semantics for distributed UML activities. In: Hatcliff, J., Zucca, E. (eds.) FMOODS/FORTE 2010, Part II. LNCS, vol. 6117, pp. 17–31. Springer, Heidelberg (2010)
16. Mekki, A., Ghazel, M., Toguyeni, A.: Validating time-constrained systems using UML statecharts patterns and timed automata observers. In: VECoS, pp. 112–124. British Computer Society (2009)
17. Peixoto, D.C., Batista, V.A., Atayde, A.P., Pereira, E.B., Resende, R.F., Pádua, C.I.: A comparison of BPMN and UML 2.0 activity diagrams. In: Simposio Brasileiro de Qualidade de Software (2008), http://homepages.dcc.ufmg.br/~cascini/

18. Reggio, G., Leotta, M., Ricca, F.: Precise is better than light: A document analysis study about quality of business process models. In: First International Workshop on Empirical Requirements Engineering (EmpiRE), pp. 61–68 (2011)
19. Reggio, G., Ricca, F., Scanniello, G., Di Cerbo, F., Dodero, G.: A precise style for business process modelling: Results from two controlled experiments. In: Whittle, J., Clark, T., Kühne, T. (eds.) MODELS 2011. LNCS, vol. 6981, pp. 138–152. Springer, Heidelberg (2011)
20. Workflow Patterns Initiative. Workflow patterns home page, `http://www.workflowpatterns.com`
21. Zhang, S.J., Liu, Y.: An automatic approach to model checking UML state machines. In: SSIRI (Companion), pp. 1–6. IEEE Computer Society (2010)

S-TunExSPEM: Towards an Extension of SPEM 2.0 to Model and Exchange Tunable Safety-Oriented Processes

Barbara Gallina, Karthik Raja Pitchai, and Kristina Lundqvist

Abstract. Prescriptive process-based safety standards (e.g. EN 50128, DO-178B, etc.) incorporate best practices to be adopted to develop safety-critical systems or software. In some domains, compliance with the standards is required to get the certificate from the certification authorities. Thus, a well-defined interpretation of the processes to be adopted is essential for certification purposes. Currently, no satisfying means allows process engineers and safety managers to model and exchange safety-oriented processes. To overcome this limitation, this paper proposes S-TunExSPEM, an extension of Software & Systems Process Engineering Meta-Model 2.0 (SPEM 2.0) to allow users to specify safety-oriented processes for the development of safety-critical systems in the context of safety standards according to the required safety level. Moreover, to enable exchange for simulation, monitoring, execution purposes, S-TunExSPEM concepts are mapped onto XML Process Definition Language 2.2 (XPDL 2.2) concepts. Finally, a case-study from the avionics domain illustrates the usage and effectiveness of the proposed extension.

Keywords: DO-178B, safety-oriented processes, process modelling, SPEM 2.0, process exchange, XPDL 2.2, process reuse.

1 Introduction

The increasing awareness of software development being a complex task has since the 1980's received increased attention from the research community working on engineering software processes [11]. Software processes can be defined as coherent sets of policies, organizational structures, technologies, procedures, and artefacts that are needed to conceive, develop, deploy, and maintain a software product [11].

Barbara Gallina · Karthik Raja Pitchai · Kristina Lundqvist
Mälardalen University, P.O. Box 883, SE-72123 Västerås, Sweden
e-mail: {barbara.gallina,kristina.lundqvist}@mdh.se,
kpi10001@student.mdh.se

R. Lee (Ed.): *SERA*, SCI 496, pp. 215–230.
DOI: 10.1007/978-3-319-00948-3_14

The research motivation surrounding software processes is based on the assumption that there is a correlation between the quality of the process and the quality of the software developed. According to what is stated in [15], each life-cycle phase may represent a source of faults that if not handled lead to system failures causing serious incidents. To avoid such failures, processes must be enhanced by preventing or removing potential faults. To enhance processes, in the context of safety standards, best systems and software development practices have been collected and prescriptive processes have been defined. More specifically, these processes mandate the activities to be executed, the work-products to be provided, the qualifications needed to execute the activities, the tools to be used to execute the activities, and the guidelines to be followed. DO-178B [22], for instance, is the de facto standard for software development in civilian aircraft and its adoption is considered to be beneficial in contributing to the excellent record with remarkably few failures of avionics software [24]. Even though no strong correlation between the process and the product can be claimed in the context of dependable (safety-critical) systems, the enhancement of the processes permits the development of a deeper safety culture, leading the development team to act cautiously [16].

For certification purposes, in some domains, compliance with the processes defined within safety standards is mandatory. As investigated in [5], DO-178B leaves room for interpretation. In some cases, due to its lack of specificity in the guidelines, different users may come to different conclusions regarding the acceptability of a particular document or artefact based on their particular experience. Thus, as nicely put in [6] "for companies seeking a first-time certification, preparation for DO-178B can be a daunting challenge".

To ensure process understanding and thus eliminate inconsistencies in the process specification, a Process Modelling Language (PML) is necessary. Besides understanding, an adequate PML should permit users to document and exchange process models. In the literature, several PMLs are at disposal, e.g. Software & Systems Process Engineering Meta-Model 2.0 (SPEM 2.0). As recently reviewed in [23], SPEM 2.0 has obtained a significant acceptance and the research community is very active to propose extensions towards SPEM 3.0 in order to enhance its modelling capabilities, its executability, and its tool support. Thus, we decide to join this active research community to propose an extension, called S-TunExSPEM, to support the modelling as well as the exchange of safety-oriented processes. Our focus is limited to core process elements since our goal is to ease the adoption of S-TunExSPEM by providing an easy-to-digest PML. To define the set of core elements we have analyzed DO-178B to extract a list of key safety-related concepts. For these concepts we have defined: the abstract syntax by extending the SPEM 2.0 meta-model, the concrete syntax by providing new safety-oriented icons. Then, to enable process models interchange towards the usage of existing execution as well as monitoring and simulation engines, we have provided a mapping between S-TunExSPEM concepts and corresponding concepts of XML Process Definition Language 2.2 (XPDL 2.2). Finally, we have used S-TuneExSPEM to model processes for the development of avionics software.

The rest of the paper is organized as follows. In Sect. 2, we provide essential background information. In Sect. 3, we present S-TunExSPEM the proposed SPEM 2.0 extension that targets safety-oriented processes. In Sect. 4, we illustrate the usage and effectiveness of S-TunExSPEM by modelling a process taken from the avionics domain. In Sect. 5, we discuss related work. Finally, in Sect. 6 we present some concluding remarks and future work.

2 Background

In this section, we present the background information on which we base our work. In particular, in Sect. 2.1 we provide general information on safety-oriented processes and their role in the certification process. In Sect. 2.2 we provide essential information concerning the software development process defined in DO-178B. In Sect. 2.3, we briefly present SPEM 2.0, the process modelling language from which stems our extension. Finally, in Sect. 2.4, we briefly present XPDL 2.2, the process definition language onto which we map our SPEM 2.0 extension.

2.1 Safety-Oriented Processes and Their Role in Certification

Prescriptive safety-oriented processes also known as safety life-cycles are systems (or software) development processes that prescribe best practices to be followed to achieve systems capable of managing safety by addressing the causes of accidents, namely hazards. Generally, a safety process requires safety analysts to identify and categorize the hazards according to domain-specific levels and risk assessment procedures. These levels, whose determination is not straightforward due to potential misconception/misuse/abuse [21], are called Design Assurance Levels (DALs) in the avionics domain in the context of DO-178B, Automotive Safety Integrity Levels (ASILs) in the automotive domain in the context of ISO 26262, and Safety Integrity Levels (SILs) in other domains that inherit the levels from IEC 61508. These levels (four or five depending on the specific standard) span from negligible to catastrophic hazards and they determine the number of objectives to be satisfied (eventually with independence) during the system (or software) development. Once hazards are classified, safety managers elicit safety requirements aimed at reducing risk. Then, they verify and validate the correct implementation and deployment of the elicited safety requirements throughout the safety life-cycle. It must be noted that it is not always possible to show that the systems developed meet the safety requirements.

Certification refers to the "process of assuring that a product or process has certain stated properties, which are then recorded in a certificate" [16]. Thus, for safety certification purposes, product and process-based arguments are needed to claim an acceptable level of safety. Process-based arguments are of particular value whenever confidence in product-based arguments is limited.

To provide convincing process-based arguments claiming for compliance, first of all it is necessary to achieve a well-defined and agreed-upon interpretation of the processes mandated within the standards [19]. Thus, adequate process modelling means are necessary and should be developed.

2.2 DO-178B

DO-178B [22] has been the de facto standard in the avionics domain. Currently, it is being replaced by a revised version (DO-178C), which addresses the inconsistencies of the previous document but preserves its basic and valuable principles. DO-178B provides guidance for the development of software for airborne systems and equipment. Its purpose is to guarantee a level of confidence in the correct functioning of the software developed in compliance with airworthiness requirements.

In this subsection, we provide a brief description of the software development process. This description is then used in Sect. 3 to extract process models. The software development process is constituted of four phases (requirements, design, coding and integration), which can be chained, if a waterfall process model is selected. The standard, however, does not impose a specific process model. In what follows, for each phase we provide its characteristics in terms of input/output, guidelines and roles. For sake of clarity, it must be noted that no role is explicitly assigned in DO-178B. Roles, however, can be inferred from the skills that are required and mentioned in the standard.

The **requirements phase** is characterized by:

Input: System requirements, hardware interface, system architecture, Software Development Plan, Software Requirements Standards.

Output: Software Requirements Data that include functional as well as non functional requirements.

Roles: requirement engineers in charge of functional requirements and quality (safety) experts in charge of non-functional requirements.

Guidelines: guidelines, defined in Section 5.1.2 of the standard, contain general as well as safety specific information.

The **design phase** is characterized by:

Input: Software development plan, Software Requirements Data, Software Design Standards.

Output: Design description.

Roles: designers in charge of the design decision related to functional requirements and quality (safety) experts in charge of the design decision related to non-functional requirements.

Guidelines: guidelines, defined in Section 5.2.2 of the standard, contain general as well as safety specific information.

The **coding phase** is characterized by:

Input: Software development plan, Design description, Software Code Standards.

Output: Source code and object code.

Roles: programmers in charge of the implementation decisions related to functional aspects of the design and quality (safety) experts in charge of the implementation decision related to non-functional aspects of the design.

Guidelines: guidelines, defined in Section 5.3.2 of the standard, contain general as well as safety specific information.

The **integration phase** is characterized by:

Input: Source code and object code, target computer, linking and loading data.

Output: Executable Object Code.

Roles: integration experts.

Guidelines: guidelines, defined in Section 5.4.2 of the standard, contain general as well as safety specific information.

With respect to the outputs that characterize the phases, a general remark is that for reuse purposes, outputs (e.g. Software Requirements Data) should be split to take into consideration the different views.

Details concerning how to break down the work within each phase are not provided in the standard. For sake of simplicity, we consider that each phase is constituted by a single task. Similarly, no specific tool is mentioned. However, at the organization-specific level, tools have to be planned (indeed a specific section called *Software development environment* is expected within the *Software Development Plan*) and used. The standard however recommends to guarantee traceability among the phases thus an additional task aimed at checking traceability can be considered.

2.3 SPEM 2.0

As recalled in the introduction, a software process can be defined as a coherent set of policies, organizational structures, technologies, procedures, and artefacts that are needed to conceive, develop, deploy, and maintain a software product. From this definition, it emerges that the core conceptual elements that are necessary to define a process are: guidelines, roles, tools, artefacts, and finally the breakdown structure to define the work to be executed.

In the literature, several PMLs that support those concepts are available [28, 1, 4]. SPEM 2.0 (Software & Systems Process Engineering Meta-Model 2.0) [18] is one of them and since it has appealing features in terms of standardization, reuse, tool-support, etc. (as surveyed in [4]) as well as in terms of an active community working towards its enhancement [23], it answers our expectations. SPEM 2.0 is the OMG's standard for systems and software process modelling and it is defined as a MOF-based meta-model. SPEM 2.0 meta-model is composed of seven main packages, which are briefly recalled in what follows.

The *Core* package defines concepts allowing for the foundation of the other packages. The *Method Content* package defines concepts allowing for the specification of a knowledge base of reusable process elements, as partially depicted in Fig. 1. The *Process Structure* package defines concepts allowing for the representation of process models composed of inter-related activities, roles (actual performers, called RoleUse), work-products (actual data, called WorkProductUse). The

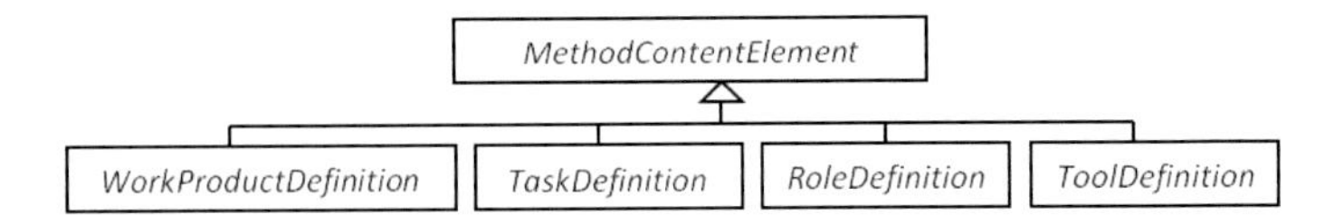

Fig. 1 Taxonomy of MethodContentElement

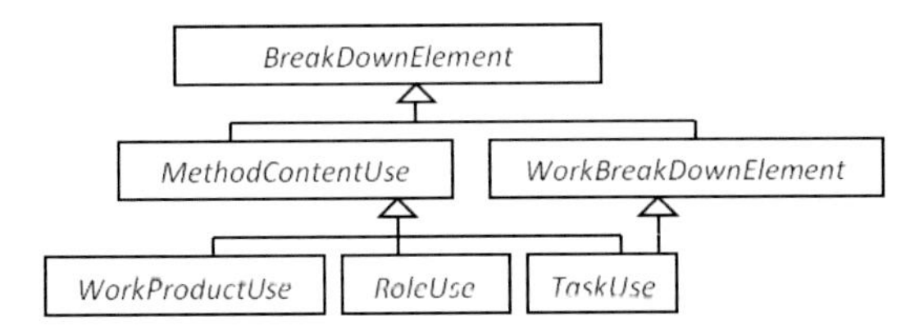

Fig. 2 Taxonomy of BreakDownElement

Managed Content package defines concepts such as Guidance allowing for the addition of descriptions in natural language to be attached to other process elements defined in other packages. The *Process with Method* package defines concepts such as Method Content Use elements for the integration of processes defined by using the concepts available in Process Structure with the instances of concepts available in Method Content. Fig. 2 depicts a sub-set of these concepts. The *Method Plugin* package defines mechanisms allowing for the reuse and management of method content and processes. The *Process Behaviour* package defines mechanisms and concepts (i.e. proxy meta-classes) allowing process elements to be linked to external models (e.g. UML 2.0 Activity Diagrams) for behavioural specification.

For a subset of the concepts that belong to the meta-model, graphical modelling elements (icons) are at disposal. In Table 1, we recall those elements for which we propose a safety-oriented decoration in Sect. 3.1. Tasks, roles and work-products (shortened as WP in Table 1) are commonly considered as process core elements [4]. Beside these elements, since we are focusing our work on safety-oriented processes, tools and guidances are also considered being core elements.

Table 1 Icons denoting Method Content (MC) and Method Content Use (MCU) elements

MC Elements					MCU Elements		
Task Definition	**Role Definition**	**Tool**	**WP Definition**	**Guidance**	**TaskUse**	**RoleUse**	**WPUse**

2.4 XPDL 2.2

XML Process Definition Language 2.2 (XPDL 2.2) [27] is the current version of the XPDL specification recently issued by the Workflow Management Coalition (WfMC). XPDL 2.2 is a standard that defines an interchange format for process models. XPDL 2.2 syntax is specified by an XML schema. A process description in XPDL 2.2 is an XML document, which includes core modelling elements such as: Process, Activity, Transition, Participant, DataObject, and DataAssociation, and Application, Annotation. Below we recall the informal semantics of these elements and in Fig. 3 we provide the cut of XPDL 2.2 meta-model that includes them.

Annotation represents a piece of textual information that can be attached to activities or lanes. *Application* is used to specify the applications/tools invoked by the process. *Activity* represents a logical, self-contained unit of work. *Transition* represents the sequence-flow that relates two activities. Each individual transition has three elementary properties: the from-activity, the to-activity and the condition under which the transition is made. Activities and transitions are the key elements that form the *process*, which consists of an oriented graph composed of nodes (activities) and edges (transitions). *Participant* is used to specify the participants in the workflow, i.e., the entities that can execute work. There are six types of participant: ResourceSet, Resource, Role, OrganizationalUnit, Human, and System. *Pool* acts as the container for activities and transitions. *Lane* represents a performer information at the activity level. A lane is used to subdivide a pool and thus model who does what. *DataObject* (and related concepts such as DataInputs and DataOutputs) belongs to the set of new concepts, which have been introduced in XPDL 2.2. DataObject represents the primary construct for modelling data within a process and opens the possibility to model global as well as local variables and to model that data objects are transformed during the process execution [25]. DataInputs and DataOutputs are used to specify the I/O parameters needed by e.g. activities. *DataAssociation* represents a mapping between a data object on one end and a data input or data output on the other end. XPDL also offers extensibility mechanisms supported by the *extended attribute* modelling element. This element can be used to customize all the other XPDL 2.2 concepts.

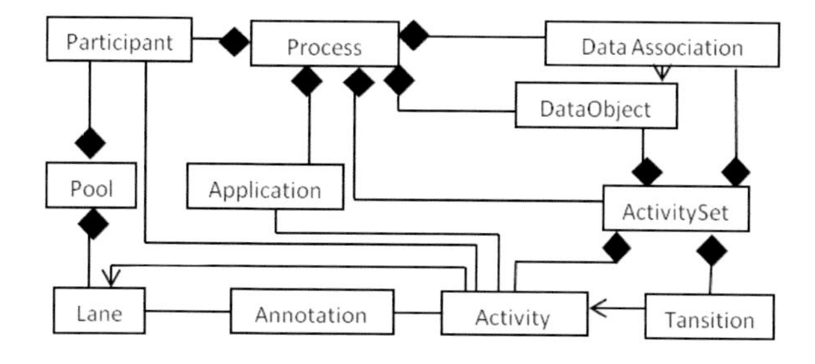

Fig. 3 Cut of XPDL 2.2 meta-model

Currently, several commercial and open-source tools (e.g. process execution / monitoring / simulation engines) take XPDL descriptions in input and it is likely that soon new releases will be provided to support XPDL 2.2. This is why in Sect. 3 we propose a mapping onto XPDL 2.2 and not onto older versions. Moreover, we select XPDL 2.2 and not one specific execution language (e.g. Business Process Execution Language - BPEL) because by focusing on the exchangeability we can take advantage of the existing and various engines.

3 S-TunExSPEM

As discussed in the previous sections, development processes defined within safety standards exhibit safety-related concepts, which should be better supported by PMLs in order to allow process engineers and assessors to better communicate and easily identify process-based evidence. Thus, in this section we introduce S-TunExSPEM, the SPEM 2.0 extension aimed at supporting the modelling as well as the exchange of safety-oriented processes. In particular, in Sect. 3.1 we focus on the modelling aspect and in Sect. 3.2 we focus on the exchangeability aspect.

3.1 Modelling Safety-Oriented Information

In this subsection, we focus on one aspect of our SPEM 2.0 extension: its safety-tunability (recalled in the first part of its name *S-Tune*). Our extension involves mainly four SPEM 2.0 packages, namely Method Content, Process with Method, Managed Content and Process Structure.

To provide safety-tunability, we add an attribute to the Activity meta-class to allow process engineers to set the safety level. We only consider four levels since in case of negligible (e.g. no effect, level E in D0-178B) consequences related to the hazards, no specific safety-related process elements are needed. Moreover, we extend each meta-class pertaining to the definition of the core process elements (namely, RoleDefinition/etc. as depicted in Fig. 1) with a corresponding safety-related meta-class (SafetyRole/etc.). Similarly to what proposed for the core process elements-related meta-classes, we extend the Method Content Use-related meta-classes (recalled in Fig. 2) with corresponding safety-related meta-classes, as partially depicted in Fig. 4 (e.g. SafetyWorkProductUse, SafetyRoleUse, etc.). Finally, the extension of the WorkSequence meta-class permits process engineers to highlight safety-related flows within the process.

Whenever DO-178B provides information to further classify the core process elements, we add an attribute to the corresponding meta-classes to allow the kind to be set. For sake of clarity, in what follows we provide some examples. According to DO-178B, a safety activity (task) can be further characterized by setting its kind (check, review, or audit). Thus, as shown in Fig. 4, we add an attribute called *S-ActivityKind* to the *SafetyActivity* class and an appropriate enumeration to allow the kind to be set. As seen in Sect. 2.2, workproducts that flow through the tasks belong to different kinds (Plans e.g. Software Development Plan, Standards e.g Software

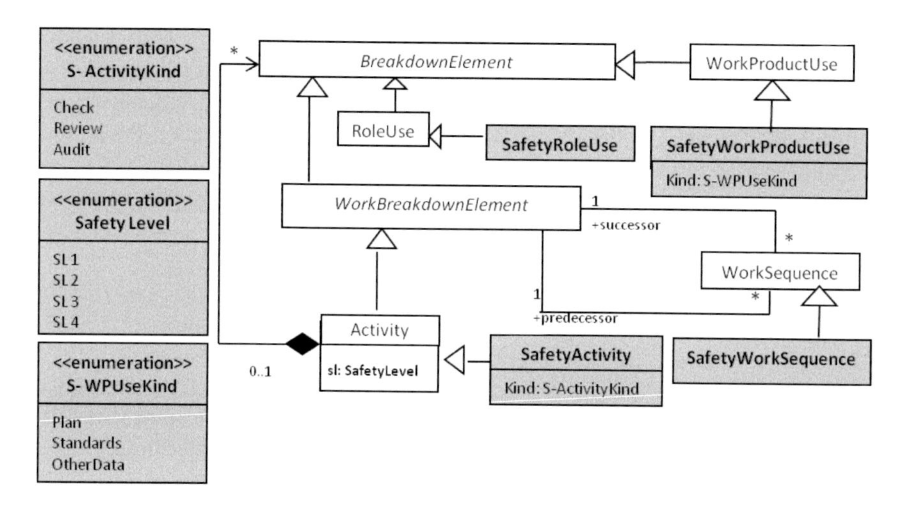

Fig. 4 Cut of S-TunExSPEM meta-model

Design Standards, or other software life-cycle data e.g. code). Thus, an attribute is added to the SafetyWorkProduct meta-class and an additional enumeration is available to allow the kind to be set (Plan, Standards, OtherData). This characterization is possible also for Method Content Use-related meta-classes, as shown in Fig. 4 for SafetyWorkProductUse. DO-178B also allows guidances to be further characterized (namely, checklists to guide for example reviews, guidelines and additional supporting material). Thus, also in this case, even if not shown in Fig. 4, we add an attribute to the SafetyGuidance meta-class and an enumeration.

To the meta-classes, we associate intuitive icons. Table 2 shows some of the S-TunExSPEM icons to be used to model safety-related tasks, roles, tools, workproducts and guidelines. Except for the Safety Work Sequence, which is represented as a yellow/black line, the remaining elements are obtained by adding a safety hat to the original Method Content SPEM 2.0 icons presented in Table 1. Similarly, a safety hat is added for the Method Content Use SPEM 2.0 icons. According to the safety level, a different colour for the hat can be used (i.e. red for the most critical safety level, followed by orange, yellow and bitter lemon). In case of sub-processes related to non-safety functions, no hat is needed.

Table 2 Graphical core elements of S-TunExSPEM

Task Definition	Role Definition	Tool Definition	Work Product Definition	Guidance
Safety Work Sequence				

3.2 Exchangeability of Safety-Related Processes

In this subsection, we focus on the other aspect of our SPEM 2.0 extension: its exchangeability (recalled in the second part of its name *Ex*). In particular, we present the mapping between some S-TunExSPEM concepts and corresponding XPDL 2.2 concepts. We focus our attention on the safety-related concepts. The interested reader may find details concerning the entire mapping as well as a pseudo-code version of the transformation algorithm in [20]. The aim of this mapping is to support exchangeability of process models and thus enable the exploitation of engines (available off the shelf) for execution, simulation, monitoring purposes.

Table 3 shows our rather self-explanatory mapping which further develops what was presented in [10] to take into consideration the beneficial changes (introduced in XPDL 2.2), which allow for a better semantic mapping. As mentioned in Sect. 2.4, XPDL 2.2 provides modelling elements for the data/artefacts that flow within a process, thus instead of mapping a work-product onto an extended attribute as authors did in [10], we are able to map a work-product onto a closer semantic element. Similarly, we map the concept of guidance onto the concept of textual annotation. Moreover, we also preserve the distinction between RoleDefinition and RoleUse, by mapping these elements onto more appropriate XPDL 2.2 elements. We indeed map the reusable method content element role onto the concept of participant and we map the process-specific task-role (method content use element) onto the concept of lane. Then, to model the safety concern, we make an extensive usage of the extensibility mechanisms of XPDL.

Table 3 Concepts mapping

S-TunExSPEM	XPDL 2.2
SafetyRoleDefinition	Participant +extended attribute
SafetyTaskUse	Activity +extended attribute
SafetyWorkProductUse	DataObject+ extended attribute
SafetyRoleUse	Lane in a pool + extended attribute
SafetyGuidance	Annotation +extended attribute
SafetyTool	Application+extended attribute
SafetyWorkSequence	Transition + extended attribute. Remark: from-activity or to-activity must be an activity representing a SafetyTaskUse

4 Case Study

In this section, we show the usage of S-TunExSPEM by modelling the software development process defined in DO-178B, which was briefly recalled in natural language in Sect. 2.2. The purpose is not to provide a detailed model but to provide evidence with respect to the richer expressiveness of the language as well as its potential in terms of exchangeability. For sake of clarity, it must be highlighted that S-TunExSPEM only aims at offering usable and expressive modelling capabilities targeting safety-oriented processes. Its usage should allow process engineers to model safety concerns in a more straightforward way and to communicate with safety assessors more easily. S-TunExSPEM does not contribute to safer code directly. If the process mandated by the standard contributes to safer code and if this process is properly understood, S-TunExSPEM may help in spreading and formalizing its understanding as well as graphically recalling what should be done.

Fig. 5 shows the design phase modelled by using S-TunExSPEM. From the figure it is straightforward to grasp that this phase is dealing with some design decisions related to some safety concerns of major (yellow hat) relevance. Moreover, in case of need, it is straightforward to detect the roles that are responsible of safety related decisions. Hanna is the only human being in charge of the design. Hanna however has all the skills that are needed since she acts as safety expert as well as designer. Hanna is in charge of: checking that all the work-products in input are available, following the guidances and using the appropriate tools to provide all the work-products in output. It is also straightforward to identify safety-related work-products and thus be aware about the deliverables that are involved in the certification process.

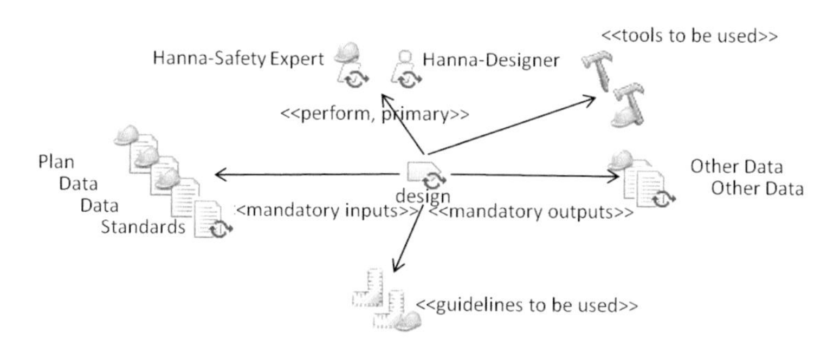

Fig. 5 DO-178B design phase in S-TunExSPEM

Fig. 6 shows the dynamics of the entire software development process. For space reasons, however, in Fig. 6 we do not provide in S-TunExSPEM all the characteristics of the phases as done textually in Sect. 2.2 and graphically in Fig. 5 for the design phase. For the same reason, we do not show the usage of the safety-oriented flow that takes place whenever an output from the traceability check tasks is available as a feedback to the preceding task. Fig. 6 is simply aimed at showing that S-TunExSPEM permits process engineers to intuitively separate safety concerns from functional concerns.

In what follows, we provide the essential XPDL 2.2 snippets corresponding to some S-TunExSPEM process elements, depicted in Fig. 5. We do not provide the entire code but only significant parts needed to highlight our mapping related to safety concerns and our timely and pertinent exploitation of the current release of XPDL. In bold, we highlight the first-class entities for readability purposes.

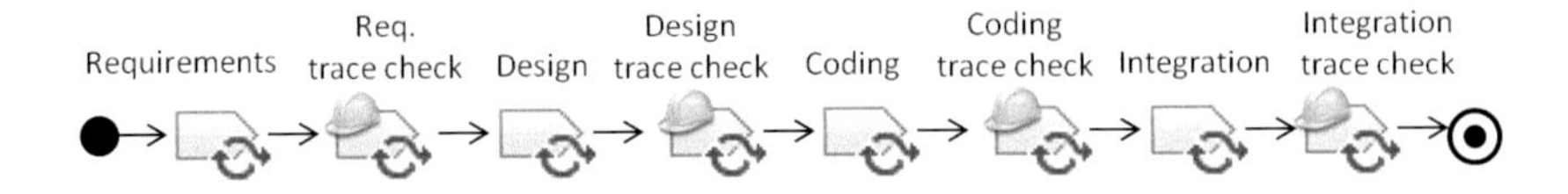

Fig. 6 DO-178B software development process in S-TunExSPEM

```
<!-Input data of Activity (TaskUse) "Design" -->
<xpdl:Artifact ArtifactType="DataObject" FormalParameterRef="IN"
Id="DO1" Name="SW development Plan"></xpdl:Artifact>
<xpdl:Artifact ArtifactType="DataObject" FormalParameterRef="IN"
Id="DO2" Name="SW Requirements Data (functional)"></xpdl:Artifact>
<xpdl:Artifact ArtifactType="DataObject" FormalParameterRef="IN"
Id="DO3" Name="SW Requirements Data (Safety-related)"> </xpdl:Artifact>
<xpdl:Artifact ArtifactType="DataObject" FormalParameterRef="IN"
Id="DO4" Name="SW Design Standards"> </xpdl:Artifact>

<!-Output data of Activity (TaskUse) "Design" -->
<xpdl:Artifact ArtifactType="DataObject" FormalParameterRef="OUT"
Id="DO5" Name="SW design Description (Functional)"> </xpdl:Artifact>
<xpdl:Artifact ArtifactType="DataObject" FormalParameterRef="OUT"
Id="DO6" Name="SW design Description (Safety-related)"></xpdl:Artifact>
```

As the above snippets show, work-products involved in the design phase are defined as Artifacts of type DataObject. Moreover, if artefacts are provided in input (output, respectively), the attribute FormalParameterRef must be set to ”IN” (”OUT” respectively).

```
<!-Guidance attached to Activity (TaskUse) "Design" -->
<xpdl:Artifact ArtifactType="Annotation" Id="AN1" Name="Safety guidance">
</xpdl:Artifact>
<xpdl:Artifact ArtifactType="Annotation" Id="AN2" Name="guidance">
</xpdl:Artifact>
```

As the above snippets show, guidances involved in the design phase are defined as Artifacts of type Annotation.

```
<!-Participants  of Activity (TaskUse) "Design" -->
<xpdl:Participants>
<xpdl:Participant Id="RO1" Name="Designer"><xpdl:ParticipantType Type=""/)
<xpdl:Description>In charge of design decision related to functional
requirements </xpdl:Description></xpdl:Participant>
<xpdl:Participant Id="RO2" Name="Safety Expert"><xpdl:ParticipantType Type=""/)
<xpdl:Description>In charge of design decision
related to non-functional (safety) requirements
</xpdl:Description> </xpdl:Participant></xpdl:Participants>
```

As the above snippets show, the involved roles are defined as Participants.

```
<!-Pool and Lane containing TaskUse "Design" -->
<xpdl:Pools>
<xpdl:Pool BoundaryVisible="true" Id="RO1" MainPool="true"
Name="PARTICIPANT NAME" Orientation="HORIZONTAL" Process="SW Life Cycle">
<xpdl:Lanes>
<xpdl:Lane Id="" Name="Hanna">
<xpdl:Performers>
<xpdl:Performer>RO1</xpdl:Performer><xpdl:Performer>RO2</xpdl:Performer>
</xpdl:Performers>
</xpdl:Lane></xpdl:Lanes></xpdl:Pool></xpdl:Pools>
```

As the snippets concerning the pool specification states, Hanna, consistently with what modelled in Fig. 5, is the only human being in charge of the design. Hanna is the actual role responsible of acting as designer as well as safety expert.

```
<!-extended attributes for the safety-oriented customization  -->
<xpdl:ExtendedAttributes>
<xpdl:ExtendedAttribute Name="Safety Role" Value="Safety Expert">
<xpdl:ExtendedAttribute Name="Safety Data object" Value="SW Development Plan">
<xpdl:ExtendedAttribute Name="Safety Data object"
Value= "SW Requirements Data (Safety-related)">
<xpdl:ExtendedAttribute Name="Safety Data object" Value="SW Design Standards">
<xpdl:ExtendedAttribute Name="Safety Data object"
Value= "SW Design Description (Safety-related)">
<xpdl:ExtendedAttribute Name="Safety Guidelines" Value="Safety Guidance">
</xpdl:ExtendedAttributes>
```

As the above snippets show, extended attributes customize/specialize the XPDL 2.2 concepts towards safety. As presented in Table 3, an extended attribute is used to customize each safety-related process element.

5 Related Work

In this section, we discuss those related works that contribute to either provide modelling capabilities for safety-oriented processes or transform process models into other models for execution purposes. To support the modelling of safety-oriented processes, a new meta-model, called Repository-Centric Process Metamodel is provided in [13, 29, 14]. Besides, meta-classes aimed at representing generic process concepts (e.g. activity), RCPM includes one safety related meta-class (check point), which specializes a generic meta-class. RCPM also includes one meta-class to represent safety-related relationships (safety relationship). Finally, a safety level can be specified for a process. Thus, in principle, safety process engineers are enabled to model safety-related activities and how these activities are related from a safety-related flow point of view.

Similarly to what is proposed in [13, 29, 14], we also provide a meta-class to represent safety-related activities as well as a meta-class to represent safety-related flows. However, our work highly differs from [13, 29, 14] since we do not introduce a new meta-model but propose to extend an existing one. Moreover, we broaden our focus on other conceptual elements that are crucial in the context of safety critical systems development. The concept of role, for instance is of paramount importance to stress that every piece of information produced during the development process requires the appropriate set of skills. Similarly, the way in which an activity is performed is of paramount importance. Thus guidelines represent first-class modelling elements. Finally, we also propose a rather intuitive safety-oriented concrete syntax.

Another related work which was aimed at modelling DO178B processes by using OpenUp is presented in [6]. This work is of interest for its pioneering intention of exploiting existing process modelling capabilities to document safety-related processes. Authors conclude that customization of the existing capabilities is needed.

When quality attributes (e.g. safety) are crucial for the systems development, it becomes relevant to model the techniques that target that attribute. In [7], authors investigate how safety analysis techniques could be modelled in SPEM. They explore two alternatives: the usage of *step* eventually combined with *guidance* or the usage of *task* eventually combined with *guidance*. In our case, we also model the techniques but we only use *guidance* since we model the remaining and conceptually different information onto other modelling elements.

Concerning process models interchange or simulation/execution/monitoring, several works exist. Some of these works have investigated approaches for mapping process models onto interchangeable models others have provided SPEM 2.0 extensions to enhance its support for executability.

In [10], authors provide a mapping as well as a transformation algorithm to transform SPEM1.0 models into XPDL (draft 1.0) models. As a running example they use a review process. As mentioned in Sect. 3.2, our approach borrows from this one and goes beyond it since we transform S-TunExSPEM models into XPDL 2.2 models and thus we provide support for safety concerns and a more suitable semantic mapping. In [3], authors make a critical analysis of SPEM 2.0 support for executability and then propose a SPEM 2.0 extension, called xSPEM. Their extension includes a set of concepts and behavioural semantics aimed at enhancing SPEM 2.0 executability. Similarly, in [8, 9], authors propose a tool-supported SPEM 2.0 extension, called eSPEM to enhance the support for executability. eSPEM is defined as CMOF meta-model and is based on both SPEM 2.0 and UML Superstructure. Authors replace the Process Behaviour package recalled in Sect. 2.3 with a new one defining fine-grain concepts for behaviour specification.

To provide our contribution, we have focused our attention on the textual descriptions of safety-related processes available in safety standards. We have not yet tried to model real processes and thus the mechanisms for behavioural specification, provided within the SPEM 2.0 Process Behaviour package, were enough for our purposes. So, we have not integrated the above extensions within our proposal.

6 Conclusion and Future Work

To ensure the safety of safety-critical systems, compulsory as well as advisory safety standards have been issued. Some of these standards define (prescriptive) safety-oriented processes. Modelling processes in compliance with the standards is relevant to provide process-based evidence for certification purposes. To support safety-oriented process engineers in these activities, in this paper we have proposed a PML, called S-TunExSPEM, obtained by extending SPEM 2.0 with safety-specific constructs extracted by examing safety standards (mainly DO-178B). Moreover, besides offering modelling capabilities for safety-related concepts, S-TunExSPEM provides the first tile to pave the road towards process models exchangeability aimed at exploiting existing simulation, monitoring and execution engines.

In the immediate future, first of all, we aim at validating the effectiveness of our proposal in supporting process modelling activities in industrial settings. We are currently in contact with some military as well as civil organizations responsible for software development of avionics software. Then, we aim at investigating model transformation approaches to automatize the generation of XPDL 2.2 models from S-TunExSPEM models. In a long-term future, we plan to provide a tool-chain support for modelling and monitoring / executing / etc. safety-processes.

Finally, since safety-oriented processes can be considered as a process line [12], safety-related process elements of S-TunExSPEM could be considered as variability elements and divided into commonalities, partial commonalities, and variabilities either by reusing the current SPEM 2.0 support for variability modelling or by adopting the in-progress SPEM 2.0 extension for process lines, called vSPEM [17]. The intention would be to contribute to pushing towards a SPEM 3.0 version allowing for richer modelling support as well as exchangeability/execution targeting safety.

Acknowledgements. This work has been partially supported by the European Project ARTEMIS SafeCer [2] and by the Swedish SSF SYNOPSIS project [26].

References

1. Acuña, S.T., Ferré, X.: Software Process Modelling. In: Proceedings of the World Multiconference on Systemics, Cybernetics and Informatics, Orlando, FL, pp. 237–242 (2001)
2. ARTEMIS-JU-269265: SafeCer-Safety Certification of Software-Intensive Systems with Reusable Components (2013), http://www.safecer.eu/
3. Bendraou, R., Combemale, B., Cregut, X., Gervais, M.P.: Definition of an Executable SPEM 2.0. In: Proceedings of the 14th Asia-Pacific Software Engineering Conference, APSEC, Nagoya, Japan, pp. 390–397 (2007)
4. Bendraou, R., Jezequel, J., Gervais, M.P., Blanc, X.: A Comparison of Six UML-Based Languages for Software Process Modeling. IEEE Transactions Software Engineering 36, 662–675 (2010)
5. Berk, R.H.: An Analysis of Current Guidance in Certification of Airborne Software. Master's thesis, Massachusetts Institute of Technology, Cambridge, USA (2009)
6. Bertrand, C., Fuhrman, C.P.: Towards Defining Software Development Processes in DO-178B with Openup. In: Proceedings of 21st IEEE Canadian Conference on Electrical and Computer Engineering, CCECE, Niagara Falls, Ontario, Canada, pp. 851–854 (2008)
7. Chiam, Y.K., Staples, M., Zhu, L.: Representation of Quality Attribute Techniques Using SPEM and EPF Composer. In: European Software Process Improvement, EuroSPI, Spain. Springer (2009)
8. Ellner, R., Al-Hilank, S., Drexler, J., Jung, M., Kips, D., Philippsen, M.: eSPEM – A SPEM extension for enactable behavior modeling. In: Kühne, T., Selic, B., Gervais, M.-P., Terrier, F. (eds.) ECMFA 2010. LNCS, vol. 6138, pp. 116–131. Springer, Heidelberg (2010)
9. Ellner, R., Al-Hilank, S., Jung, M., Kips, D., Philippsen, M.: Integrated Tool Chain for Meta-model-based Process Modelling and Execution. In: Proceedings of First Workshop on Academics Modeling with Eclipse, ACME, Lyngby, Denmark (2012)
10. Feng, Y., Mingshu, L., Zhigang, W.: SPEM2XPDL-Towards SPEM Model Enactment. In: Software Engineering. Front. Comput. Sci. China, pp. 1–11. Higher Education Press, Bejing (2008); Co-published with Springer-Verlag GmbH
11. Fuggetta, A.: Software Process: A Roadmap. In: Proceedings of the International Conference on Software Engineering, ICSE, New York, USA, pp. 25–34 (2000)

12. Gallina, B., Sljivo, I., Jaradat, O.: Towards a Safety-oriented Process Line for Enabling Reuse in Safety Critical Systems Development and Certification. In: Post-proceedings of the 35th IEEE Software Engineering Workshop, SEW-35, Greece (2012)
13. Hamid, B., Geisel, J., Ziani, A., Gonzalez, D.: Safety lifecycle development process modeling for embedded systems - example of railway domain. In: Avgeriou, P. (ed.) SERENE 2012. LNCS, vol. 7527, pp. 63–75. Springer, Heidelberg (2012)
14. Hamid, B., Zhang, Y., Geisel, J., Gonzalez, D.: First Experiment on Modeling Safety LifeCycle Process in Railway Systems. International Journal of Dependable and Trustworthy Information Systems 2, 17–39 (2011)
15. Health and Safety Executive (HSE): Out of Control. Why Control Systems Go Wrong and How to Prevent Failure (2003)
16. Jackson, D., Thomas, M., Limmet, L.I.: Software for Dependable Systems: Sufficient Evidence? National Academy Press, Washington DC (2007)
17. Martínez-Ruiz, T., García, F., Piattini, M., Münch, J.: Modeling Software Process Variability: An Empirical Study. IET Software 5, 172–187 (2011)
18. Object Management Group: Software & Systems Process Engineering Meta-Model (SPEM), v2.0. Full Specification formal/08-04-01 (2008)
19. Panesar-Walawege, R.K., Sabetzadeh, M., Briand, L.: Using Model-Driven Engineering for Managing Safety Evidence: Challenges, Vision and Experience. In: Proceedings of the 1st International Workshop on Software Certification, WoSoCER, Hiroshima, Japan, pp. 7–12 (2011)
20. Pitchai, K.R.: An Executable Meta-model for Safety-oriented Software and Systems Development Processes within the Avionics Domain in Compliance with RTCA DO-178B. Master's thesis, Mälardalen University, School of Innovation, Design and Engineering, Sweden (2013)
21. Redmill, F.: Safety Integrity Levels - Theory and Problems. Lessons in System Safety. In: Proceedings of the Eighth Safety-critical Systems Symposium, Southampton (2000)
22. RTCA Inc.: Software Considerations in Airborne Systems and Equipment Certification, RTCA DO-178B (EUROCAE ED-12B), Washington DC (1992)
23. Ruiz-Rube, I., Dodero, J.M., Palomo-Duarte, M., Ruiz, M., Gawn, D.: Uses and Applications of SPEM Process Models. A Systematic Mapping Study. Journal of Software Maintenance and Evolution: Research and Practice, 1–32 (2012)
24. Rushby, J.: New Challenges in Certification for Aircraft Software. In: Proceedings of the Ninth ACM International Conference on Embedded Software, EMSOFT, New York, USA, pp. 211–218 (2011)
25. Shapiro, R.M.: XPDL 2.2: Incorporating BPMN2.0 Process Modeling Extensions. Extracted from BPM and Workflow Handbook, Future Strategies (2010)
26. SYNOPSIS-SSF-RIT10-0070: Safety Analysis for Predictable Software Intensive Systems. Swedish Foundation for Strategic Research
27. Workflow Management Coalition: Workflow Management Coalition Workflow Standard- Process Definition Interface - XML Process Definition Language, WfMC-TC-1025, v2.2 (2012)
28. Zamli, K.Z., Lee, P.A.: Taxonomy of Process Modeling Languages. In: Proceedings of the ACS/IEEE International Conference on Computer Systems and Applications, AICCSA, Beirut, Lebanon, pp. 435–437 (2001)
29. Zhang, Y., Hamid, B., Gouteux, D.: A metamodel for representing safety lifecycle development process. In: Proceedings of the Sixth International Conference on Software Engineering Advances (ICSEA), pp. 550–556. IEEE Computer Society Press, Barcelona (2011)

Solving SMT Problems with a Costly Decision Procedure by Finding Minimum Satisfying Assignments of Boolean Formulas

Martin Babka, Tomáš Balyo, and Jaroslav Keznikl

Abstract. An SMT-solving procedure can be implemented by using a SAT solver to find a satisfying assignment of the propositional skeleton of the predicate formula and then deciding the feasibility of the assignment using a particular decision procedure. The complexity of the decision procedure depends on the size of the assignment. In case that the runtime of the solving is dominated by the decision procedure it is convenient to find short satisfying assignments in the SAT solving phase. Unfortunately most of the modern state-of-the-art SAT solvers always output a complete assignment of variables for satisfiable formulas even if they can be satisfied by assigning truth values to only a fraction of the variables. In this paper, we first describe an application in the code performance modeling domain, which requires SMT-solving with a costly decision procedure. Then we focus on the problem of finding minimum-size satisfying partial truth assignments. We describe and experimentally evaluate several methods how to solve this problem. These include reduction to partial maximum satisfiability – PMAXSAT, PMINSAT, pseudo-Boolean optimization and iterated SAT solving. We examine the methods experimentally on existing benchmark formulas as well as on a new benchmark set based on the performance modeling scenario.

Martin Babka · Tomáš Balyo
Department of Theoretical Computer Science and Mathematical Logic, Faculty of Mathematics and Physics, Charles University, Malostranské nám. 2/25, 118 00 Prague, Czech Republic
e-mail: {babka,balyo}@ktiml.mff.cuni.cz

Jaroslav Keznikl
Department of Distributed and Dependable Systems, Faculty of Mathematics and Physics, Charles University, Malostranské nám. 2/25, 118 00 Prague, Czech Republic
e-mail: keznikl@d3s.mff.cuni.cz

Jaroslav Keznikl
Institute of Computer Science, Academy of Sciences of the Czech Republic, Pod Vodárenskou veží 2, 182 07 Prague, Czech Republic
e-mail: keznikl@cs.cas.cz

R. Lee (Ed.): *SERA*, SCI 496, pp. 231–246.
DOI: 10.1007/978-3-319-00948-3_15

1 Introduction

Boolean satisfiability (SAT) is one of the most important and most studied problems of computer science. It is important in theoretical computer science, it was the first NP-complete problem [12], as well as in practical applications. SAT has a lot of successful applications in many fields such as A.I. planning [19], automated reasoning [26] and hardware verification [29]. This is possible because of the high practical efficiency of modern SAT solvers.

An important extension of the SAT problem is the SMT (Sat Modulo Theories) problem [4, 24]. SMT is a combination of SAT and some theories, for example arithmetic, arrays, or uninterpreted functions. Like SAT, SMT has numerous applications for example bounded model checking [1] or performance modeling of software [10, 11]. SMT solving can be done by using a SAT solver to evaluate the propositional skeleton of the SMT formula and then checking the result of the SAT solver using the theory evaluation procedures. It might be the case, that the evaluation of the theory is very time consuming and therefore it is beneficial to try to find minimum satisfying assignments in the SAT solving phase.

Unfortunately, most of the current state-of-the-art SAT solvers always output a complete satisfying truth assignment even for formulas that can be satisfied by small partial truth assignments. It is because these solvers implement the conflict-driven clause learning (CDCL) DPLL algorithm [7] in a very efficient manner. The search for a satisfying assignment in these implementations is continued until all variables are assigned or an empty clause is learned. Therefore the output of the solvers is a complete truth assignment for satisfiable instances.

In this paper we first give a brief description and example of the challenge of solving SMT problems with a theory that has a very costly decision procedure. We show how it can be addressed using a special SAT solver, that gives minimum partial satisfying assignments. The rest of the paper is then dedicated to finding such assignments.

In the theory of Boolean functions the problem of finding a partial satisfying truth assignment with the minimal number of assigned variables is called the shortest implicant problem. The decision version of this problem has been shown to be Σ_2^P – complete for general formulas [28]. However, for CNF formulas, it is in NP (see below), thus, theoretically, it is not harder than SAT.

This problem is also referred to as finding minimum-size implicants. It is sometimes confused with the problem of finding minimal-size implicants (implicants that cannot be shortened, i.e., prime implicants). A minimum-size implicant is always a minimal-size (prime) implicant but not vice versa. The problem of finding prime implicants is well studied and there are many papers devoted to this topic, see e.g. [25]. On the other hand, methods for finding minimum-size implicants are often hidden inside papers dealing with other problems, where they are only briefly mentioned as a possible application. There are however some papers dealing directly with minimum-size implicants such as [23] and [22].

Our goal is to give an overview of several methods for the minimum satisfying assignment problem based on reducing this problem to other well known problems.

Two of the described reductions (PMaxSAT and PMinSAT) have not been described elsewhere. The others are mentioned in the literature. For more information please see Section Related Work. In the paper we also do experimental comparison of the described methods using relevant benchmark problems and state-of-the-art solvers.

2 Motivation

2.1 SMT *Solving with a Costly Decision Procedure*

In general, the main motivation for short satisfying assignments is the case of an SMT-solving [4, 24] algorithm with a costly decision procedure. SMT-solving is a technique for finding satisfying assignments of predicate-logic formulas. The basic idea of one of the approaches to SMT-solving is to employ a SAT solver for finding a satisfying assignment of the propositional skeleton of a given predicate formula. Having such a satisfying assignment, a decision procedure (specific to the particular predicate logic) is employed in order to decide the feasibility of the assignment with respect to the predicates. If the assignment is not feasible, the SAT solver is (incrementally) asked for another satisfying skeleton assignment until the assignment is feasible or there are no undecided assignments left. As an aside, the state-of-the-art SMT solvers operate incrementally; i.e., they call the decision procedure already for partial skeleton assignments. Nevertheless, since this potentially increases the number of expensive decision procedure calls, we will consider the non-incremental case. Note, that the unsatisfiability of a propositional skeleton implies unsatisfiability of the associated predicate formula (the opposite does not hold). Additionally, the satisfiability of a predicate formula implies the satisfiability of its propositional skeleton.

A typical decision procedure of an SMT-solving algorithm is designed to work with the conjunctive fragment of the predicate logic (i.e., conjunctions of predicates and their negations). A formula in the conjunctive fragment can be easily obtained from a satisfying skeleton assignment. Therefore, while deciding feasibility of a skeleton assignment, it is necessary to evaluate some of the associated predicates; in the case of a feasible assignment all of them.

Taking into account a decision procedure where an evaluation of a predicate is a costly operation [11], it is beneficial to minimize the number of evaluated predicates while deciding feasibility of a skeleton assignment. However, this minimization has to be performed by the SAT solver by providing small satisfying assignments (as the decision procedure works with the conjunctive fragment and thus has to evaluate all the corresponding predicates).

2.2 Stochastic Performance Logic

To illustrate this problem, we describe the Stochastic Performance Logic (SPL) [10, 11], for which evaluating the predicates is a very time-consuming operation and which will thus greatly benefit from minimization of the satisfying skeleton assignments during SMT-solving. Specifically, it is a predicate logic designed for expressing assumptions about performance of code and is motivated by the challenges in the performance modeling domain. In particular, according to [11], it is beneficial to provide means for performance testing similar to functional unit-testing approaches – that is, being able to express performance-related developer assumptions or intended usage in code in a platform-independent way and test or verify them automatically.

The main goal of SPL is thus to capture performance conditions that should be met by software (expressing performance-related developer assumptions or intended usage) in a form of predicate formulas, semantics of which is platform-independent. Specifically, the approach of SPL is based on capturing performance conditions on a given function relatively to performance of a baseline function (rather than on absolute metrics); e.g., in case of an encryption function, the baseline can be the memory-copying function (i.e., no encryption). In practice, SPL formulas are inserted into code (e.g., as Java annotations) and automatically validated [18].

The semantics of the predicates expressing the relative performance is based on instrumentation and monitoring of the execution times of both the tested and baseline function and performing a statistical test in order to validate or invalidate the statistical hypothesis determined by the predicate. Therefore, the decision procedure in SPL has to perform (expensive) execution-time measurements and a statistical test in order to evaluate a single performance predicate. Thus, it is an extremely time-consuming operation.

To provide a clearer perspective on this issue, we present a brief summary of the SPL-solving algorithm (Fig. 1). Before going into detail, we first describe the notation. For a given SPL formula F, the *MakeSkeleton* function returns its propositional skeleton F_S. A_P is a partial assignment of F_S enforcing the results of the previous decision-procedure runs. The $ApplyAssignment(F,A)$ function returns formula F after applying the assignment A ; i.e., with all variables from A replaced by their assigned values. The *PartSAT* function returns for the given formula a satisfying assignment with only some variables assigned (i.e., a partial satisfying assignment). The tuple (var, val) denotes a variable and its value in an assignment. The $FilterAssigned$ function returns the assigned variables in the given assignment. *MeasureAndTest* is the very expensive decision procedure deciding validity of a single performance predicate associated with the given skeleton variable. Finally, m is the result of the procedure (i.e., true or false).

After the propositional skeleton F_S is created and the partial assignment A_P is initialized (lines 1-2), a partial satisfying truth assignment A_{temp} of F_S after applying A_P is obtained via the *PartSAT* function (line 3). If *PartSAT* indicates that F_S after applying A_P is unsatisfiable, the algorithm returns "false" (lines 4-6), because it implies that the original SPL formula is unsatisfiable with respect to measurements dictating A_P. Otherwise, the algorithm sequentially processes assignments of all assigned variables; i.e., those which were not yet checked by the decision procedure (line 7). Note that the order in which the variables are processed may depend on further optimization; e.g., the variable corresponding to the "cheapest to measure"

```
1: F_S ← MakeSkeleton(F)
2: A_P ← ∅
3: A_temp ← PartSAT(ApplyAssignment(F_S, A_P))
4: if A_temp = false then
5:    return false
6: end if
7: for all (var, val) ∈ FilterAssigned(A_temp) do
8:    m ← MeasureAndTest(var)
9:    A_P ← A_P ∪ {(var, m)}
10:   if val ≠ m then
11:      goto line 3
12:   end if
13: end for
14: return true
```

Fig. 1 SPL-solving algorithm

performance predicate will be processed first. For each assigned variable, it is necessary to call the decision procedure *MeasureAndTest* (line 8). The result of the decision procedure is added to A_P to be enforced in the subsequent *PartSAT* runs (line 9). If the stored result conforms to the current skeleton valuation A_{temp}, the next variable is processed. Otherwise (lines 10-12), A_{temp} is infeasible with respect to the measurements and a new skeleton valuation has to be obtained from *PartSAT*.

It is important to stress, that each call of the decision procedure *MeasureAndTest* for a typical performance predicate usually takes a non-trivial amount of time; i.e., hundreds of milliseconds. Thus, it is obvious that employing a *PartSAT* function that supports partial satisfying assignments with the minimum number of assigned variables (and thus minimizes the number of performance predicates to be evaluated) would significantly reduce the execution time of the whole SPL-solving algorithm.

The rest of the paper is devoted to the computation of the *PartSAT* function i.e. finding minimum satisfying truth assignments of Boolean formulas.

3 Preliminaries

A *Boolean variable* is variable with two possible values *True (1)* and *False (0)*. A *literal* of a Boolean variable x is either x or $\bar{x}$ (*positive* or *negative literal*). A *clause* is a disjunction (OR) of literals. A *conjunctive normal form (*CNF*) formula* is a conjunction (AND) of clauses. The number of variables of a formula will be denoted by n. A (partial) truth assignment ϕ of a formula F assigns a truth value to (some of) its variables. The assignment ϕ satisfies a positive(negative) literal if it assigns the value true (false) to its variable and ϕ satisfies a clause if it satisfies any of its literals. Finally, ϕ satisfies a CNF formula if it satisfies all of its clauses. A formula F is said to be satisfiable if there is a (partial) truth assignment ϕ that satisfies F. Such an assignment is called a *satisfying assignment*. The satisfiability problem (SAT) is to find a satisfying assignment of a given CNF formula. We will call ϕ_{min} a *minimum-size satisfying assignment* of a formula F if there is no other satisfying assignment ϕ of F, such that ϕ assigns truth values to fewer variables than ϕ_{min}.

A conjunction (AND) of literals is called a *term*. An *implicant* I of a formula F is a term, such that any truth assignment that satisfies I also satisfies F. I is a *shortest implicant* of a formula F if there is no other implicant I' of F such that I' contains fewer literals than I. I is called a prime implicant if there is no other implicant I' such that $I' \subset I$. The *shortest implicant problem* is to find the shortest implicant of a given formula. It is easy to observe that an implicant corresponds to a satisfying partial truth assignment of its formula and the shortest implicant corresponds to a minimum-size satisfying assignment.

4 Related Work

In the Boolean functions community the problem of shortest implicants is studied mostly in the context of Boolean function minimization [28], which is the problem of finding a minimal representation of Boolean functions [13]. The function is often given in form of a CNF formula and the desired output is an equivalent CNF or DNF formula of minimum size. In this context, finding shortest implicants is Σ_2^P – complete for general formulas, [28].

Some papers about enumerating prime implicants also describe methods for finding the shortest implicants. One such paper is by Bieganowky and Karatkevich, which presents a heuristic for Thelen‘s method [5]. Thelen‘s method is an algorithm for enumerating all prime implicants of a CNF formula. The proposed heuristic should lead to a minimal prime implicant, but it is not guaranteed to find an optimal solution.

In [25] a 0-1 programming scheme is used to encode the formula and additional constraints which allow selective enumeration. The constraint can, of course, be the length of the implicant, therefore this method is suitable for our purposes. Considering the efficiency of state-of-the-art pseudo-Boolean optimization (PBO) solvers, this approach appears to be a promising one.

In [23] and [22] the authors describe some methods based on integer linear programming (ILP) and binary decision diagrams (BDD).

Finally, in [8] there is a suggestion, that the problem could be solved by incremental SAT solving. This requires us to encode cardinality constraints into SAT. There are several available methods to do this, a survey of such methods is given in [2].

5 Solving the Shortest Implicant Problem

We shall start this section by describing a technique called *dual rail encoding* [9], which will be used in all of the following methods.

5.1 Dual Rail Encoding

The first step of the dual rail encoding of a CNF formula F is introducing new *dual rail variables* representing possible positive and negative assignments to the original variables of F.

Definition 1 (Dual rail variables). Let $X = \{x_1, \ldots, x_n\}$ be a set of Boolean variables. Then the Boolean variables $X_{DR} = \{px_1, nx_1, px_2, nx_2, \ldots, px_n, nx_n\}$ are the dual rail variables for X.

Let ϕ be a partial truth assignment of X. Then we define ϕ_{DR} as a truth assignment of X_{DR} so that $\phi_{DR}(px_i) = 1 \Leftrightarrow \phi(x_i) = 1$ and $\phi_{DR}(nx_i) = 1 \Leftrightarrow \phi(x_i) = 0$.

Notice that px_i and nx_i are both negative under ϕ_{DR} iff x_i is unassigned under ϕ. This implies that the number of assigned variables under ϕ is equal to the number of dual rail variables that are assigned 1 by ϕ_{DR}. Also observe that given ϕ_{DR} we can easily construct ϕ and vice versa.

Definition 2 (Dual rail encoding). Let F be a CNF SAT formula with variables $X = \{x_1, \ldots, x_n\}$ and clauses C. Let C_{DR} be the clauses obtained from the clauses C by replacing all occurrences of the literal x_i by px_i and literal $\overline{x}_i$ by nx_i for all $i \in \{1 \ldots n\}$. The dual rail encoding of F is a CNF formula

$$F_{DR} = C_{DR} \wedge \bigwedge_{i \in \{1 \ldots n\}} (\overline{px}_i \vee \overline{nx}_i)$$

Example 1 (Dual rail encoding). $(x_1 \vee \overline{x}_2) \wedge (x_3 \vee \overline{x}_1) \wedge (\overline{x}_2 \vee \overline{x}_3)$ would be encoded as $(px_1 \vee nx_2) \wedge (px_3 \vee nx_1) \wedge (nx_2 \vee nx_3) \wedge (\overline{px}_1 \vee \overline{nx}_1) \wedge (\overline{px}_2 \vee \overline{nx}_2) \wedge (\overline{px}_3 \vee \overline{nx}_3)$.

Lemma 1. *Let F be a* CNF *formula. Then ϕ is a satisfying assignment of F iff ϕ_{DR} is a satisfying assignments of F_{DR}.*

Proof. Let ϕ satisfy F. Let C be an arbitrary clause of F, then there is a literal x (or $\overline{x}$) in C that is satisfied under ϕ. It implies by definition that px (or nx) is $True$ under ϕ_{DR}. Hence the clause corresponding to C in F_{DR} is satisfied by px (or nx). The clauses $(\overline{nx}_i \vee \overline{px}_i)$ of F_{DR} are satisfied under any ϕ_{DR} since a Boolean variable cannot be assigned both values $True$ and $False$.

On the other hand, let ϕ_{DR} satisfy F_{DR}. The $(\overline{nx}_i \vee \overline{px}_i)$ clauses ensure that either px_i or nx_i is $False$ under ϕ_{DR} and thus ϕ is a valid partial truth assignment. Let C be an arbitrary clause in F. The corresponding clause to C in F_{DR} is satisfied by a literal px (or nx), surely is then C satisfied by x (or $\overline{x}$) under ϕ.

5.2 Solving via Pseudo-Boolean Optimization

In this section we describe a method for solving the shortest implicant problem by reducing it to the pseudo-Boolean optimization problem [7]. We start by its definition.

A *PB-constraint* is an inequality $C_0 \times x_0 + C_1 \times x_1 + \cdots + C_{k-1} \times x_{k-1} \geq C_k$, where C_i are integer coefficients and x_i are literals. The integer value of a Boolean variable is defined as 1 (0) if it is $True$ ($False$). Positive (negative) literals of a variable x are expressed as x ($(1-x)$) in the inequality. A (partial) truth assignment ϕ *satisfies a PB-constraint* if the inequality holds. An *objective function* is a sum $C_0 \times x_0 + C_1 \times x_1 + \cdots + C_l \times x_l$, where C_i are integer coefficients and x_i are literals. The pseudo-Boolean optimization problem is to find a satisfying assignment to a set of PB-constraints that minimizes a given objective function.

Now, we describe how a CNF formula F can be reduced into a PB optimization problem. For a clause $C = (l_1 \vee l_2 \vee \cdots \vee l_k)$ we define its PB-constraint $\mathrm{PB}(C) = (1 \times l_1 + 1 \times l_2 + \cdots + 1 \times l_k \geq 1)$. It is easy to see that a partial assignment ϕ satisfies the clause C iff it satisfies its PB-constraint $\mathrm{PB}(C)$. For a CNF formula F we denote its PB-constraints $\mathrm{PB}(F) = \{\mathrm{PB}(C) \mid C \in F\}$.

Example 2 (Reducing a clause into a PB-constraint). $(x_1 \vee \overline{x}_2 \vee x_3)$ would yield $1 \times x_1 + 1 \times (1 - x_2) + 1 \times x_3 \geq 1$.

For a given CNF formula F we encode the instance of the shortest implicant problem as the pseudo-Boolean optimization problem $\mathrm{PBO}(F)$ as follows. First we construct the dual rail encoding F_{DR} of F. Then we translate it into its PB-constraints $\mathrm{PB}(F_{DR})$. Finally, we define the objective function $\mathrm{O}(F_{DR})$ as $\mathrm{O}(F_{DR}) = \sum_{i=1}^{n}(1 \times px_i + 1 \times nx_i)$. Let us denote $\mathrm{PBO}(F) = (\mathrm{PB}(F_{DR}), \mathrm{O}(F_{DR}))$ the pseudo-Boolean optimization problem with the constraints $\mathrm{PB}(F_{DR})$ and the objective function $\mathrm{O}(F_{DR})$.

The optimal solution of $\mathrm{PBO}(F)$ is a truth assignment that satisfies all the constraints and minimizes the objective function. Now, we can use a PB solver to find an optimal solution of $\mathrm{PBO}(F)$ and from the optimal solution we can extract the shortest implicant in the following way.

Definition 3. For a truth assignment ψ of the dual rail variables we define the term I_ψ as

$$I_\psi = \bigwedge_{i:\ \psi(px_i)=1} x_i \wedge \bigwedge_{i:\ \psi(nx_i)=1} \overline{x}_i$$

Theorem 1. *Let F be a* CNF *formula and ψ the optimal solution of* $\mathrm{PBO}(F)$*. Then I_ψ is the shortest implicant of F.*

Proof. From Lemma 1 and the correspondence of satisfying assignments and implicants we get that I_ψ is an implicant of F. By contradiction we show that there is no shorter implicant. Let I' be a shorter implicant than I_ψ. Then I' defines a satisfying assignment ϕ of F. Realize the fact that the length of the implicant is exactly the number of the variables assigned by ϕ which equals the value of the objective function $\mathrm{O}(F_{DR})$ for ϕ_{DR}. Thus ϕ allows us to construct a better solution for $\mathrm{PBO}(F)$ than ψ. That is contradictory with ψ being an optimal solution of $\mathrm{PB}(F)$.

5.3 *Solving via Partial Maximum Satisfiability*

In this section we describe a reduction of shortest implicant problem into a partial maximum satisfiability (PMAXSAT) problem [7]. The reduction is again based on dual rail encoding, therefore it is very similar to the PB optimization approach. First we define the PMAXSAT problem.

A PMAXSAT *formula* is a tuple of two sets of clauses called *soft clauses* and *hard clauses*. A *solution* of a PMAXSAT problem is a truth assignment that satisfies all hard clauses and some soft clauses. An *optimal solution* of a PMAXSAT problem is a solution ϕ that there is no other solution that satisfies more soft clauses than ϕ.

To reduce shortest implicant problem given by a CNF formula F to a PMAXSAT problem PMAX(F) we first apply dual rail encoding on F. The clauses of F_{DR} are the hard clauses of PMAX(F). The soft clauses of PMAX(F) are defined as the unit clauses $\overline{px_i}$ and $\overline{nx_i}$ for each i. The shortest implicant from the optimal solution of PMAX(F) is extracted in the same way as in the case of PB optimization. A precise formulation and proof follows.

Theorem 2. *Let F be a* CNF *formula and ψ an optimal solution of* PMAX(F). *Then I_ψ is a shortest implicant of F.*

Proof. Let ψ be an optimal solution of PMAX(F). All hard clauses of PMAX(F) are satisfied under ψ and thus by Lemma 1 I_ψ is an implicant of F. The implicant I_ψ is also the shortest possible. The existence of a shorter one would allow a partial truth assignment ϕ of F such that ϕ_{DR} satisfies more soft clauses than ψ. Indeed, the number of unsatisfied soft clauses is equal to the number dual rail variables assigned the value True.

5.4 *Solving via Partial Minimum Satisfiability*

The partial minimum satisfiability (PMINSAT) problem [7] is analogous to the PMAXSAT problem with the only difference being, that the goal is to minimize the number of satisfied soft clauses. The reduction of the shortest implicant problem to PMINSAT is a straightforward modification of the PMAXSAT reduction. Instead of using the unit soft clauses $\overline{px_i}$ and $\overline{nx_i}$ for each i we use px_i and nx_i (e.g. the soft clauses of PMAX(F) are negated).

5.5 *Solving via Iterative* SAT *Solving*

The method described in this section is in a way similar to planning as satisfiability [19]. For a given CNF formula F we construct another CNF formula $G(F,k)$ which will be satisfiable iff F has an implicant of size k or shorter. We construct and test $G(F,k)$ for various k iteratively until we find the smallest k such that $G(F,k)$ is satisfiable. From the satisfying assignment of $G(F,k)$ we extract a shortest implicant of size k.

To construct $G(F,k)$ we again start by dual rail encoding F into F_{DR} and then we add a cardinality constraint $\leq_k (px_1, nx_1, \ldots, px_n, nx_n)$ meaning $(\sum_{i=1}^{n} px_i + nx_i) \leq k$. There are several methods of encoding cardinality constraints into SAT. A survey on these methods is given in [2]. Many of these encodings are polynomial (relative to n and k) in size and time required to construct them. There is even a linear encoding [14]. The resulting formula $G(F,k)$ is a conjunction of the cardinality constraint and the dual rail encoding of the original formula.

The reduction can be improved by adding a set of n new variables sx_i which encode if the variable x_i is assigned: $(px_i \vee nx_i) \rightarrow sx_i$. Then we encode the cardinality constraint over sx_i instead of px_i and nx_i. The improved reduction $G_s(F,k)$ is then

$$G_s(F,k) = F_{DR} \wedge \bigwedge_{i=1}^{n} [(\overline{px}_i \vee sx_i) \wedge (\overline{nx}_i \vee sx_i)] \\ \wedge \leq_k (sx_1, sx_2, \ldots, sx_n)$$

Why is this an improvement? In fact most encodings of cardinality constraints add a lot of new variables and clauses to the formula. Therefore it is good to use the cardinality constraint on fewer variables. Overall, $G_s(F,k)$ has fewer variables than $G(F,k)$ for almost every known cardinality encoding. Also, in our experiments $G_s(F,k)$ vastly outperformed $G(F,k)$ in terms of time required to solve them by a SAT solver. A theorem of this approach's validity follows.

Theorem 3. *Let F be a* CNF *formula, k be the smallest integer such that $G_s(F,k)$ is satisfiable. If ψ is a partial truth assignment satisfying $G_s(F,k)$, then the term $I_{\psi|\{px_1,nx_1,\ldots,px_n,nx_n\}}$*[1] *is the shortest implicant of F.*

Proof. Lemma 1 implies that $I = I_{\psi|\{px_1,nx_1,\ldots,px_n,nx_n\}}$ is an implicant of F. Observe that I has length exactly k. For the sake of contradiction assume that there is a shorter implicant I' with length $k' < k$. Then $G_s(F,k')$ must be satisfiable which is contradictory with the choice of k.

The proper k can be found for example by iteratively solving $G_s(F,k-1)$ for $k = n, n-1, \ldots, 1$ until $G_s(F,k-1)$ is unsatisfiable. A better way is to use binary search to find the proper k, which we used in our experiments.

What we described above is actually a polynomial reduction of the decision version of the shortest implicant problem into SAT. Since SAT is in NP, the shortest implicant problem for CNF formulas is also in NP.

5.6 Solving via Incomplete Methods

In the previous sections we incorporated various complete methods which find the optimal solution for certain optimization problems. However it is often the case that there are also incomplete methods based on local search which solve the same

[1] By $\psi|\{px_1, nx_1, \ldots, px_n, nx_n\}$ we mean the restriction of ψ to the variables $px_1, nx_1, \ldots, px_n, nx_n$.

Table 1 Results for the tested algorithms and instances

Benchmark Set		maxsat	minsat	iter. sat	pbo	local	sat
Random	no. solved (opt. no./approx.)	79	80	80	**81**	79 (75/1.0003)	81
	total time [s]	8767	6516	5887	**5037**	**4005**	20.15
SPL	no. solved (opt. no./approx.)	24	35	**98**	19	**98** (20/1.0045)	98
	total time [s]	139835	133047	**4680**	143827	18447	1.51
BMC	no. solved (opt. no./approx.)	0	3	**9**	3	0 (0/∞)	13
	total time [s]	23400	18302	**12096**	18502	11714	14

The bold value indicates the best result. In the case of local search we also give the number of optimal solutions and the average approximation ratio.

problems. The general advantage of incomplete solvers is that they run fast and are able to quickly produce a first but rough estimate of the objective function. The unpleasant price is that they are not guaranteed to find the optimal solution.

Several incomplete methods have been already designed for the PMAXSAT problem. They can, of course, be used for solving the shortest implicant instances using the same encoding. Thus the only difference is that the produced implicant cannot be proven to be of minimum size. When using incomplete methods we have to consider the quality of the solutions together with the running time of the algorithms in order to compare the algorithms correctly.

6 Experiment Setup

To compare the practical usability of the above described methods, we conducted experiments on various benchmark problems. We implemented the reductions in Java, particularly, to encode cardinality constraints for iterative SAT we employed the BoolVar/PB Java library [3]. BoolVar/PB implements several methods; we concretely used the "linear" encoding, which implements a sorter based encoding introduced by Eén and Sörensson [14].

For PMAXSAT solving we used Akmaxsat by Adrian Kügel [20] and for PMINSAT minsat [15]. For PB-optimization we selected bsolo [21]. The SAT solver used for iterative SAT solving was PrecoSAT by Armin Biere [6]. As for the incomplete solver we chose UBCSAT [27] particularly the g2wsat algorithm.

Our focus was on our own new benchmark set – SPL – but we also used benchmark formulas from SATLIB [17].

As for the SPL benchmark, we have exploited several SPL use-cases [10, 11], As a simple example, consider the following: if a method M uses two implementations A and B of a library function, we may want to express that performance of M depends on performance of the fastest one of A and B. In the propositional skeleton of the corresponding SPL formula, this could be expressed (after simplification) by the following conjunction:

$$
\begin{aligned}
&(V_{A\_is\_faster\_than\_B} \implies V_{M\_depends\_on\_A}) \\
\wedge\ &(V_{B\_is\_faster\_than\_A} \implies V_{M\_depends\_on\_B}) \\
\wedge\ &(V_{A\_is\_faster\_than\_B} \vee V_{B\_is\_faster\_than\_A}) \\
\wedge\ &(\neg V_{A\_is\_faster\_than\_B} \vee \neg V_{B\_is\_faster\_than\_A})
\end{aligned}
$$

Note, that in SPL such a formula corresponds to particular values of performance parameters (e.g., size of an input array). A scenario typically covers several such values (e.g., array sizes 100, 200, and 500). We have always considered several scenarios to generate each benchmark CNF formula. In particular, the formula comprises a conjunction of sub-formulas encoding the individual scenarios. Since the scenarios are independent, all the sub-formulas use disjoint sets of variables. Basically, the scenarios cover different forms of selection of a suitable variant of a function implementation, based on the relative performance of the implementation for the given performance parameters (e.g. size of an input array). In general, the main parameters determining the produced sub-formula for each scenario are: (i) the number of alternative implementation variants, and (ii) the range of performance parameter values to be covered. The former case increases the size of the clauses of the generated sub-formulas, while the latter increases the number of the sub-formulas. Overall, the SPL benchmark uses randomization while generating the sub-formulas. The final formula is produced by repeating the randomized generation process until reaching the required number of clauses and/or variables. For our experiments, we have generated formulas in a range of sizes, starting with hundreds of clauses and variables, and ending with tens of thousands. As an aside, in SPL, the relation "faster than" has a slightly different semantics to "slower or equal", therefore there are two different variables in the example – $V_{A\_is\_faster\_than\_B}$ and $V_{B\_is\_faster\_than\_A}$ – rather the just one and its negation. Moreover, because of SPL semantics, the variable $V_{M\_depends\_on\_A}$ has to be in the benchmark formula actually represented as a conjunction of variables "*M is at most c1% slower than A*" and "*M is at most c2% faster than A*", where c1 and c2 express the level of dependency of *M* on *A*.

The other input data are chosen from the SATLIB benchmarks [16]. For our experiments we selected the "bmc" and "Uniform Random-3-SAT" formulas. The BMC formulas arise from bounded model checking problem instances which are modelled as SAT. And the random formulas are from phase transition region with number of variables ranging from 50 to 250. For further explanation of the formulas consult the SATLIB benchmark site [16].

The experiments were run for each input type on a computer with Intel i7 920 CPU @ 2.67 GHz processor and 6 GB of memory. The timelimit for a single instance was 1800 seconds. The instances were sorted by the number of variables and if the solver timed out eight times in a row we stopped running it on that input set.

7 Experiment Results

In Table 1 and Figure 3 we compared the running times required to solve the SAT formulas (using Precosat [6]) with the running times required to find the shortest

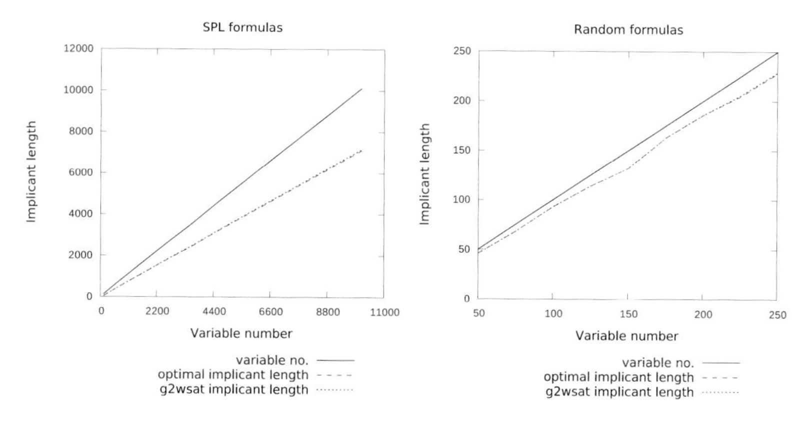

Fig. 2 Comparison of the length of the shortest implicant to the number of variables for SPL and BMC formulas.

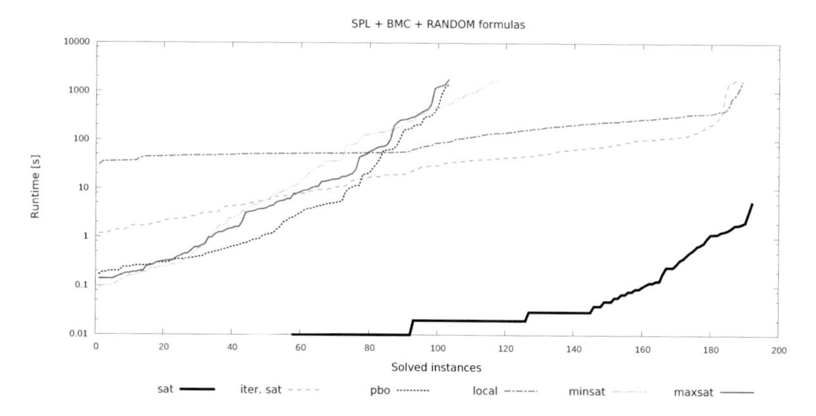

Fig. 3 Number of solved instances and runtimes of all the algorithms for all the input instances

implicants by the described methods. In the case of the local search, g2wsat algorithm, (named **local**) we also provide the number of optimal solutions found and the average approximation ratio – the length of the found implicant divided by the optimal length. The *total time* is the sum of the running times on all the instances. If the solver did not terminate within the time limit we used the time limit as the running time. Let us note that the SAT solver was able to solve all the instances.

Verifying if the input formula is satisfiable turns out to be by orders of magnitude faster than finding the minimum satisfying assignment. We think that this is also the reason why the iterative SAT is the fastest method.

To support this idea we also experimented with various modifications of iterative SAT. First we chose different initial lower and upper bounds on the the length of the shortest implicant. We obtained them using the Akmaxsat solver or set them to 1 and n respecitvely. The other modification is just a simple linear search, i.e., we used the iterative SAT reduction with the limit u, then $u-1$ and so on until the optimum length was found.

Out of all the possible modifications binary search with the initial bounds set to 1 and n performed the best when considering the number of solved instances. It also always ran faster and solved more instances than the binary search with the bounds initialized by Akmxasat solver. Thus we think that Akmaxsat spends a nonnegligible amount of time by deriving the bounds, especially the upper one. This fact is based on the observation that the linear search starting from the lower bound achieves comparable results to the binary search.

All the iterative methods solve more instances than the Akmaxsat solver, especially the method approaching the optimum from below. We also observed that the iterative methods based on the linear search are less stable than the binary search with bounds 1 and n, i.e. they never solved more instances. However on some inputs the linear search approaching the optimum from below performed faster.

The performance of linear search does not substantially depend on the fact if the bound is derived by akmaxsat – the methods have roughly the same performance. For the methods starting with lower bounds, which seems to be easy to obtain, we think that sat solving dominates the running time. For the methods starting with the upper bound the number of iterations is certainly lower see Figure 2. On the other hand sat solving is harder since the upper limit on the implicant length prunes less of the search space than the lower bound. For the lower bound methods the fact that solving unsatisfiable formulas does seem to have a detrimal effect.

The other complete methods (PMAXSAT, PMINSAT, and PBO) give very similar results relative to each other but are considerably weaker than iterative SAT. It is interesting that there is a relatively big gap between PMAXSAT (worst of the 3) and PMINSAT (best of the 3) since these problems and our encodings for them are very similar. The difference is probably caused by the different heuristics and implementation of the solvers.

Let us note that the performance of incomplete methods, especially the quality of the solution, crucially depends on a proper choice of the parameters of the algorithm such as the number of steps, number or restarts and overall iteration count. When these parameters are well chosen the quality of the solution is comparable to the optimal solution as observed in Figure 2.

Altogether we can conclude that the best strategy is iterative SAT followed by iterative SAT using simple linear search. For the hard formulas incomplete methods could also be useful but one has to tweak their parameters.

8 Conclusion

In this paper we have shown that finding minimum-size satisfying assignments is both useful and can be computed relatively efficiently for many relevant formulas.

The usefulness was demonstrated by describing a class of SMT problems with a costly decision procedure and an application of this kind – the SPL framework.

We described five possible methods to solve this problem from which the reductions to PMINSAT and PMAXSAT are novel to our best knowledge. Although the other three already appeared in the literature, there is no published comparison of these methods.

We did exhaustive experiments using modern state-of-the-art solvers and relevant benchmark problems to measure the performance of the methods we described. One of the benchmark sets was generated according to ideas of the SPL framework. Unfortunately, we were unable to do direct experiments to measure the usefulness of the methods for the SPL framework, since it is still under development and the number of its large-scale case studies is limited.

As for future work, we plan to improve the methods with support for assignment costs. Finding optimal short assignments with respect to a given assignment cost function would be beneficial in the cases presented in the motivation section, SPL in particular. Here, the cost of a SAT assignment could be determined by the execution times of the measurements to be performed by the SMT decision procedure in order to decide the feasibility of the skeleton assignment. In consequence, this would allow preferring the fast measurements to the slower ones while solving the SPL formulas.

Acknowledgements. This research was partially supported by the SVV project number 267314, the Grant agency of the Charles University under contracts no. 266111 and 600112, and the Charles University institutional funding SVV-2013-267312. This work was also partially supported by the Grant Agency of the Czech Republic project GACR P202/10/J042.

References

1. Armando, A., Mantovani, J., Platania, L.: Bounded model checking of software using smt solvers instead of sat solvers. International Journal on Software Tools for Technology Transfer (STTT) 11(1), 69–83 (2009)
2. Bailleux, O.: On the cnf encoding of cardinality constraints and beyond. CoRR abs/1012.3853 (2010)
3. Bailleux, O.: Boolvar/pb v1.0, a java library for translating pseudo-boolean constraints into cnf formulae. CoRR abs/1103.3954 (2011)
4. Barrett, C.W., Dill, D.L., Stump, A.: Checking satisfiability of first-order formulas by incremental translation to SAT. In: Brinksma, E., Larsen, K.G. (eds.) CAV 2002. LNCS, vol. 2404, pp. 236–249. Springer, Heidelberg (2002)
5. Bieganowski, J., Karatkevich, A.: Heuristics for thelen's prime implicant method. Schedae Informaticae 14, 125–125 (2005)
6. Biere, A.: Precosat home page (2013), `http://fmv.jku.at/precosat/`
7. Biere, A., Heule, M.J.H., van Maaren, H., Walsh, T. (eds.): Handbook of Satisfiability. Frontiers in Artificial Intelligence and Applications, vol. 185. IOS Press (2009)
8. Brauer, J., King, A., Kriener, J.: Existential quantification as incremental SAT. In: Gopalakrishnan, G., Qadeer, S. (eds.) CAV 2011. LNCS, vol. 6806, pp. 191–207. Springer, Heidelberg (2011)

9. Bryant, R.E.: Boolean analysis of mos circuits. IEEE Trans. on CAD of Integrated Circuits and Systems 6(4), 634–649 (1987)
10. Bulej, L., Bures, T., Horky, V., Keznikl, J., Tuma, P.: Performance Awareness in Component Systems: Vision Paper. In: Proceedings of COMPSAC 2012 (2012)
11. Bulej, L., Bures, T., Keznikl, J., Koubkova, A., Podzimek, A., Tuma, P.: Capturing performance assumptions using stochastic performance logic. In: Proceedings of ICPE 2012 (2012)
12. Cook, S.A.: The complexity of theorem-proving procedures. In: STOC, pp. 151–158 (1971)
13. Crama, Y., Hammer, P.L.: Boolean Functions - Theory, Algorithms, and Applications. In: Encyclopedia of Mathematics and its Applications, vol. 142. Cambridge University Press (2011)
14. Eén, N., Sörensson, N.: Translating pseudo-boolean constraints into sat. JSAT 2(1-4), 1–26 (2006)
15. Heras, F., Morgado, A., Planes, J., Silva, J.P.M.: Iterative sat solving for minimum satisfiability. In: ICTAI, pp. 922–927 (2012)
16. Hoos, H., Stutzle, T.: Satlib benchmark site (2013), http://www.cs.ubc.ca/~hoos/SATLIB/benchm.html
17. Hoos, H.H., Stutzle, T.: Satlib: An online resource for research on sat, pp. 283–292. IOS Press (2000)
18. Horky, V.: Stochastic Performance Logic (SPL) Home Page (2013), http://d3s.mff.cuni.cz/projects/performance_evaluation/spl/
19. Kautz, H.A., Selman, B.: Planning as satisfiability. In: ECAI 1992: Tenth European Conference on Artificial Intelligence, Vienna, Austria, pp. 359–363 (1992)
20. Kügel, A.: Homepage of Adrian Kügel (2012), http://www.uni-ulm.de/en/in/institute-of-theoretical-computer-science/m/kuegel.html
21. Manquinho, V.: bsolo home page (2012), http://sat.inesc-id.pt/~vmm/research/index.html
22. Manquinho, V., Oliveira, A., Marques-Silva, J.: Models and algorithms for computing minimum-size prime implicants. In: Proceedings of the International Workshop on Boolean Problems (1998)
23. Manquinho, V.M., Flores, P.F., Silva, J.P.M., Oliveira, A.L.: Prime implicant computation using satisfiability algorithms. In: ICTAI, pp. 232–239 (1997)
24. de Moura, L., Bjørner, N.: Satisfiability modulo theories: An appetizer. In: Oliveira, M.V.M., Woodcock, J. (eds.) SBMF 2009. LNCS, vol. 5902, pp. 23–36. Springer, Heidelberg (2009)
25. Palopoli, L., Pirri, F., Pizzuti, C.: Algorithms for selective enumeration of prime implicants. Artificial Intelligence 111(1), 41–72 (1999)
26. Robinson, J.A., Voronkov, A. (eds.): Handbook of Automated Reasoning (in 2 volumes). Elsevier and MIT Press (2001)
27. Tompkins, D.: Ubcsat home page (2012), http://www.satlib.org/ubcsat/
28. Umans, C.: The minimum equivalent dnf problem and shortest implicants. In: FOCS, pp. 556–563 (1998)
29. Velev, M.N., Bryant, R.E.: Effective use of boolean satisfiability procedures in the formal verification of superscalar and vliw microprocessors. J. Symb. Comput. 35(2), 73–106 (2003)

Repository-Centric Process Modeling – Example of a Pattern Based Development Process

Jacob Geisel, Brahim Hamid, and Jean-Michel Bruel

Abstract. Repositories of modeling artefacts have gained more attention recently to enforce reuse in software engineering. In fact, repository-centric development processes are more adopted in software/system development, such as architecture-centric or pattern-centric development processes.

In our work, we deal with a specification language for development methodologies centered around a model-based repository, by defining both a metamodel enabling process engineers to represent repository management and interaction and an architecture for development tools.

The modeling language we propose, has been successfully evaluated by the TERESA project for specifying development processes for trusted applications centered around a model-based repository of security and dependability (S&D) patterns.

Keywords: Metamodel, Model-Driven Engineering, Process, Security, Dependability, Repository, Pattern.

1 Introduction

Non-functional requirements such as Security and Dependability (S&D) [12] become more and more important as well as more and more difficult to achieve, particularly in embedded systems development [17]. Such systems come with a large number of common characteristics, including real-time and temperature constraints, security and dependability as well as efficiency requirements. In particular, the development of Resource Constrained Embedded Systems (RCES) has to address constraints regarding memory,

Jacob Geisel · Brahim Hamid · Jean-Michel Bruel
IRIT, University of Toulouse
118 Route de Narbonne, 31062 Toulouse Cedex 9, France
e-mail: {geisel,hamid,bruel}@irit.fr

R. Lee (Ed.): *SERA*, SCI 496, pp. 247–261.
DOI: 10.1007/978-3-319-00948-3_16

computational processing power and/or energy consumption. The integration of S&D features requires the availability of both application domain specific knowledge and S&D expertise at the same time. Hence capturing and providing this expertise by means of a repository of S&D patterns and models can enhance embedded systems development. We seek mechanisms which allow a safer, easier and faster RCES development processes.

Modeling software and system process is fundamental in order to improve the quality of applications. The main goal of these processes is to provide to organizations with the means to define a conceptual framework. For this reason, several tentatives (including those developed by the OMG[1]) have been proposed to model software process. For instance, the SPEM [10] specification is used for describing a concrete software development process or a family of related software development processes. It conforms to the OMG MOF metametamodel and is defined as a UML profile.

In this paper, we study the RCPM metamodel which defines a new formalism for system development processes. This formalism is centered around a repository of modeling artefacts, providing new concepts related to repository management and interaction. The paper also presents the design environment for process modeling, supporting reuse in form of predefined libraries of process element types. These libraries may be used to facilitate process modeling from scratch or to adapt existing process models for certain domains. Furthermore, the design environment offers the ability to build new type libraries based on the recommendations of a targeted domain.

The rest of this paper is organized as follows. In Section 2, we introduce the context and background related to this work. Then, Section 3 details the specification of the repository-centric process modeling language. Section 4 describes our proposed tool implementation through an example of a process model from Railway domain targeting RCES applications. In Section 5, we present an extract from a process enactment to develop an RCES application. In Section 6, we review some principal existing process metamodels close to our work. Finally, Section 7 concludes and draws future work directions.

2 Development Context and Background

2.1 Development Context

The proposed methodology promotes a model-based approach coupled with a repository of modeling artefacts. In this vision, the modeling artefacts derived from (resp. associated with) domain specific models aim at helping the application developer to integrate these artefacts as building blocks. The repository presented here is a model-based repository of modeling artefacts. Concretely, the repository is a structure that stores specification languages and the modeling artefacts coupled with a set of tools to

[1] Organization normalizing the UML language.

manage/visualize/export/instantiate these artefact in order to use them in engineering processes. For instance, to define an engineering discipline for S&D that is adapted to RCES, a repository-centric engineering process model will have to recognize the need to separate expertise on applications (represented by an application designer), expertise on security and dependability (represented by an S&D engineer), and expertise on repository-based development (represented by a model-driven and pattern engineer).

2.2 Process Models and Artefacts

Models are used to denote some abstract representation of system engineering processes. Specifically, we need models to represent the process activities, models to encode the artefacts and software platforms to test, to simulate and to validate the proposed solutions. Accordingly, comprehension, study and analysis of system engineering processes require the seek of models which make it as easy as possible to express and to encode them with the following characteristics:

- Intuitive: to develop them and teach them,
- Practical: to test and validate them by a simple implementation.
- Formal: to prove their correctness using formal method tools,

As a benefit, the study of problems on high-level models allows deducing properties on other less abstract models. Here, we deal with the two first characteristics through metamodeling technique and its associated implementation environment.

2.3 DSL Buildung Process

Domain Specific Modeling Languages (DSML) [2] have recently increased in popularity to cover a wider spectrum of concerns. A process defining those DSMLs reuses many practices from Model-Driven Engineering. For instance, metamodeling and transformation techniques. SEMCO[2] is a set of federated DSLs working as a group, each one relevant to the key concern. A DSL process [3] is divided into several kinds of activities: DSL definition, transformations and consistency and relationships rules as well as design with DSLs and qualification. The first three activities are achieved by the DSL designer and the two last ones are used by the final DSL user.

There are several DSML environments available. In our context, we use the Eclipse Modeling Framework (EMF) [15] open-source platform to support such a building process and to create our tool suite. Note, however, that our vision is not limited to the EMF platform.

[2] `http://www.semcomdt.org`

[3] DSL process defines how development projects based on DSL are achieved.

2.4 Working Example

The illustrating example is a simple variant of the well-know V-Model. In this process model, the developer starts by requirements engineering/ specification, followed by system specification. In a traditional approach (non repository-of-pattern-based approach) the developer would continue with the architecture design, module design, implementation and test.

In our vision, instead of following this phases and performing their related activities, which usually are time and efforts consuming as well as errors prone, the system developer merely needs to select appropriate patterns from the repository and integrate them into the system under development (Figure 1 shows the process and points out the phases with repository interactions). For each phase, the system developer executes the search/select from the repository to instantiate appropriate patterns in his modeling environment and then integrates them in his models following an incremental process. The downside of this approach is that in a very early stage of the development, mainly during the requirements and design phases, the requirements engineers and the system architects have to be aware of existing patterns.

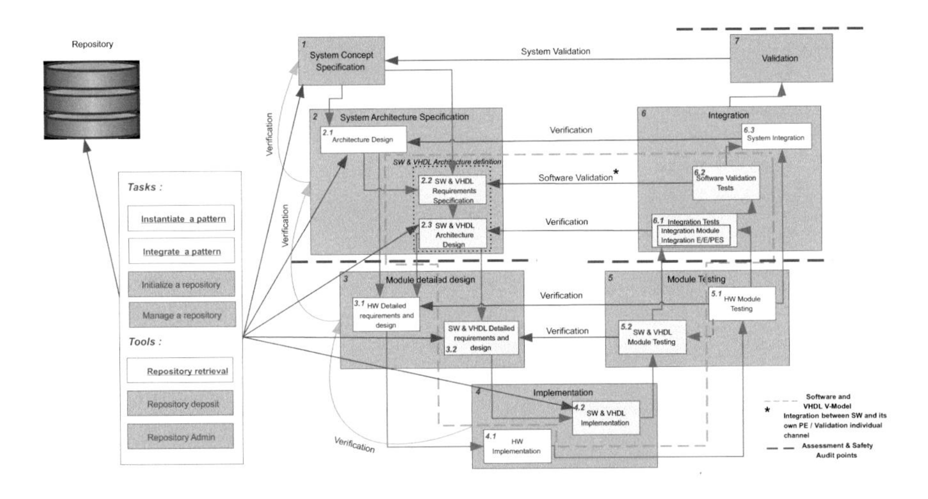

Fig. 1 Railway Engineering Process Lifecycle

3 Repository Interaction Metamodel

In the following subsection, we highlight the sub-metamodel architecture of the Repository- Centric Process Metamodel (RCPM), while the next subsections concentrate on the presentation of the repository interaction part of the RCPM metamodel.

3.1 RCPM

The RCPM is a metamodel defining a new formalism for system development process modeling based on a repository of modeling artefacts. The RCPM metamodel contains different sub-metamodels, as shown in Fig. 2, which offer different capabilities. RCPM is oriented to support:

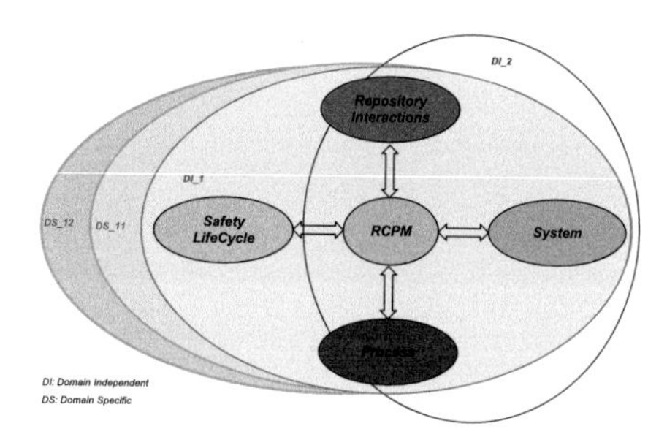

Fig. 2 Design principles of RCPM

- *The development of embedded systems.* The metamodel orients to facilitate the modeling the development of embedded systems, including the concepts of partitions which are popular in embedded system development.
- *Reuse of existing solutions.* The metamodel enables to model existing modeling artefacts and their integration process. For instance, the metamodel supports the repository-centric design methodology, introducing new concepts on repository management and interactions with the traditional process metamodel.
- *A safety process lifecycle.* As we can find in standards as IEC 61508 [7], there are more and more requirements for transforming traditional processes to safety processes to meet specific safety requirements of systems or software. This metamodel adds the concepts used in the safety lifecycle to support this kind of process model, such as verification and validation [4].

In this paper, we concentrate on presenting the repository part of the RCPM metamodel. For a general description and other referenced metamodel concepts see [3].

3.2 Repository Interaction Sub-Metamodel

Our specification language is described by a metamodel that we call Repository Interaction Sub-Metamodel, as depicted in Figure 3. It constitutes the base of our process modeling language, describing all the concepts (and their relations) required to capture all the facets of Repository Interactions.

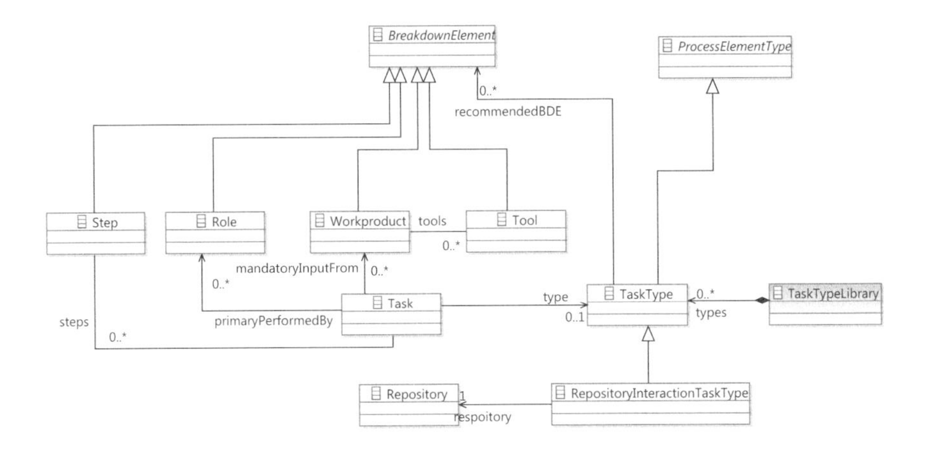

Fig. 3 Overview of the Repository Interaction Sub Metamodel

The principal classes of the metamodel are described with the Ecore notations of the Eclipse Modeling Framework[4] in Figure 3 as well as the link with the libraries models [5]. As we shall see, we define a set of libraries with a set of tasks and steps dedicated to specify the repository interaction tasks and steps during the process model enactment. These libraries will be used as external models to type the process tasks. The meaning of the main elements of the metamodel with the working example are described in the sequel.

- BREAKDOWNELEMENT. A BreakdownElement is an abstract generalization for any Process Element that is part of a breakdown structure. Any of its concrete sub-classes can be used to compose an Activity*[6].
- TASK. A Task is a WorkBreakdownElement* that represents the work that should be done in an Activity*. The Task should be related to a Role, a WorkProduct and, if necessary, a Tool. In our example, as visualized in Figure 1 we define a set of Tasks, which are related to repository management (*initialize a repository, manage a repository*) and those related to repository interactions (*instantiate a pattern, deposit a pattern, integrate a pattern*). The later Task is not strictly related to repository interaction, but may lead to some repository interactions. A Task is decomposed into Steps, which detail what exactly is done in which order. A Task has normally WorkProducts (mandatory or optional) as input and output.

[4] `http://www.eclipse.org/modeling/emf/`

[5] We use *gray* to label concepts imported from the library model

[6] Elements marked with * are not shown in Figure 3, please refer to [3] for more details

- STEP. A Step is a detailed description of the work to be done. It is the smallest entity in the decomposition of Process*, Phase*, Activity* and Task. It describes the elementary step, which leads to the realization of a WorkProduct. For instance, *instantiate a pattern* task may be decomposed into three steps: *search a pattern in the repository*, *select the appropriate one from the search results list* and finally *import the selected one into the development environment.*
- ROLE. A Role describes the role of an actor in a Process/Phase/Activity/Task. It is generally linked to the realization of a WorkProduct for a specific Task using a specific Tool. In our example, we can associate *repository manager* role to the actor responsible of the *manage a repository* task and *system engineer* role to actor responsible of the *instantiate a pattern* task.
- WORKPRODUCT. A WorkProduct is a special BreakdownElement that represents an input and/or output for a Task. The WorkProduct is related to a Task and a Role. A *pattern* is a key workproduct of the proposed process model.
- TOOL. A Tool represents the tool used to fulfill a Task and to realize a WorkProduct. Here, we deal with a set of tools supporting to the repository management (*Repository Admin*) and repository interactions (*Repository Retrieval*).
- PROCESSELEMENTTYPE. The ProcessElementType allows to type a ProcessElements*, adding mandatory or optional properties to a ProcessElements*, as well as references to different Phases*, Roles, Tools, WorkProducts or Activities*.
- TASKTYPE. A TaskType allows to type a Task to reuse capitalize knowledge about Roles, Tools, WorkProducts and Steps. This Type links these information.
- REPOSITORYINTERACTIONTASKTYPE. A RepositoryInteractionTaskType is a specialization of a TaskType introducing the idea of Repository. Thess TaskTypes can be linked to a Repository. In our example, we could define *instantiate a pattern* as REPOSITORYINTERACTIONTASKTYPE instance.
- TASKTYPELIBRARY. A Library containing TaskTypes which are common to an application domain or standard recurring TaskTypes and can be reused to type recurring Tasks in a process or Tasks in different processes. The repository specific interaction tasks may be grouped into one or multiple libraries to foster reuse.
- REPOSITORY. It describes the repositories used in development process. As the repository-centric development processes are more and more adopted in software/system development, such as architecture-centric or pattern-centric development processes. In our example, we use a repository of S&D patterns.

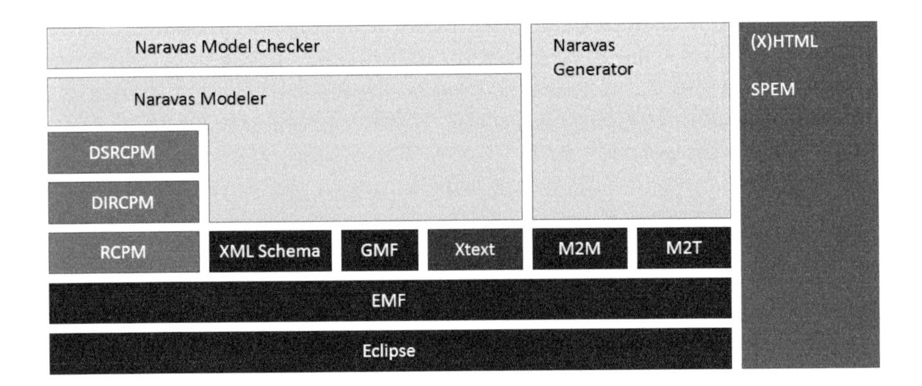

Fig. 4 Overview of the Naravas Architecture

4 Tool Architecture and Implementation

Using the proposed metamodels, ongoing experimental work with *SEMCOMDT*[7] (SEMCO Model Development Tools, IRIT's editors and platform as Eclipse plugins) is realized, testing the features of *Naravas*, a tool for formalizing process models and documentation generation. In the following subsections, we present our tooling. Figure 4 depicts the architecture of the development framework based on Eclipse Technologies.

4.1 How the Process Model Editor Is Built?

We used the Eclipse EMF based Ecore editor to model our Repository-centric Process Metamodel (RCPM), creating one Ecore file containing the three packages needed for the process model, the core package, the type package and the process package. Minor modifications have been applied on the metamodel to support an EMF based editor and HTML documentation generation. The generated editor code was modified to limit the user actions on the ones needed and to enhance user experience (e.g. modifying the process model creation workflow).

The second part of the project was to create the HTML code generator based on Acceleo[8], a Model-to-Text (M2T) component of the Eclipse Modeling Framework. We developed modularized code transformation templates, generating one HTML file per process model object and type and managing the links among them.

---

[7] http://www.semcomdt.org

[8] http://www.eclipse.org/acceleo/

4.2 *Process Model Designer: Naravas*

Naravas is an EMF tree-based editor for specifying models of processes, libraries of types and generation of documentation. Naravas implements several facilities conforming to the RCPM metamodel.

4.2.1 Library Design

The design environment of the type libraries is presented in Figure 5. The figure represents a Task Type Library for Repository Task. The Task Types presented here are identically to the ones presented in Figure 1. For instance, the second Repository Interaction Task Type (Artefact Instantiation) show the mandatory steps (*Search repository*, *Select Patterns in Repository* and *Import Patterns to IDE*), the optional Roles, the Tool and the output Work Product for a Task typed by this type. The other Task Types in this library represent Tasks with Repository Interactions, encountered multiple times in the shown process (e.g. Repository Management, Artefact Publishing, Artefact Retrieval).

4.2.2 Process Model Design

The design environment is presented in Figure 6. Naravas enables the user to model processes in a tree-based manner. There is a design palette on the right (enabled by a right click on an element), a tree view of the project on the left and the main design view in the middle. As we shall see, the design palette is updated regarding the targeted process element. The used example shows the Railway Application Process built by Ikerlan. It represents the Repository, the Phases, Activities, Tasks, Steps, Roles, Tools, Work Products and Flows among the Elements, such as Control, Retrieve, Verification and Validation Flows. The Process model editor allows to add, delete, move and modify the elements, as well as conformance validation. It also allows the import of external resources, such as Process Element Type Libraries. The usage of the Task Types is shown in Figure 7. When creating a Task, it is possible to type it from the Task Types already defined in a Library. By choosing a Type (*Repository Interaction*), the mandatory Steps, Work Products, Roles and Tools are filled in automatically (*Search Repository*, *Select Artefact in Repository*, *Import Artefact to IDE*), and the mandatory ones are proposed in addition to the standard items when creating new entities.

4.2.3 Conformance Validation

Further, using EMF features, we added the metamodel conformance validation to the editor. The process validation tool is used to guarantee design validity conforming to the process metamodel. Process model validation starts by right clicking on Process Core and pressing the *Validation* tool. In our

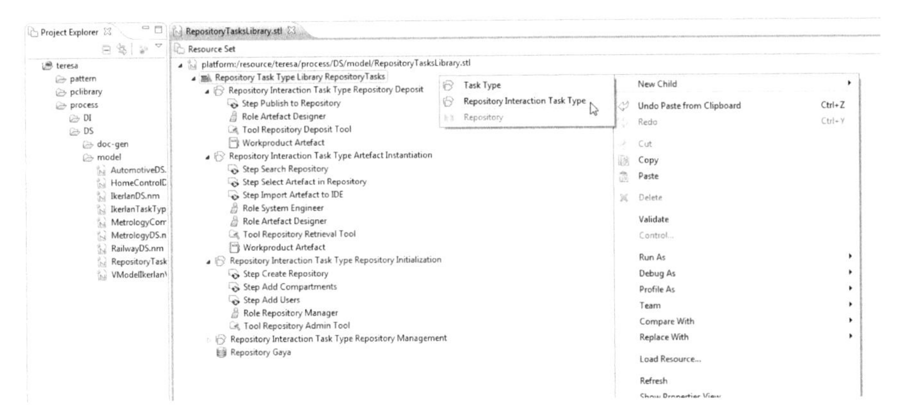

Fig. 5 Naravas for Library Design Environment

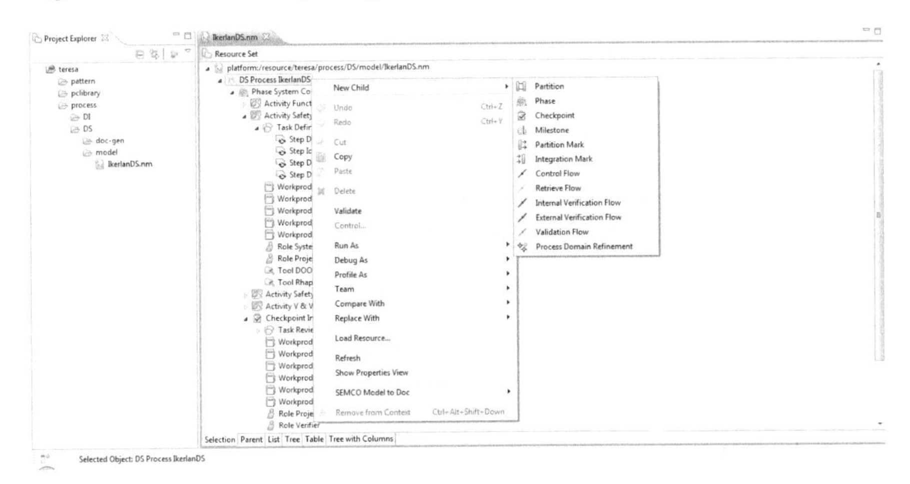

Fig. 6 Naravas Process Design Environment

example, the process model built by Ikerlan for the railway domain can be validated, where a violation of a metamodel construct will yield an error message (see Figure 8).

4.2.4 Documentation Generation

Documentation generation of a process model is triggered by running the *SEMCO Model to Doc* tool. Our implementation allows so far to generate HTML documentation using M2T transformations through Acceleo.

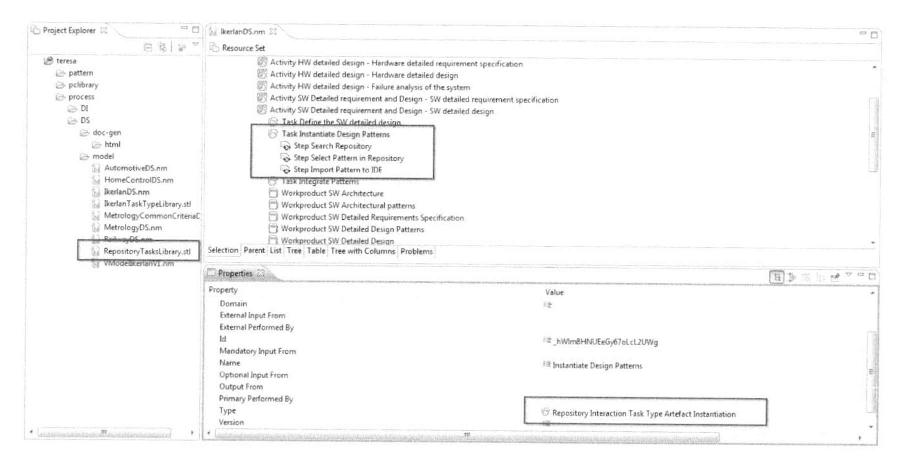

Fig. 7 Example of the Usage of a Library - Repository Interaction Task Types

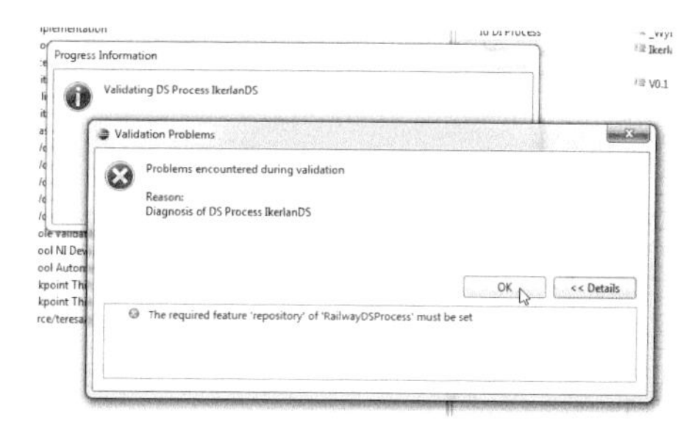

Fig. 8 Process Validation

5 Process Model Enactment

In this section we will present an extract of the process model and its enactment to build an industry control application from the railway called *Safe4Rail* acting as a TERESA case study. In this case, SIL4 level is targeted. A repository of patterns for TERESA called Gaya was built. Gaya contains so far (as of March 2013):

- Users. 5 organizations and 10 users.
- Patterns. 59 S&D patterns.

The following table depicts a subset of inputs and outputs consumed and produced during the chosen activities of the process enactment, mainly those related to the repository. Repository Interactions are highlighted, as well as results from Repository Interaction.

Table 1 Description of the Railway Process Enactment (Extract from the Module Detailed Design)

Phase	Activity	Task	Step	Role	Tool	WP in	WP out
Module Detailed Design	SW Detailed requirement and Design - SW detailed requirement specification	Define the SW detailed requirements	Analysis and Definition	SW Architect, SW Designer	Rhapsody, DOORS	SW Requirements Specification, SW Architecture	SW Detailed Requirements Specification
	SW Detailed requirement and Design - SW detailed design	Define the SW detailed design	Define Internal Description	SW Designer	Rhapsody	SW Architecture, SW Detailed Requirements Specification	SW Detailed Design
			Define Components	SW Designer	Rhapsody		
			Define Interfaces	SW Designer	Rhapsody		
			Define Communication	SW Designer	Rhapsody		
			Generate SW Detailed Design	SW Designer	Rhapsody		
		Instantiate Design Patterns	**Search Repository**	SW Designer	**Repository Retrieval Tool**	SW Architecture, **SW Architectural patterns**, SW Detailed Requirements Specification, SW Detailed Design	**SW Detailed Design Patterns**
			Select Patterns in Repository	SW Designer	**Repository Retrieval Tool**		
			Import Pattern to IDE	SW Designer	**Repository Retrieval Tool**, Rhapsody		
		Integrate Patterns	Elicitation	SW Designer	Rhapsody	SW Detailed Design, **SW Detailed Design Patterns**	SW Detailed Design with integrated Patterns
			Binding	SW Designer	Rhapsody		
			Consolidation	SW Designer	Rhapsody		

6 State of the Art

State of the Art of process metamodels have been analyzed from a perspective of repository interactions, embedded systems and safety lifecycles support. Process metamodels can be modeled from different types of views: activity-oriented, product-oriented and decision-oriented views [13, 6]. Most process metamodels and process frameworks based on metamodels adopt the activity-oriented views, such as SPEM, RUP and OPF.

The SPEM (Software & Systems Process Engineering Metamodel) [10] is a *de facto*, high-level standard for process modeling used in object-oriented software development. The scope of SPEM is intentionally limited to the minimal elements necessary to define any software and systems development process, without adding specific features for particular development domains or disciplines. The goal is to accommodate a large range of development methods and processes of different styles, cultural backgrounds, levels of formalism, lifecycle models, and communities.

The RUP (IBM's Rational Unified Process Framework) and its extension RUP SE (SE stands for System Engineering) are derived from the Unified Process Framework [8]. Both metamodels are, like SPEM, described by a UML profile and define a Process Modeling Language (PML). The OPEN Process Framework (OPF) [11] is a componentized OO development methodology underpinned by a full metamodel, encapsulating business as well as quality and modeling issues.

In addition to the above mentioned process metamodels, exist other activity-based metamodels like OOSPICE [5] and SMSDM [14]. Other types of process metamodels such as decision based etc., do not orient to safety critical system development. As far as we know, the studied process metamodels unfortunately do not support safety related development processes explicitly or facilitate the modeling of safety lifecycles. Many safety critical systems use Safety Instrument Systems (SIS) to manage the safety lifecycle, however, these SIS do not have process metamodels. Works like [1] propose to model different standards and try to give recommendations during the application development.

[16] presents a survey of business process model repositories and their related frameworks. This work deals with the management of a large collections of business processes using repository structures and providing common repository functions such as storage, search and version management. It targets the process model designer allowing the reuse of process model artefacts. A comparison of process model repositories is presented to highlight the degree of reusability of artefacts. For example, the repository for process models described in [9], supports activity, control-flow and monitoring aspects. The metamodel described in this paper may be used to specify the management and the use of this kind of process models. In fact, a process model aspect or the process model as a whole of the aforementioned process models can be seen as artefacts supported by our metamodel. In return, the vision of the

business process model repositories may be used in our work to manage the process element type libraries.

7 Conclusion

In our work, we target the development of a modeling framework built around a model-based repository of modeling artefact in order to be used in an MDE approach for trusted RCES applications in several domains. In this paper, we have proposed a modeling language to specify repository-centric process models, providing new appropriate concepts related to repository management and interaction. The design environment supports reuse in form of libraries of types, facilitating process modeling. The later may be used to specialize process models for a certain domain. In this case, the library is build on the recommendations of the targeted domain. Furthermore, we walk through a prototype with EMF editors supporting the metamodel. Currently the tool is provided as part of a tool-suite named *SEMCOMDT* as Eclipse plugins.

The design environment presented here has been evaluated in two use studies from TERESA industrial partners mainly for a repository of S&D and resource property and pattern modeling artefacts. By this illustration, we can validate the feasibility and effectiveness of the proposed specification and design frameworks.

As future work, we plan to study new libraries for additional process elements. Also, we will seek new opportunities to apply the framework to other domains.

References

1. Cheung, L.Y.C., Chung, P.W.H., Dawson, R.J.: Managing process compliance, pp. 48–62. IGI Publishing, Hershey (2003)
2. Gray, J., Tolvanen, J.-P., Kelly, S., Gokhale, A., Neema, S., Sprinkle, J.: Domain-Specific Modeling. Chapman & Hall/CRC (2007)
3. Hamid, B., Zhang, Y.: D3.2 - Common Engineering Metamodels. Technical report, TERESA-Project (2012), `http://www.teresa-project.org/`
4. Hamid, B., Geisel, J., Ziani, A., Gonzalez, D.: Safety lifecycle development process modeling for embedded systems - example of railway domain. In: Avgeriou, P. (ed.) SERENE 2012. LNCS, vol. 7527, pp. 63–75. Springer, Heidelberg (2012)
5. Henderson-Sellers, B., Gonzalez-Perez, C.: A comparison of four process metamodels and the creation of a new generic standard. Information & Software Technology 47(1), 49–65 (2005)
6. Hug, C., Front, A., Rieu, D., Henderson-Sellers, B.: A method to build information systems engineering process metamodels. J. Syst. Softw. 82, 1730–1742 (2009)
7. I. S. IEC 61508. Functional safety of electrical/electronic/programmable electronic safety-related systems (2000)

8. Kruchten, P.: The Rational Unified Process: An Introduction. Addison-Wesley Longman Publishing Co., Inc., Boston (2003)
9. Liu, C., Lin, X., Zhou, X., Orlowska, M.E.: Building a repository for workflow systems. In: TOOLS (31), pp. 348–357. IEEE Computer Society (1999)
10. OMG. Software & Systems Process Engineering Meta-Model Specification (2008)
11. OPF Repository Organization. OPEN Process Framework (OPF) (2009), http://www.opfro.org/
12. Ravi, S., Raghunathan, A., Kocher, P., Hattangady, S.: Security in embedded systems: Design challenges. ACM Trans. Embed. Comput. Syst. 3(3), 461–491 (2004)
13. Rolland, C.: A comprehensive view of process engineering. In: Pernici, B., Thanos, C. (eds.) CAiSE 1998. LNCS, vol. 1413, pp. 1–24. Springer, Heidelberg (1998)
14. Standards Australia. Standard Metamodel for Software Development Methodologies (2004)
15. Steinberg, D., Budinsky, F., Paternostro, M., Merks, E.: EMF: Eclipse Modeling Framework 2.0, 2nd edn. Addison-Wesley Professional (2009)
16. Yan, Z., Dijkman, R.M., Grefen, P.: Business process model repositories - framework and survey. Information & Software Technology 54(4), 380–395 (2012)
17. Zurawski, R.: Embedded systems. CRC Press Inc. (2005)

Applying CBD to Build Mobile Service Applications

Haeng-Kon Kim and Roger Lee

Abstract. Mobile service applications must be developed following component based and object oriented principles of encapsulation, abstraction and code reusability. Future changes to a particular functionality developed in this context will be paid only as per the individual instance of change according to its single-instance complexity. This has to be taken into account from the very beginning in the mobile service applications design. For development of mobile service applications, the use of appropriate existing tools is generally supported. Specifically, open source software should be used where possible. The set of tools in use must be kept to a minimum. The tools / external libraries / external dependencies that have to remain available to the software after development is completed must be approved in writing.

In this paper, we discuss some of the problems of the current mobile service applications development and show how the introduction of CBD (Component Based Development) provides flexible and extensible solutions to it. Mobile service applications resources become encapsulated as components, with well-defined interfaces through which all interactions occur. Builders of components can inherit the interfaces and their implementations, and methods (operations) can be redefined to better suit the component. New characteristics, such as concurrency control and persistence, can be obtained by inheriting from suitable base classes, without necessarily requiring any changes to users of these resources. We describe the MSA (Mobile Service Applications) component model, which we have developed, based upon these ideas, and show, through a prototype implementation, how we have used the model to address the problems of referential integrity and transparent component (resource) migration. We also give indications of future work.

Haeng-Kon Kim
School of Information Technology, Catholic University of Daegu, Korea
e-mail: hangkon@cu.ac.kr

Roger Lee
Dept. of Computer Science, Central Michigan Univ. Mt.Pleasant, MI 48859, U.S.A
e-mail: lee@cps.cmich.edu

R. Lee (Ed.): *SERA*, SCI 496, pp. 263–277.
DOI: 10.1007/978-3-319-00948-3_17

Keywords: Mobile Devices, Mobile Application Development, User Interface Design, Mobile Service Applications, Component-Based Development, Referential Integrity, Mobility, Distributed Systems, Mobile Application Model.

1 Introduction

Mobile applications are a term used to describe Internet applications that run on smart phones and other mobile devices. Mobile applications usually help users by connecting them to Internet services more commonly accessed on desktop or notebook computers, or help them by making it easier to use the Internet on their portable devices. A mobile app may be a mobile site bookmarking utility, a mobile-based instant messaging client, Gmail for mobile, and many other applications. Mobile apps are add-on software for handheld devices, such as smart phones and personal digital assistants (PDA). Among the most popular are games, social networking, maps, news, business, and weather and travel information. All of these leverage at least one of the device's technical features: communications interfaces (Wi-Fi, WiBro/mobile WiMAX, GSM/EDGE, W-CDMA/ UMTS/HSPA and Bluetooth), audio and video processors, camera, sensors or GPS module [1].

In this paper, we will show how the current mobile service applications development is component-based, with a single interface. Although extensions have been implemented to allow the incorporation of nonstandard resources, We will show how making the change to an Component-Based Development system can yield an extensible infrastructure that is capable of supporting existing functionality and allows the seamless integration of more complex resources and services. We aim to use proven technical solutions from the distributed Component-Based Development community to show how many of the current problems with the mobile service applications development can be addressed within the proposed model. In the next section, a critique of the current mobile service applications development is presented, highlighting existing problems in serving standard resources and the current approach for incorporating nonstandard resources. The section entitled MSA (Mobile Service Applications) Component describes the MSA (Mobile Service Applications) component design, its aims, component model, and system architecture. The Illustrations section gives an example, describing how particular mobile shortcomings can be addressed within the proposed architecture.

2 Related Works

2.1 Characteristics of Mobile Application Development Model

These powerful development tools and frameworks greatly simplify the task of implementing a mobile application. However, they are predominantly focused on the individual developer who is trying to create an application as quickly as possible. For small and medium-sized mobile applications that can be built (and easily updated) by a single developer, they represent a vast improvement on the previous generations of tools, and encourage developers to adhere to the important principles of abstraction and modularity that are built into the platform architectures. However, as mobile applications become more complex, moving beyond inexpensive recreational applications to more business critical uses, it will be essential to apply software engineering processes to assure the development of secure, high-quality mobile applications. While many "classic" software engineering techniques will transfer easily to the mobile application domain, there are other areas for new research and development. In many respects, developing mobile applications is similar to software engineering for other embedded applications. Common issues include integration with device hardware, as well as traditional issues of security, performance, reliability, and storage limitations. However, mobile applications present some additional requirements that are less commonly found with traditional software applications, including [2,3]:

1) Potential interaction with other applications – most embedded devices only have factory-installed software, but mobile devices may have numerous applications from varied sources, with the possibility of interactions among them;
2) Sensor handling – most modern mobile devices, e.g., "smart phones", include an accelerometer that responds to device movement, a touch screen that responds to numerous gestures, along with real and/or virtual keyboards, a global positioning system, a microphone usable by applications other than voice calls, one or more cameras, and multiple networking protocols;
3) Native and hybrid (mobile Mobile) applications – most embedded devices use only software installed directly on the device, but mobile devices often include applications that invoke services over the telephone network or the Internet via a Mobile browser and affect data and displays on the device;
4) Families of hardware and software platforms – most embedded devices execute code that is custom-built for the properties of that device, but mobile devices may have to support applications that were written for all of the varied devices supporting the operating system, and also for different versions of the operating system. An Android developer, for example, must decide whether to build a single application or multiple versions to run on the broad range of Android devices and operating system releases
5) Security – most embedded devices are "closed", in the sense that there is no straightforward way to attack the embedded software and affect its operation, but mobile platforms are open, allowing the installation of new "malware" applications that can affect the overall operation of the device, including the surreptitious transmission of local data by such an application.
6) User interfaces – with a custom-built embedded application, the developer can control all aspects of the user experience, but a mobile application must share common elements of the user interface with other applications and must adhere to externally developed user interface guidelines, many of which are

implemented in the software development kits (SDKs) that are part of the platform.

7) Complexity of testing – while native applications can be tested in a traditional manner or via a PC-based emulator, mobile applications are particularly challenging to test. Not only do they have many of the same issues found in testing mobile applications, but they have the added issues associated with transmission through gateways and the telephone network
8) Power consumption – many aspects of an application affect its use of the device's power and thus the battery life of the device. Dedicated devices can be optimized for maximum battery life, but mobile applications may inadvertently make extensive use of battery-draining resources.

2.2 Mobile Applications Referential Integrity

A system supports mobile applications referential integrity if it guarantees that resources will continue to exist as long as there are outstanding references to the resources. The mobile application does not support this property and cannot do so since the system is unaware of the number of references that exist to a particular resource. It is impractical to maintain every resource that has ever been published on a particular server forever, this simply does not scale. Mobile applications resources that are no longer of value, for whatever reason, become garbage and need to be collected. This may involve moving the resources to backing storage, or in some cases, deleting the resources entirely. Access pattern information, which is currently available through examination of server logs, is not a sufficient basis to decide whether a component is safe to garbage collect as important though rarely used references to a resource may exist. Safe garbage collection can only be performed if referencing information is available [4]. The consequences of deleting resources that are still referenced affect both the user and the information provider. In mobile application environment, deleting a resource is referenced by another resource results in a broken hypertext link. Such broken links are the single most annoying problem faced by browsing users in the current mobile. Broken links result in a tarnished reputation for the provider of the document containing the link, annoyance for the document user, and possible lost opportunity for the owner of the resource pointed to by the link.

2.3 Component Based Mobile Applications Development

Well-defined models can substantially improve the development and evolution of complex, multi-platform and long-running software systems. Software models play a pivotal role particularly for component-, framework-, and product line-based development. Modeling expertise requires both domain knowledge and software knowledge. Software modeling disciplines are rapidly accumulating in terms of languages, codified expertise, reference models, and automated tools as in figure 1. The areas where such technologies are extensively practiced, the quality features re neither of main concern nor adequately tackled. It is a well-known truth that CBD is important for large and complex systems but why it is important for mobile device applications. It tackles vital concerns such as productivity, high level of abstraction, partitioning of the system development process from the component development process and reusability [7]. Reusability offers a number of advantages to a software development team. An assembly of component assembly leads to a 70 percent reduction in development cycle time; an 84% reduction in project cost, and a productivity index of 26.2, compared to an industry norm of 16.9. For the development of mass mobile examination system, CBD is a smart method, but due to its explicit requirements such as real time, safety, reliability, minimum memory and CPU consumption, standard component models cannot be used [7]. Rather than, a new CBD methodology is very much needed for the development of mobile mass examination system to deal with its specific requirements.

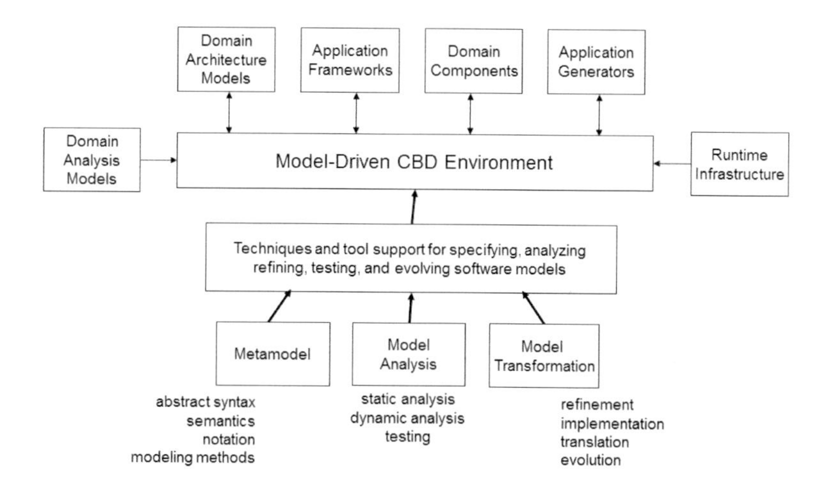

Fig. 1 CBD Driven Mobile Applications Development

MSA(Mobile Service Applications) development model in this paper is based on component based software development. One of the principles of computer science field to solve a problem is divide and conquer i.e., divide the bigger problem into smaller chunks. This principle fits into component based development. The aim is to build large computer systems from small pieces called a component that has already been built instead of building complete system from scratch. Reuse of software components concept has been taken from manufacturing industry and civil engineering field [8]. Manufacturing of vehicles from parts and construction of buildings from bricks are the examples. Car manufacturers would have not been so successful if they had not used standardized parts/components. Software companies have used the same concept to develop software in standardized parts/components. Software components are shipped with the libraries available with software.

2.4 Quality of Service of Mobile Applications

The perceived quality of service (QoS) of the mobile is influenced by many factors, including the broken link problems already mentioned. Even if a user holds a correct reference to an existing mobile resource, it may still be unavailable due to a number of reasons, including unavailability of the machine serving the resource, and partitions in the network between the client and server. Partitions may either be real, caused by breaks in the physical network, or virtual, due to excessive network or server load making communications between the client and server impossible. Even if communication is possible, very poor response characteristics may effectively make the resource unusable. QoS will become more of an issue as the Mobile continues its transformation into a commercially Based Development system. Technical solutions for improving QoS are fairly well understood, including caching for responsiveness, replication for availability, and migration for load balancing. Caching in the mobile is reasonably common, both through the use of browser memory and disc caches, and also through the use of caching servers [5]. Current caching servers use a heuristic approach for consistency management, where resources can only apply coarse-grained tuning based on expiry dates.

3 Design of MSA(Mobile Service Application) Components

A new component-based development (CBD) model has been proposed for a mobile sever applications system. A MSA model is a process model that provides a framework to develop software from previously developed components. The primary componentive of our research is to develop an extensible Mobile infrastructure which is able to support a wide range of resources and services. Our model makes extensive use of the concepts of component-orientation to achieve the necessary extensibility characteristics. Within this component-Based Development framework, proven concepts from the distributed component-Based Development community will be applied to the problems currently facing the mobile. The next section introduces our component model, describing how the principles of component-orientation are applied to the Mobile domain. The interactions between the system components are described in the section entitled "System Architecture," which is followed by a section entitled "MSA Properties" which classifies and describes a collection of properties applicable to different classes of MSAComponent.

3.1 Component Model for Mobile Applications

In the proposed model, Mobile resources are transformed from file-based resources into components, *MSAComponent*. MSAComponent are *encapsulated* resources possessing internal state and a well-defined behavior. The components themselves are responsible for managing their own state transitions and properties, in response to method invocations. This model supports *abstraction* since clients only interact with MSAComponent through the published interfaces; the implementation of a particular operation is not externally visible. Different classes of MSAsupport different operational interfaces, which are obtained through the use of *interface inheritance*. Abstract classes are used to define an interface to a particular component abstraction, without specifying any particular implementation of the operations. Different classes of MSAComponent may share conformance to a particular abstract interface, but may implement the operations differently, in a manner appropriate to the particular class. The use of interface inheritance provides *polymorphism*; that is, all derived classes that conform to an interface provided by some base class may be treated as instances of that base class, without regard for any other aspects of that class' behavior. Continuing with the previous example, consider a dedicated GUI-based Mobile site management tool, providing a graphical interface for performing management-style operations on the components; one such operation may be component migration. The management tool is able to abstract away from other features of the different components (supported through various other interfaces) and simply address all of the different components as instances of the Manageable interface. In addition to inheritance of interface, the model also supports *behavioral inheritance*, thereby supporting code reuse. For example, mixinbase classes to be inherited as required may provide component properties such as persistence and concurrency control. Mix-in classes are not designed to be instantiated themselves. They are used to augment the functionality of the derived class, by providing some particular behavior, usually orthogonal to the primary function of the class. The diagram in Figure 2 illustrates the key points of our component model by showing how two example MSAclasses, backend, SAP web application server and mobile devices are composed using both interface and behavioral inheritance. The abstract class, Manageable provides the interface description for management-style operations (only a single operation, migrate, is shown). Both of the derived classes inherit this interface, providing their own implementations. Also shown in the diagram are three different clients, which manipulate instances of SAP and mobile device. A Mobile site management tool, previously mentioned, is solely concerned with the operations provided through the manageable interface. The tool is able to invoke the migrate operation on instances of either derived class without knowledge of the nature of the class. Two further clients are shown, a theatre booking application and a spreadsheet tool, which manipulate instances of backend, SAP web application server and mobile device respectively. The fact that these classes also conform to the manageable interface is of no consequence to the clients who only interact with the components via the interfaces supporting the classes' primary function.

3.2 Mobile Component Architecture

In common with the current Mobile Applications, the proposed MSA architecture consists of three basic entity types, namely, clients, servers, and published components.

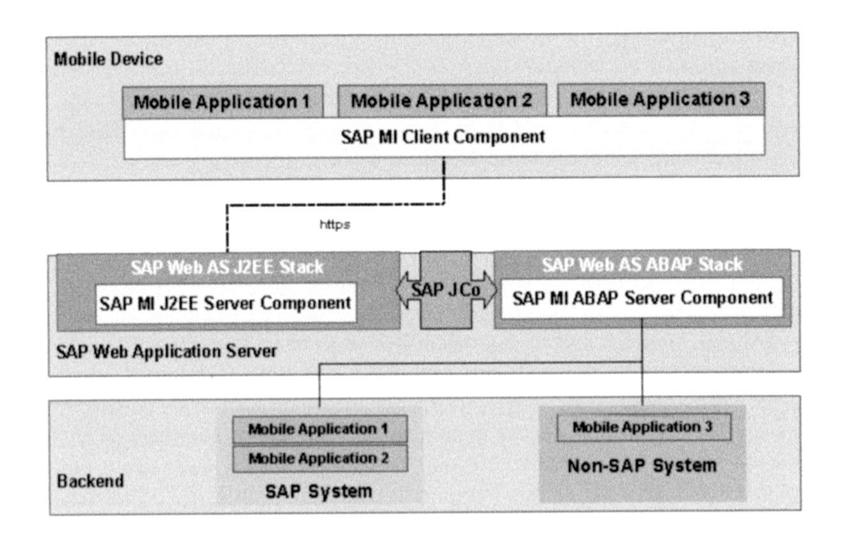

Fig. 2 Basic Component Model for Mobile Applications

In the current Mobile environment, these three types correspond to Mobile browsers (e.g., mosaic), Mobile daemons (e.g., CERN HTTPD), and documentation resources (e.g., HTML documents) respectively. Our architecture supports both client-component (client-server), and intercomponent (peer-to-peer) communication.

3.2.1 Client-Component Interactions

Figure 3 shows the logical view of client-component interactions within the MSA architecture. A single server process is shown, managing a single MSA(although servers are capable of managing multiple components of different types), which is being accessed via two different clients, a standard Mobile browser, and a dedicated bespoke application. This diagram highlights *interoperability* as one of the key concepts of the architecture, that is, the support for component accessibility via different applications using multiple protocols. As stated earlier, MSAComponent are encapsulated, meaning that they are responsible for managing their own properties (e.g., security, persistence, concurrency control etc.) rather than the application accessing the component.

For example, in the case of concurrency, the component manages its own access control, based upon its internal policy, irrespective of which application method invocations originate from. The local representation of a component, together with the available operations, may vary depending upon the particular type of client accessing it. The Mobile browser uses a URL to bind to the particular component in the server. The operations that are permitted on the component, via the URL,

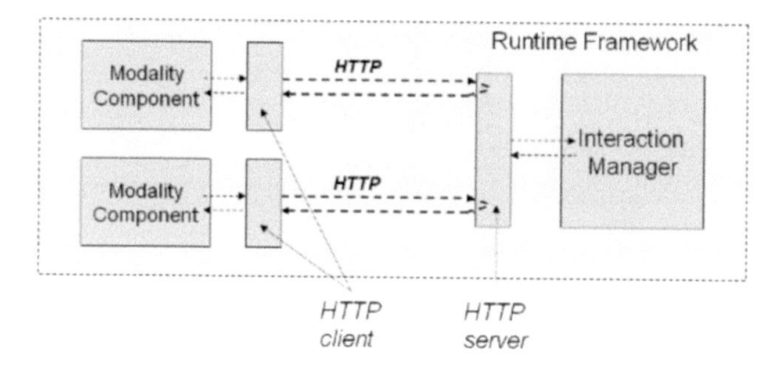

Fig. 3 Client-Component Interactions of MSA

are defined by the HTTP protocol.. The HTTP communication end-point of the server may perform name mapping between URL and the internal name for the component and may also map HTTP requests to appropriate method invocations on the component. From the point of view of the application, this stub component presents the illusion that the remote component is actually within the address space of the client. Like any other component, the stub presents a well-defined interface describing the supported operations. This interface has the potential to be much richer than that provided through HTTP, including application specific operations. Operation invocations on the stub are passed to the component using the remote procedure call (RPC) protocol. Client-stub components may be automatically generated from a description of a component interface. Our implementation uses C++ as the definition language and we provide stub-generation support for creating client and server side stubs which handle operation invocation and parameter. Other possible interface definition languages are possible, including CORBA IDL. The common gateway interface (CGI) could be used to provide a richer client-side interface than is readily available through HTTP. Although, it has been already stated that we believe CGI to be too low-level for direct programming, CGI interfaces to remote components can be automatically created using stub-generation tools. We have implemented a basic stub-generator, which uses an abstract definition of the remote component, and ANSA have recently released a more complete tool based on CORBA IDL. Recent developments using interpreted languages within the Mobile, including Java and SafeTcl are potentially very useful for developing client-side interfaces to MSAComponent. Using such languages, complex, architecture-neutral, front-ends dedicated to a particular MSAclass can be developed, supporting rich sets of operations.

3.2.2 Inter-Component Interactions

In addition to client-component communication, our architecture also supports inter-component communication, regardless of the components' location. In effect, the architecture may be viewed as a single distributed service, partitioned over different hosts as illustrated in Figure 4. Inter-component communication is used for a variety of purposes, including referencing, migration, caching, and replication.

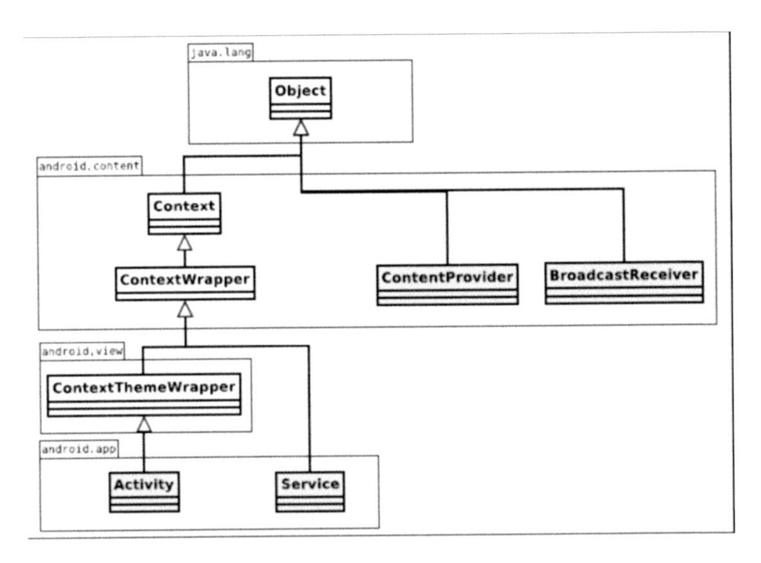

Fig. 4 Inter-Component Interactions as Android

In addition to MSAComponent, servers may contain MSAstubs, or aliases, which are named components that simply forward operation invocations to another component, transparently to clients. One particular use of aliases is in implementation of name-servers, since a name-server may be viewed simply as a collection of named components which alias other components with alternative names (activity, view and contents in diagram). Components may also contain stubs to other components. This feature is used in our implementation of referencing.

3.2.3 Inter-Component Interactions

One method of interfacing with multiple servers is to make use of an HTTP Gateway, which uses stub components to forward component invocations through to the appropriate server. The gateway is transparent to clients accessing the components; incoming requests are simply forwarded to the destination component, which parses the request and replies accordingly. This is illustrated in Figure 5, in which backend server manages a number of different types of component middle server SAP and mobile devices as mobile applications manages components of a single type. As the processing of operations is entirely the responsibility of the individual component, the introduction of new component types is transparent to the gateway.

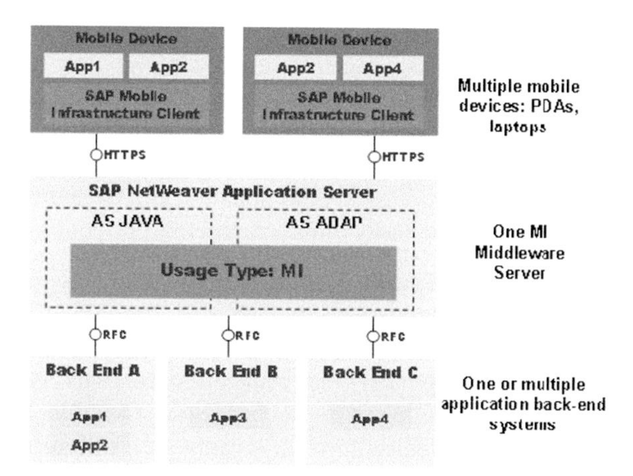

Fig. 5 Client-Component Communication Through Gateway

3.3 MSAProperties

Based on critiques of the current Mobile by ourselves and others [19], and also our experience with distributed systems in general, we have attempted to identify the set of properties that are required by MSAComponent. We have classified these properties into three categories: core properties, common properties, and class-specific properties. In this section we shall present what we believe to be the core properties required by all MSAComponent and give examples of some common properties.

3.3.1 Core Properties

Four properties have been identified as being the core requirements for MSAComponent: Naming, Sharing, Mobility, and Referencing. The implementation of these properties is divided between the components themselves and the supporting infrastructure, which manages the components. Each property will be considered in turn.

***Naming*:** One of the fundamental concepts of the component-Based Development paradigm is identity. The ability to name a component is required in order to unambiguously communicate with and about it. Context-relative naming is an essential feature of our environment so as to support interoperability and scalability. As mentioned previously, different clients may use different local representations of a remote component (URLs, client-stub components, etc.). Since it is impractical to impose new naming conventions on existing systems, we require the ability to translate names between system-boundaries. Furthermore, for extensibility, we need to be able to incorporate new naming systems. Within our design, naming is provided via the component infrastructure.

Sharing**:** Implicit within the Mobile domain is the requirement that components can be shared. Although the basic function of allowing multiple users to interact with components is simple to achieve, there are a number of other associated mechanisms that require interaction with the base sharing functionality. Access control, either user and group based, or access restriction based on the location of the client, are both likely requirements. Additionally, with components supporting a rich set of interfaces, the granularity of the control must be configurable.

Mobility**:** One of the lessons learned from the current Mobile is that support for component mobility is a necessary requirement for MSAComponent. At component creation time, migration of the component may not be envisaged, but it is virtually impossible to predict the future requirements of a particular component. Mobility may be required for many reasons, including load balancing, caching, and improved performance through locality etc., with different forms of migration, including intra- or inter-host.

Referencing**:** In order to address what may be viewed as the primary problem with the current Mobile, namely referential integrity, we believe that low-level referencing support is required by all components. A range of schemes is possible, including forward referencing, call-backs, and redirection through a location server (as in the URN approach). Referencing is closely related to mobility, since referencing schemes may be used to locate components even in the event of component migration.. There are a potentially large number of common properties for MSAComponent, which can be encapsulated within appropriate base classes.

Replication**:** There is a range of replication protocols from active to passive, and strong consistency to weak consistency. There is no single replication protocol which is suitable for every component which may need to be replicated and at the same time can satisfy a user's required quality of service. As such, it is our intention to implement a suitable base class for component providers, which will enable them to select the appropriate replication protocol on a per component basis. In addition component providers will also be able to select the optimum number and location of these replicas, and modify this as required.

Concurrency control**:** By enabling users to share arbitrary components it may be necessary for these component state transitions to be managed through an appropriate concurrency control mechanism. Consider the theatre booking example earlier: if user A wishes to examine the seats which are available while user B is in the process of booking a seat, it would be desirable for B to lock the seat in order to prevent conflicting updates. There are a number of concurrency control mechanisms available, but our initial implementation will be based upon the familiar multiple reader, single writer policy.

Caching**:** The caching of component states, either at or close to users, can help alleviate problems of network congestion and latency. However, as with replication, there is a need for a range of caching policies based upon user requirements and component properties.

Fault tolerance: In a large-scale distributed system, fault tolerance is an important property. One way of addressing the issues of fault tolerance is by using atomic actions to control method invocations. Components inherit necessary persistence and concurrency control characteristics, and application programmers then manipulate these components within atomic actions, which guarantee the usual ACID properties.

3.4 Implementation of MSA

Having described our model in the previous sections, we shall now illustrate how two of the core properties, referencing and mobility, are implemented within the model. Our aim is to address the current problem of broken links and provide transparent component migration. Figure 6 shows our implementation model for mobile component referencing architecture which consists of service component accounting and service component accounting business rule. Referencing is closely related to mobility, since referencing schemes may be used to locate components even in the event of component migration. There are a potentially large number of common properties for MSAComponent, which can be encapsulated within appropriate base classes.

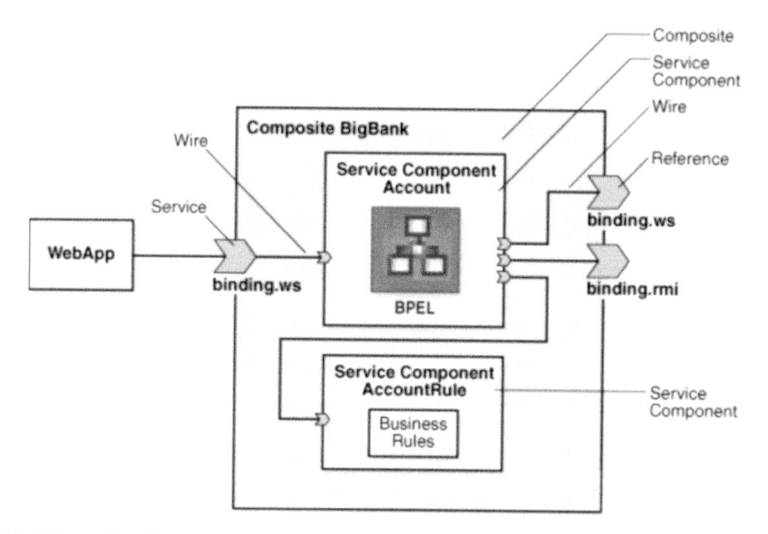

Fig. 6 Mobile applications Component Referencing

4 Conclusions

As a way of serving standard resources, the Mobile has proven extremely successful but still suffers from a number of shortcomings. Furthermore, in order to cope with new resource types the Mobile needs improved flexibility and extensibility characteristics. We have illustrated how the application of the concepts of component-orientation can achieve these extensibility requirements and how problems, such as the lack of referential integrity, can be addressed through the application of techniques developed by the distributed component research community. The MSAmodel, presented in this paper, is intended to provide a flexible and extensible way of building Mobile applications, where Mobile resources are encapsulated as components with well-defined interfaces. Components inherit desirable characteristics, redefining operations as is appropriate; users interact with these components in a uniform manner. We have identified three categories of component properties: core, common, and specific, and have described an implementation using the core properties which addresses what we believe to be one of the most significant problems facing the current Mobile --that of referential integrity. A key feature of our design is support for interoperability; for example, in addition to sophisticated clients which may use the rich component inter faces that our model provides, our implementation will also allow MSAComponent to continue to be accessed using existing Mobile browsers.

When developing mobile based applications, it is important to remember the following factors involved:

- **Usability**: Try to keep the forms small, simple and easy to use for the mobile device being targeted. If scrolling is required, contain it in one direction (usually vertical). While newer mobile browsers approach desktop browser capabilities, it is not always wise to increase the complexity of your mobile forms, because the browser is capable of displaying large desktop forms.
- **Testing**: Not all browsers work the same, and it is important to test with the browsers that your target audience will be using. Testing can be carried out with emulators, but using the physical device is usually a better alternative, since emulators may not always be up to date with the software levels delivered on the device itself.
- **Design**: System Modeler includes a new “FormLayout” property for mobile forms. Use this property when your Mobile Browser does not support absolute positioning of controls on the form. Also add <br> tags and space labels for specific placement of elements on the form. Remember to turn the Grid on in the Painter and enable the SnapToGrid property, to assist in aligning the tope borders of controls on the form.
- **Using Translations**: You can add a new language to your application that contains different presentations to be used for Mobile devices, and the application logic remains the same for both presentation types.

Mobile browsers are constantly changing and improving in capabilities, and there is a trend which shows a convergence with desktop browsers. It is important to test new browser versions as they emerge, to ensure that your application still behaves correctly.

Acknowledgement. This work was supported by the Korea National Research Foundation (NRF) granted funded by the Korea Government (Scientist of Regional University No. 2012-0004489).

References

1. Roy, Ramanujan: Understanding Mobile services. IT Professional 3(6), 69–73 (2001)
2. Ogbuji, U.: The Past, Present and Future of Mobile Services (2004), http://www.Mobileservices.org/index.php/article/articleview/663/4/61/
3. Litoiu, M.: Migrating to Mobile Services-latency and scalability. In: Proceedings of Fourth International Workshop on Mobile Site Evolution, pp. 13–20 (October 2002), http://www.tigris.org/
4. Brown, A.: Using service-oriented architecture and component-based development to build Mobile service applications. Rational Software white paper from IBM (April 2002)
5. Soley, R., and OMG Staff Strategy Group: Model Driven Architecture, OMG White Paper Draft 3.2 (2000), http://www.omg.org/~soley/mda.html
6. Poole, J.D.: Model Driven Architecture: Vision, Standards and Emerging Technologies. In: European Conference on Object-Oriented Programming (April 2004), http://www.omg.org/mda/mda_files/Model-Driven_Architecture.pdf
7. Rizwan Jameel Qureshi, M.: Reuse and Component Based Development. In: Proc. of Int. Conf. Software Engineering Research and Practice (SERP 2006), Las Vegas, USA, June 26-29, pp. 146–150 (2006)
8. Barnawi, A., Rizwan Jameel Qureshi, M., Khan, A.I.: A Framework for Next Generation Mobile and Wireless Networks Application Development using Hybrid Component Based Development Model. International Journal of Research and Reviews in Next Generation Networks (IJRRNGN) 1(2), 51–58 (2011)
9. Champion, M., Ferris, C., Newcomer, E., Orchard, D.: Mobile Services Architecture: W3C Working Draft (2002), http://www.w3.org/TR/ws-arch/
10. OMG, Common Component Request Broker Architecture and Specification, OMG Document Number 91.12.1

Author Index

Printed in the United States
By Bookmasters